Lecture Notes in Statistics **193**

Edited by P. Bickel, P.J. Diggle, S. Fienberg, U. Gather,
I. Olkin, S. Zeger

Alexander Meister

Deconvolution Problems in Nonparametric Statistics

Dr. Alexander Meister
University of Ulm
Helmholtzstraße 22
89081 Ulm
Germany
alexander.meister@uni-ulm.de

ISBN: 978-3-540-87556-7 e-**ISBN**: 978-3-540-87557-4
DOI: 10.1007/978-3-540-87557-4

Library of Congress Control Number: 2008940961

Cover design: SPi Publishing Services

Printed on acid-free paper

springer.com

Contents

1

Introduction

Deconvolution problems occur in many fields of nonparametric statistics, for example, density estimation based on contaminated data, nonparametric regression with errors-in-variables, image and signal deblurring. During the last two decades, those topics have received more and more attention. As applications of deconvolution procedures concern many real-life problems in econometrics, biometrics, medical statistics, image reconstruction, one can realize an increasing number of applied statisticians who are interested in nonparametric deconvolution methods; on the other hand, some deep results from Fourier analysis, functional analysis, and probability theory are required to understand the construction of deconvolution techniques and their properties so that deconvolution is also particularly challenging for mathematicians.

The general deconvolution problem in statistics can be described as follows: Our goal is estimating a function f while any empirical access is restricted to some quantity

$$h = f * G = \int f(x-y)\,\mathrm{d}G(y), \tag{1.1}$$

that is, the convolution of f and some probability distribution G. Therefore, f can be estimated from some observations only indirectly. The strategy is estimating h first; this means producing an empirical version $\hat{h}$ of h and, then, applying a deconvolution procedure to $\hat{h}$ to estimate f. In the mathematical context, we have to invert the convolution operator with G where some regularization is required to guarantee that $\hat{h}$ is contained in the invertibility domain of the convolution operator. The estimator $\hat{h}$ has to be chosen with respect to the specific statistical experiment. Its shape may be quite different for the different models and data structures considered in this book; but in the deconvolution step, we notice some unifying properties. Obviously, to ensure that the specific convolution operator is known, we have to assume that the distribution G is known. Although this condition is not realistic in many practical problems, full knowledge of G is assumed in the classical (or standard) deconvolution approaches. Then, G may be used in the construction

A. Meister, *Deconvolution Problems in Nonparametric Statistics*, Lecture Notes in Statistics 193, DOI 10.1007/978-3-540-87557-4_1,

of deconvolution estimators. Some recent approaches try to relax the exact knowledge of G. Nevertheless, as one faces troubles of identifiability in problems with unknown G, either more restrictive conditions on f or additional data or repeated measurements are required. That will be made precise in this book. While there are discrete deconvolution problems where all probability mass of G is concentrated on a finite set, the large majority of problems discussed in this book deals with continuous convolution models, where G has a density function g in the Lebesgue sense. Then, g is called error density or blurring density, according to the corresponding model,

$$h = f * g = \int f(x-y)g(y)\,\mathrm{d}y. \tag{1.2}$$

Then, the integral is to be understood in the Lebesgue sense; and f and g are real-valued functions mapping into $\mathbb{R}$.

The current book intends to provide a comprehensive overview on results derived during the last 20 years and to give discussion on modern and recently solved problems; estimation methods are introduced and the underlying ideas are explained; the asymptotic theory and adaptive choice of smoothing parameters are derived; applicability is underlined by some real data examples. While the main target group of readers of the book are scientists and graduate students working in the field of mathematical statistics, we also aim at addressing to people from econometrics, biometrics, and other fields of applied statistics, who are interested in the methodology and theory of statistical deconvolution. One main goal of this book is to present theory so that the mathematical prerequisites are kept at a rather low level. Therefore, people having some knowledge as acquired in undergraduate lectures on mathematics (analysis, linear algebra) and elementary probability and measure theory shall be able to understand the proofs of the theorems. In particular, the readers should be familiar with the standard terms and results occurring in stochastics such as probability measure, Lebesgue measure, (Lebesgue) density, expectation, variance, (stochastic) independence. The other major tools of deconvolution, which lie in the field of Fourier analysis, are provided in a separate chapter in the appendix of this book. Readers who are not much familiar with Fourier techniques are strongly encouraged to read this chapter.

Let us roughly explain why Fourier methods are very popular in deconvolution problems. The Fourier transform of a distribution G, defined by

$$G^{\mathrm{ft}}(t) = \int \exp(itx)\,\mathrm{d}G(x)\,, \qquad t \in \mathbb{R}$$

and also the Fourier transform of a function f (not necessarily a density function), defined in the same way when $\mathrm{d}G(x)$ is replaced by $f(x)\,\mathrm{d}x$, are utilized. Throughout this book, the Fourier transform is denoted by G^{ft} and f^{ft}, respectively. Using the Fourier transform is motivated by the fact that it changes convolution into simple multiplication. More concretely, (1.1) is equivalent with

$$h^{\mathrm{ft}} = f^{\mathrm{ft}} \cdot G^{\mathrm{ft}}.$$

We realize that the reconstruction of f from h just becomes dividing h^{ft}, which is empirically accessible, by G^{ft} in the Fourier domain. As a rough scheme for the construction of deconvolution estimators, we give

1. Estimate h^{ft} based on direct empirical information; denote the empirical version of h^{ft} by $\hat{h}^{\mathrm{ft}}$.
2. Calculate $\hat{h}^{\mathrm{ft}}(t)$ and divide it by $G^{\mathrm{ft}}(t)$, leading to the estimator $\hat{f}^{\mathrm{ft}}(t)$. Note that G^{ft} is calculable whenever G is known (at least, theoretically).
3. Regularize $\hat{f}^{\mathrm{ft}}$ so that its inverse Fourier transform $\hat{f}$ exists. Take $\hat{f}$ as the deconvolution estimator of f.

However, this shall not sound too much of straight-forward as the mathematical effort for the regularization steps must not be underestimated.

Also, we describe real data applications to show the broad practical merit of deconvolution estimators. Nevertheless, the book focuses on methodology and theory rather than computational aspects and programming.

The book deals with nonparametric methods. In parametric statistics, the function f to be estimated is known up to finitely many real-valued parameters. As an example for a parametric approach, we could assume that f is a normal density

$$f_{\mu,\sigma^2}(x) = \frac{1}{\sqrt{2\pi}\sigma} \exp\left[-(x-\mu)^2/(2\sigma^2)\right],$$

with unknown mean μ and variance σ^2. Then we could estimate those two parameters and insert the estimated values $\hat{\mu}$ and $\hat{\sigma^2}$ so that $f_{\hat{\mu},\hat{\sigma^2}}$ is an estimator for the function f. Contrarily, nonparametric models are more generally applicable, as the shape of f need not be assumed to be known in advance. The conditions on f are mild such as integrability, boundedness, in some cases also continuity or smoothness. There are problems where f is known to be a density function, in addition (see Chap. 2).

The book is so organized that each of the most important deconvolution problems is introduced in an own chapter. Density deconvolution – maybe the most common deconvolution problem – is introduced in Chap. 2. Then, Chap. 3 focuses on errors-in-variables regression problems; and, in Chap. 4, we study problems of image deblurring. Then, the Appendix Chap. A on Fourier analysis follows.

2

Density Deconvolution

2.1 Additive Measurement Error Model

A basic problem in statistics is the estimation of the (Lebesgue) density function f of the corresponding distribution F when independent real-valued random variables $X_1, \ldots, X_n$ are observed, where each X_j has the distribution F. Those random variables are statistically interpreted as the data. One could think of several examples:

- *From econometrics.* We are interested in the distribution F of the wages of all employees within a certain country. Then we observe the random variables $X_1, \ldots, X_n$, which denote the salary per month of n arbitrarily chosen employees.
- *From biometrics.* We are interested in the distribution F of a certain quantity belonging to an animal race (e.g., size, weight, maximum age). Then, this quantity is measured where n individuals are involved; the results are reported by the data $X_1, \ldots, X_n$.
- We are interested in the traffic intensity of some road. Then, $X_1, \ldots, X_n$ denote the number of cars passing that road, observed on different days.

In all those situations, the random variables $X_1, \ldots, X_n$ are assumed to be i.i.d. (independent and identically distributed). Stochastic independence of the random variables is interpreted by the fact that the data do not influence each other. With respect to the biometrical example, that condition is violated if some individuals have been measured twice. The field of statistics where the assumption of independence is omitted is called time series analysis, but that is not the subject of this book. Identical distribution is justified if the data are measured under identical circumstances. In the third example above, the assumption of identical distribution does not hold true if some measurements are observed on working days; and others on Sundays when traffic of commuters is not taken into account. Still, we can save the i.i.d.-assumption by suitable modeling. Concretely, we define each X_j by the number of cars passing the road on an arbitrarily chosen day. Identical distribution is valid if

A. Meister, *Deconvolution Problems in Nonparametric Statistics*, Lecture Notes in Statistics 193, DOI 10.1007/978-3-540-87557-4_2,

we do not have any information about the specific day of the week when the observation was measured. Sometimes, it is advantageous to avoid too precise categorization of the data with respect to the circumstances under which they were surveyed, because that reduces the sample size that is available for each category of data. Therefore, very exact modeling is not always preferable.

The distribution F must be continuously distributed so that the corresponding density f exists; the latter is defined by

$$F(A) = \int_A f(x)\,\mathrm{d}x$$

for any Lebesgue measurable $A \subseteq \mathbb{R}$. There is a unique one-by-one correspondence between the distribution F and its density f up to changes of the values of f on a set with Lebesgue measure zero. By the Radon–Nikodym–Lemma, such a density function f exists when $F(A) = 0$ holds for any set A with the Lebesgue measure zero. In practice, this is mainly justified by the fact that the observation X_1 is equal to an arbitrary but fixed real number with probability zero. Note that there are uncountably many other possible values in each neighborhood of any single value; so why should that specific value be preferred? Of course, in problems where the X_j denote a specific amount of money, we actually have a discrete situation; however, if huge amounts are involved, a continuous model can be used as a legitimate approximation.

In many real-life situations, direct data are not available since measurement error occurs. Then, we observe the contaminated data $Y_1, \ldots, Y_n$ instead of the true data $X_1, \ldots, X_n$. The elementary model of noisy data is the additive measurement error, that is, any empirical access is restricted to the data $Y_1, \ldots, Y_n$ with

$$Y_j = X_j + \varepsilon_j, \qquad j \in \{1, \ldots, n\} \tag{2.1}$$

instead of the incorrupted i.i.d. random variables $X_1, \ldots, X_n$. Nevertheless, our goal is still to estimate the density f of the incorrupted, but unobserved random variable X_1. The i.i.d. random variables $\varepsilon_1, \ldots, \varepsilon_n$ represent the error or the contamination of the data; the density of each ε_j – consequently called error density – is denoted by g. Further, we assume that X_j and ε_j are real-valued and independent.

Consider the following examples where measurement error should not be neglected:

1. *From statistics in sports*: We are interested in the distribution density f of the time that an athlete needs for 100-m-dash. We observe some data measured at several competitions and trainings. However, as the results are manually measured, the data are corrupted by measurement error. When exact electronic measurement was introduced, the measurement error was indeed taken into account by some sports committees by not considering some former records any longer. In fact, deconvolution techniques do not allow to decide about the validity of a single record, which represents just one past observation. However, some changes of the distribution could

be detected if the measurement error density g is known for manually measured data.

2. *From medical statistics*: (a) We are interested in the density of the long-term log daily saturated fat intake where the National Health and Nutrition Examination Survey (NHANES I) provides some suitable data, which are, however, affected by considerable noise, see, for example, [117] for a deconvolution approach to this study. Indeed, NHANES I was one of the first surveys where nonparametric deconvolution techniques were used. An even more comprehensive study NHANES II has been published later.
(b) In [122] and [39], a dataset containing replicated measurements derived from CAT scans of the heads of 50 psychiatric patients is considered. The ventricule–brain ratio (VBR) is measured twice for each patient, using a hand-held planimeter. Therefore, the measurments are affected by considerable error. Deconvolution techniques allow to estimate the density of the VBR of an arbitrarily chosen patient whenever the error density g is known. In this problem g is empirically accessible due to the replicated measurements of each patient. This is studied in [29], for instance; and will be described in Sect. 2.6.3.
(c) In [3], datasets for two methods of measuring the peak expiratory flow rate (PERF) are given: In the first method, a Wright peak flow meter is used; while, in the second method, the data are measured by a mini Wright meter. Observations for both methods were made on 17 individuals. The fact that both methods lead to different observations for the same individual shows that at least one of the methods suffers from data contamination. Therefore, the additive measurement error model is applied to both methods. Under assumed knowledge of both error densities, density deconvolution is applicable. Again, in practice, the determination of the error density g of the methods is not easy; and its discussion is deferred to Sect. 2.6.
3. *From astronomy*: Bissantz et al. [2] refer to the problem of estimating the log-based ratio of iron compared to hydrogen as a measure of the fraction of heavy elements in some stars. That ratio is observed via the brightness of a star in a certain spectral band, where some difficult calibrations are used. The error made in that transformation may be interpreted as random additive noise and is empirically accessible by replicated measurements of the same star. Again, the practical determination of the error density can be viewed as a kind of replicated measurement problem as studied in Sect. 2.6.3.
4. *From econometrics*: (a) In [113], the Consumer Expenditure Survey is taken from the United States Department of Labor [100] to estimate the permanent log income of an individual. Then, some given measurements report the log expenditure of those individuals in some quarters. One may assume that the difference between the log expenditure in some quarter and the permanent log income may be seen as a random variable, which may be interpreted as the readiness of the people to spend their earnings.

That random variable represents the measurement error with the density g. When no further information about the expenditure behavior of the individuals is given, the difference shall be modeled to be independent of the log income. Again, g is estimable from the log expenditure of the same individuals in another quarter.
In this setting, we realize that the additive measurement error model is also useful for statistical experiments with an independent multiplicative error component. Quite obviously, the ratio between the expenditure of an individual and his income may be seen as a random variable, which is independent of the income, rather than the difference between those quantities. However, applying the logarithm to the data changes the link between the incorrupted random variable and the error from product into sum. Then, interest should be directed to the density of the log of the original variable, which is, in fact, often done in statistics.
(b) In [69], the following mean-regression model involving panel data is described:

$$Y_{jt} = X_{jt}\beta + U_j + \varepsilon_{jt}, \qquad j = 1, \ldots, n,\, t = 1, \ldots, T,$$

where the data (X_{jt}, Y_{jt}) are observed. Therein, Y_{jt} denotes some econometric quantity of the individual j; and X_{jt} denotes some vector-valued explanatory variable for the individual j at some time period t. Further, β is a vector consisting of some constants. Since the unobserved random variables U_j do not depend on the time period t, the differences $Y_{jt} - X_{jt}\beta - Y_{j't} + X_{j't}\beta$ for $j \neq j'$ make the density of the i.i.d. error variables ε_{jt} identifiable as long as it is symmetric (for details, see Sect. 2.6.3). Then the estimated density g of the ε_{jt} may be utilized to estimate the density of the stationary (independent of the time t) random variables U_j by density deconvolution. In this model, we assume that all U_j, ε_{jt} are independent.

Following the classical approach to this model, g is assumed to be exactly known; although in many real-life situations this condition cannot be justified as we have already noticed in the earlier examples. However, in most practical applications, we are able to estimate the error density g from replicated measurements. Discussion on the case of unknown g will be deferred to Sect. 2.6; so, in the following sections, we assume that g is perfectly known so that we can fully concentrate on the deconvolution step itself.

It is an elementary result of probability theory that the density of the sum of two independent random variables is equal to the convolution of the densities of both addends; hence,

$$h = f * g = \int f(x-y)g(y)\,\mathrm{d}y,$$

where h denotes the density of the observation Y_1. Since any direct empirical access is restricted to h, the problem is included in the general deconvolution framework as described in Chap. 1.

2.2 Estimation Procedures

Following the general deconvolution scheme in the introduction, the first step is estimating the density h of the observations Y_j. So we face a problem of direct density estimation, which can be solved by various methods. But we see that the main intention at this stage is making the Fourier transform h^{ft} empirically accessible. As the characteristic function Ψ_Y of a random variable Y is just the Fourier transform of the density of Y, we have

$$h^{\mathrm{ft}}(t) = \int \exp(\mathrm{i}tx)h(x)\mathrm{d}x = E\,\exp(\mathrm{i}tY_1) = \Psi_{Y_1}(t).$$

As a very common statistical method, we obtain an empirical version of h^{ft} by replacing the expectation by averaging with respect to the i.i.d. data. Then, the so-called empirical characteristic function is defined by

$$\hat{\Psi}_{Y_1}(t) = \frac{1}{n}\sum_{j=1}^{n}\exp(\mathrm{i}tY_{\mathrm{j}}). \tag{2.2}$$

We have a simple multiplicative link between h^{ft} and f^{ft} because

$$\begin{aligned} h^{\mathrm{ft}}(t) = \Psi_{Y_1}(t) &= E\,\exp\big(\mathrm{i}t(X_1+\varepsilon_1)\big) = E\big[\exp(\mathrm{i}tX_1)\exp(\mathrm{i}t\varepsilon_1)\big] \\ &= E\,\exp(\mathrm{i}tX_1)\cdot E\,\exp(\mathrm{i}t\varepsilon_1) = \Psi_{X_1}(t)\cdot\Psi_{\varepsilon_1}(t) = f^{\mathrm{ft}}(t)\cdot g^{\mathrm{ft}}(t), \end{aligned}$$

where the independence of X_1 and ε_1 implies independence of $\exp(\mathrm{i}tX_1)$ and $\exp(\mathrm{i}t\varepsilon_1)$. In particular, the elementary probabilistic result $E[A\cdot B] = EA\cdot EB$ for independent and square-integrable random variables A and B is used. Therefore, it is reasonable to consider

$$\hat{\Psi}_{X_1}(t) = \frac{1}{n}\sum_{j=1}^{n}\exp(\mathrm{i}tY_j)\,/\,g^{\mathrm{ft}}(t) \tag{2.3}$$

as an estimator of $f^{\mathrm{ft}}(t)$, assuming that g^{ft} vanishes nowhere. It is not hard to establish unbiasedness and consistency for this estimator (by the strong law of large numbers). Nevertheless, we are still interested in f rather than f^{ft}. Therefore, we shall apply Fourier inversion to $\hat{\Psi}_{X_1}$. A naive estimator of f is then

$$\hat{f}_{\mathrm{naive}}(x) = \frac{1}{2\pi}\int \exp(-\mathrm{i}tx)\hat{\Psi}_{X_1}(t)\,\mathrm{d}t\,,$$

with the convention that all integrals without any subscript are to be taken over the whole real line throughout this book. See Theorems A.2 and A.4 in the Appendix Chap. A for details on the inverse Fourier transform. However, the estimator $\hat{f}_{\mathrm{naive}}$ is not well-defined as $\hat{\Psi}_{X_1}(t)$ is neither integrable nor square-integrable over $\mathbb{R}$, unlike its true counterpart $f^{\mathrm{ft}}(t)$ to be estimated, which is square-integrable whenever f is square-integrable, due to Parseval's identity (*see* Theorem A.4). Apparently, for large $|t|$, $\hat{\Psi}_{X_1}(t)$ is no good estimator for $f^{\mathrm{ft}}(t)$ as the tail behavior is significantly different. Therefore, there is some necessity to regularize $\hat{\Psi}_{X_1}(t)$ before the Fourier inversion is employed. Several regularization methods are described in the following subsections.

2.2.1 Kernel Methods

In the case of density estimation based on direct data (i.e., the error-free case), [112] and [105] introduce the kernel density estimator, defined by

$$\hat{h}(x) = \frac{1}{nb}\sum_{j=1}^{n} K\Big(\frac{x-Y_j}{b}\Big), \tag{2.4}$$

with a kernel function $K:\mathbb{R}\to\mathbb{R}$ and a bandwidth parameter $b>0$. We restrict our consideration to the univariate case, that is, all data are real-valued, since that restriction is already made when introducing model (2.1). Although h is a more conventional notation for the bandwidth, we choose b to avoid any misunderstanding regarding the density h, which $\hat{h}$ claims to estimate. If $K\in L_1(\mathbb{R})\cap L_2(\mathbb{R})$, that is, the intersection of the sets of all absolutely or square integrable functions over the whole real line, respectively, in Lebesgue sense, the estimator $\hat{h}(x)$ also lies in $L_1(\mathbb{R})\cap L_2(\mathbb{R})$ almost surely so that its Fourier transform exists; it is calculable by

$$\begin{aligned}\hat{h}^{\mathrm{ft}}(t) &= \int \frac{1}{nb}\sum_{j=1}^{n}\exp(itx)K\Big(\frac{x-Y_j}{b}\Big)\,\mathrm{d}x = \frac{1}{nb}\sum_{j=1}^{n}\int\exp(itx)K\Big(\frac{x-Y_j}{b}\Big)\,\mathrm{d}x\\ &= \frac{1}{n}\sum_{j=1}^{n}\int\exp\big(it(zb+Y_j)\big)K(z)\,\mathrm{d}z = \frac{1}{n}\sum_{j=1}^{n}\exp(itY_j)\int\exp(itbz)K(z)\,\mathrm{d}z\\ &= \hat{\Psi}_{Y_1}(t)\cdot K^{\mathrm{ft}}(tb)\,, \end{aligned}\tag{2.5}$$

by substituting x by $z=(x-Y_j)/b$ in the integral, with $\hat{\Psi}_{Y_1}(t)$ as in (2.2). Then, we may apply

$$\hat{\Psi}_{Y_1}(t)K^{\mathrm{ft}}(tb)/g^{\mathrm{ft}}(t) \tag{2.6}$$

as a second empirical version for $f^{\mathrm{ft}}(t)$, which differs from (2.3) only due to the deterministic term $K^{\mathrm{ft}}(tb)$. Involving this factor may be seen as a regularization procedure for $\Psi_{X_1}(t)$. Further, having the representation (2.6), it is no longer required to assume $K\in L_1(\mathbb{R})$, but the term is still well-defined as long as $K\in L_2(\mathbb{R})$. There are kernel functions whose Fourier transforms are bounded and compactly supported, for example, the sinc-kernel $K(x)=\sin(x)/(\pi x)$ having the Fourier transform

$$K^{\mathrm{ft}}(t) = \chi_{[-1,1]}(t),$$

where χ_I denotes the indicator function of a set I. Under use of such a kernel function, the function given by (2.6) is supported on $[-1/b,1/b]$ and bounded whenever $g^{\mathrm{ft}}(t)\neq 0\,,\ \forall t\in\mathbb{R}$. Note that the latter condition suffices to ensure that $\inf_{t\in[-1/b,1/b]}|g^{\mathrm{ft}}(t)|>0$, since g^{ft} is the Fourier transform of a probability density and, hence, continuous according to Lemma A.1(d) in the appendix. That implies both integrability and square-integrability of the term in (2.6) so that we may apply the inverse Fourier transform to (2.6), leading to the density estimator

$$\hat{f}(x) = \frac{1}{2\pi} \int \exp(-itx) K^{\mathrm{ft}}(tb) \frac{1}{n} \sum_{j=1}^{n} \exp(itY_j)/g^{\mathrm{ft}}(t)\, dt, \tag{2.7}$$

which is well-defined for any nonvanishing g^{ft} whenever K^{ft} is, for example, compactly supported; unlike the naive estimator $\hat{f}_{\mathrm{naive}}(x)$ as proposed in the introduction of the current section. Estimator (2.7) has become known as the standard deconvolution kernel density estimator. It was first introduced by [117] and [14].

Note that estimator (2.7) may also take complex values. This seems strange considering that the function to be estimated is a density. However, one can easily solve that problem by just taking the real part of estimator (2.7). This is usually done for practical implementation. The theoretical results derived in the Sect. 2.4.2 and 2.4.3 still hold true for that modified estimator. On the other hand, restricting to the real part does not lead to any significant improvement of the quality (improvement of the convergence rates etc.) so that we still use (2.7) to derive our propositions and theorems.

As the first result on deconvolution kernel estimators, let us calculate their expectation and variance at some arbitrary but fixed $x \in \mathbb{R}$.

Proposition 2.1. *We assume that f is bounded and continuous; and, in addition $f^{\mathrm{ft}} \in L_1(\mathbb{R})$; also, assume that $g^{\mathrm{ft}}(t) \neq 0$, for all $t \in \mathbb{R}$. With respect to the deconvolution kernel density estimator (2.7) with a kernel which satisfies $K \in L_1(\mathbb{R}) \cap L_2(\mathbb{R})$ and K^{ft} is compactly supported, we obtain, for any $x \in \mathbb{R}$, that*

$$\text{(a) } E\,\hat{f}(x) = \big[K_b * f\big](x) \text{ with } K_b = \frac{1}{b} K(\cdot/b),$$

$$\text{(b) } \mathrm{var}\,\hat{f}(x) \le \frac{1}{2\pi} \|f * g\|_\infty \cdot \frac{1}{n} \int \big|K^{\mathrm{ft}}(tb)/g^{\mathrm{ft}}(t)\big|^2\, dt.$$

Proof. (a) Applying Fubini's theorem, which allows the exchange of integral and expectation, as well as Lemma A.1(b) and (e), we derive that

$$\begin{aligned} E\,\hat{f}(x) &= \frac{1}{2\pi} \int \exp(-itx) K^{\mathrm{ft}}(tb) \frac{1}{n} \sum_{j=1}^{n} E \exp(itY_j)/g^{\mathrm{ft}}(t)\, dt \\ &= \frac{1}{2\pi} \int \exp(-itx) K^{\mathrm{ft}}(tb)\, f^{\mathrm{ft}}(t)\, dt = \frac{1}{2\pi} \int \exp(-itx) K_b^{\mathrm{ft}}(t)\, f^{\mathrm{ft}}(t)\, dt \\ &= \frac{1}{2\pi} \int \exp(-itx) (K_b * f)^{\mathrm{ft}}(t)\, dt. \end{aligned}$$

As $K \in L_1(\mathbb{R})$, we have $\|K_b^{\mathrm{ft}}\|_\infty = \|K^{\mathrm{ft}}\|_\infty \leq \|K\|_1$ and, hence, $(K_b * f)^{\mathrm{ft}} = K_b^{\mathrm{ft}} f^{\mathrm{ft}} \in L_1(\mathbb{R})$, as, in addition, K^{ft} is compactly supported, where here and elsewhere the $L_p(\mathbb{R})$-norm of some $f \in L_p(\mathbb{R})$ is denoted by

$$\|f\|_p = \Big(\int |f(x)|^p \,\mathrm{d}x\Big)^{1/p}, \quad p > 0,$$

and

$$\|f\|_\infty = \sup_{x\in\mathbb{R}} |f(x)|.$$

Furthermore, we have

$$\begin{aligned}\|K_b * f\|_\infty &\leq \sup_{x\in\mathbb{R}} \int \big|K_b(x-y)f(y)\big| \,\mathrm{d}y \leq \|f\|_\infty \cdot \sup_{x\in\mathbb{R}} \int \big|K_b(x-y)\big| \,\mathrm{d}y \\ &\leq \|K_b\|_1 \cdot \|f\|_\infty = \|K\|_1 \cdot \|f\|_\infty,\end{aligned}$$

and

$$\Big|\int \big[f(x-z) - f(y-z)\big] K_b(z) \,\mathrm{d}z\Big| = \Big|\int \big[f(x-zb) - f(y-zb)\big] K(z) \,\mathrm{d}z\Big|,$$

so that the continuity of $K_b * f$ follows from that of f and dominated convergence when considering the limit $y \to x$. Now we may apply the Fourier inversion Theorem A.2 to see that

$$\frac{1}{2\pi}\int \exp(-itx)(K_b * f)^{\mathrm{ft}}(t) \,\mathrm{d}t = \big[K_b * f\big](x).$$

(b) As the $Y_1, \ldots, Y_n$ are i.i.d., we derive with respect to the variance

$$\begin{aligned}\operatorname{var} \hat{f}(x) &= \frac{1}{(2\pi)^2 n^2} \sum_{j=1}^n \operatorname{var} \int \exp\big(-it(x-Y_j)\big) K^{\mathrm{ft}}(tb)/g^{\mathrm{ft}}(t) \,\mathrm{d}t \\ &\leq \frac{1}{(2\pi)^2 n} E\Big|\int \exp\big(-it(x-Y_1)\big) K^{\mathrm{ft}}(tb)/g^{\mathrm{ft}}(t) \,\mathrm{d}t\Big|^2 \\ &= \frac{1}{(2\pi)^2 n} \int \Big|\int \exp\big(-itz\big) K^{\mathrm{ft}}(tb)/g^{\mathrm{ft}}(t) \,\mathrm{d}t\Big|^2 \big[f * g\big](x-z) \,\mathrm{d}z \\ &\leq \|f * g\|_\infty \cdot \frac{1}{(2\pi)^2 n} \int \Big|\int \exp\big(-itz\big) K^{\mathrm{ft}}(tb)/g^{\mathrm{ft}}(t) \,\mathrm{d}t\Big|^2 \,\mathrm{d}z \\ &= \frac{1}{2\pi}\|f * g\|_\infty \cdot \frac{1}{n} \int \big|K^{\mathrm{ft}}(tb)/g^{\mathrm{ft}}(t)\big|^2 \,\mathrm{d}t,\end{aligned}$$

where Parseval's identity (see Theorem A.4) has been used in the last step. We justify that $\|f * g\|_\infty$ is finite by

$$\big|\big[f * g\big](x)\big| = \Big|\int f(y) g(x-y) \,\mathrm{d}y\Big| \leq \|f\|_\infty < \infty,$$

by assumption, as g integrates to one. Note that, if we had used the more obvious inequality

$$\int \Big| \int \exp\big(-itz\big) K^{ft}(tb)/g^{ft}(t)\,dt\Big|^2 \big[f * g\big](x-z)\,dz \\ \leq \|f * g\|_\infty \cdot \Big(\int \big|K^{ft}(tb)/g^{ft}(t)\big|\,dt\Big)^2,$$

without using Parseval's identity, then we would have received a less sharp upper bound, which does not lead to the best attainable convergence rates, in general. ■

We mention that the expectation of the estimator in Proposition 2.1(a) is identical with the expectation of the standard kernel density estimator in the error-free case. Further, Proposition 2.1 allows us to consider the so-called MSE (mean squared error) of the deconvolution kernel estimator, defined by

$$\mathrm{MSE}(\hat{f}, f; x) = E\big|\hat{f}(x) - f(x)\big|^2. \tag{2.8}$$

For details about the MSE, see Sect. 2.3. We may split the MSE into variance and bias term, leading to

$$\begin{aligned} \mathrm{MSE}(\hat{f}, f; x) &= \big|E\hat{f}(x) - f(x)\big|^2 + \mathrm{var}\,\hat{f}(x) \\ &\leq \big|\big[K_b * f\big](x) - f(x)\big|^2 + \frac{1}{2\pi}\|f * g\|_\infty \cdot \frac{1}{n}\int \big|K^{ft}(tb)/g^{ft}(t)\big|^2\,dt. \end{aligned} \tag{2.9}$$

Therefore, Proposition 2.1 gives us an upper bound on the MSE, which is significant for the quality of the estimation of a density function f at a fixed site $x \in \mathbb{R}$.

To evaluate the mean quality of a density estimator on the whole real line, we may consider the MISE (mean integrated squared error), that is,

$$\mathrm{MISE}(\hat{f}, f) = E\int \big|\hat{f}(x) - f(x)\big|^2\,dx. \tag{2.10}$$

In the case of no contamination, there exists an equivalent representation of the MISE of the standard kernel density estimator, involving methods from Fourier analysis; see [126], who give equivalent representations rather than inequalities. This formula is extendable to the deconvolution context, see the following proposition.

Proposition 2.2. *Assume that the density f is contained in $L_2(\mathbb{R})$ and that $g^{ft}(t) \neq 0$, $\forall t \in \mathbb{R}$. With respect to the deconvolution kernel density estimator (2.7) with a kernel that satisfies $K \in L_2(\mathbb{R})$ and K^{ft} is compactly supported, we obtain that*

$$\mathit{MISE}(\hat{f}, f) = \frac{1}{2\pi n}\int \big|K^{ft}(tb)\big|^2\big[|g^{ft}(t)|^{-2} - |f^{ft}(t)|^2\big]\,dt \\ + \frac{1}{2\pi}\int \big|K^{ft}(tb) - 1\big|^2 |f^{ft}(t)|^2\,dt.$$

Proof. By Fourier inversion (see Theorem A.4), we derive that

$$\hat{f}^{\mathrm{ft}}(t) = K^{\mathrm{ft}}(tb)\frac{1}{n}\sum_{j=1}^{n}\exp(itY_j)/g^{\mathrm{ft}}(t).$$

In fact, that term is equivalent with (2.6), from which the deconvolution kernel density estimator has been constructed. Utilizing Fubini's theorem and Parseval's identity (see Theorem A.4) as well as the basic calculation rules for variances and expectations for i.i.d. data, we obtain that

$$\begin{aligned}
\mathrm{MISE}(\hat{f}, f) &= \frac{1}{2\pi}\int E\big|\hat{f}^{\mathrm{ft}}(t) - f^{\mathrm{ft}}(t)\big|^2\,\mathrm{d}t \\
&= \frac{1}{2\pi}\int \mathrm{var}\,\hat{f}^{\mathrm{ft}}(t)\,\mathrm{d}t + \frac{1}{2\pi}\int \big|E\hat{f}^{\mathrm{ft}}(t) - f^{\mathrm{ft}}(t)\big|^2\,\mathrm{d}t \\
&= \frac{1}{2\pi}\int |K^{\mathrm{ft}}(tb)|^2\left[\mathrm{var}\,\exp(itY_1)\right] / \left[n|g^{\mathrm{ft}}(t)|^2\right]\mathrm{d}t \\
&\qquad + \frac{1}{2\pi}\int \big|K^{\mathrm{ft}}(tb)E\,\exp(itY_1)/g^{\mathrm{ft}}(t) - f^{\mathrm{ft}}(t)\big|^2\,\mathrm{d}t \\
&= \frac{1}{2\pi}\int |K^{\mathrm{ft}}(tb)|^2\left(E|\exp(itY_1)|^2 - |E\,\exp(itY_1)|^2\right) / \left[n|g^{\mathrm{ft}}(t)|^2\right]\mathrm{d}t \\
&\qquad + \frac{1}{2\pi}\int \big|K^{\mathrm{ft}}(tb)f^{\mathrm{ft}}(t) - f^{\mathrm{ft}}(t)\big|^2\,\mathrm{d}t \\
&= \frac{1}{2\pi n}\int |K^{\mathrm{ft}}(tb)|^2\left(|g^{\mathrm{ft}}(t)|^{-2} - |f^{\mathrm{ft}}(t)|^2\right)\mathrm{d}t \\
&\qquad + \frac{1}{2\pi}\int \big|K^{\mathrm{ft}}(tb) - 1\big|^2|f^{\mathrm{ft}}(t)|^2\,\mathrm{d}t\,.
\end{aligned}$$

■

In Sect. 2.4, the Propositions 2.1 and 2.2 will be applied to derive convergence rates of the deconvolution kernel estimator as the sample size tends to infinity.

2.2.2 Wavelet-based Methods

Another widely used type of nonparametric curve estimators are known as orthogonal series estimators. As used in the previous subsection for $p \in \{1, 2\}$, the linear spaces $L_p(\mathbb{R})$ are defined by all functions, which are integrable in pth power over the whole real line, that is

$$L_p(\mathbb{R}) = \Big\{f : \mathbb{R} \to \mathbb{C} : \int |f(x)|^p\,\mathrm{d}x < \infty\Big\}. \tag{2.11}$$

By definition, any density lies in $L_1(\mathbb{R})$; however, not necessarily in $L_2(\mathbb{R})$. Nevertheless, under mild assumptions such as boundedness of the density function, we can verify its membership in $L_2(\mathbb{R})$, as shown in the following.

Lemma 2.3. *Any bounded density function f is contained in $L_2(\mathbb{R})$.*

Proof. As $f \in L_1(\mathbb{R})$, all the sets $G(c) = \{x \in \mathbb{R} : |f(x)| \leq c\}$ are measurable (in Lebesgue sense). Therefore, measurability also holds for the sets $\{x \in \mathbb{R} : |f(x)|^2 \leq c\} = G(\sqrt{c})$, from what follows that the function $|f|^2$ is measurable. Hence, it suffices to show that $|f|^2$ possesses an integrable upper bound. Consider

$$\begin{aligned} |f(x)|^2 &= |f(x)|^2 \cdot \chi_{G(1)}(x) + |f(x)|^2 \cdot \chi_{\mathbb{R}\backslash G(1)}(x) \\ &\leq |f(x)| + \big(\sup_{y\in\mathbb{R}} |f(y)|\big)^2 \cdot \chi_{\mathbb{R}\backslash G(1)}(x) \,. \end{aligned}$$

Since $|f|$ is assumed to be bounded and integrable, it remains to be shown that the Lebesgue measure of $\mathbb{R}\backslash G(1)$, denoted by $\lambda(\mathbb{R}\backslash G(1))$, is finite. We have

$$1 = \int f(x)\,\mathrm{d}x \geq \int_{\mathbb{R}\backslash G(1)} \underbrace{f(x)}_{>1}\,\mathrm{d}x \geq \lambda(\mathbb{R}\backslash G(1)) \,.$$

■

All the sets $L_p(\mathbb{R})$, $p \geq 1$, are Banach spaces, that is, normed linear spaces, which are complete with respect to their specific norm

$$\|f\|_p = \Big(\int |f(x)|^p \,\mathrm{d}x \Big)^{1/p} .$$

However, $L_2(\mathbb{R})$ is the only Hilbert space among those sets, that is, its norm is induced by an inner product $\langle \cdot, \cdot \rangle$; more concretely, we have

$$\|f\|_2^2 = \langle f, f \rangle \,.$$

So the inner product in $L_2(\mathbb{R})$ is defined by

$$\langle f, g \rangle = \int f(x)\overline{g(x)}\,\mathrm{d}x .$$

Note that $L_2(\mathbb{R})$ also contains functions mapping from $\mathbb{R}$ to $\mathbb{C}$, the set of all complex numbers. To keep nonnegativity of the norm for those complex-valued functions, we have to apply the complex-conjugated number $\overline{g(x)} = \mathrm{Re}\, g(x) - \mathrm{i} \cdot \mathrm{Im}\, g(x)$ in the definition of the inner product, where Re and Im denote the real and the imaginary part of a complex number, respectively. For more details on that, see books on functional analysis such as [130]. Two functions $f, g \in L_2(\mathbb{R})$ are called orthogonal if $\langle f, g \rangle = 0$. We learn from functional analysis that there are countable orthonormal bases of $L_2(\mathbb{R})$, that is, some sets $\{f_j : j \text{ integer}\}$ so that

$$\langle f_j, f_k \rangle = \delta_{j,k} = \begin{cases} 1, & \text{if } j = k \\ 0, & \text{otherwise,} \end{cases}$$

and, for any $f \in L_2(\mathbb{R})$, we have

$$\inf \Big\{ \Big\| f - \sum_{|j|\le J} \mu_j f_j \Big\|_2 : \mu_{-J}, \ldots, \mu_J \in \mathbb{C} \Big\} \stackrel{J\to\infty}{\longrightarrow} 0 .$$

That criterion is also referred to as separability of the space $L_2(\mathbb{R})$. The latter condition means that any $f \in L_2(\mathbb{R})$ is asymptotically representable by the orthonormal basis. That infimum, for finite J, can be replaced by a minimum where the minimizing coefficients are $\mu_j = \langle f, f_j \rangle$; they are explicitly calculable. The orthogonal projection of f onto the linear hull of $\{f_j : |j| \le J\}$ leads to the coefficients $\mu_j = \langle f, f_j\rangle$. Since

$$\Big\langle f - \sum_{|j|\le J} \langle f, f_j\rangle f_j, f_k \Big\rangle = \langle f, f_k\rangle - \sum_{|j|\le J} \langle f, f_j\rangle \cdot \underbrace{\langle f_j, f_k\rangle}_{=\delta_{j,k}} = 0 ,$$

for all $k \in \{-J, \ldots, J\}$, we have orthogonality of $f - \sum_{|j|\le J}\langle f, f_j\rangle f_j$ and $\sum_{|j|\le J} \big(\mu_j - \langle f, f_j\rangle\big) f_j$ so that, for all $\mu_{-J}, \ldots, \mu_j \in \mathbb{C}$, we obtain

$$\begin{aligned} \Big\| f - \sum_{|j|\le J} \mu_j f_j \Big\|_2^2 &= \Big\| f - \sum_{|j|\le J} \langle f, f_j\rangle f_j + \sum_{|j|\le J} \langle f, f_j\rangle f_j - \sum_{|j|\le J} \mu_j f_j \Big\|_2^2 \\ = \Big\| f - \sum_{|j|\le J} \langle f, f_j\rangle f_j \Big\|_2^2 &+ \Big\| \sum_{|j|\le J} \big(\langle f, f_j\rangle - \mu_j\big) f_j \Big\|_2^2 \ge \Big\| f - \sum_{|j|\le J} \langle f, f_j\rangle f_j \Big\|_2^2, \end{aligned} \tag{2.12}$$

implying that $\sum_{|j|\le J}\langle f, f_j\rangle f_j$ minimizes the $L_2(\mathbb{R})$-distance between any element of the linear space generated by $f_{-J}, \ldots, f_J$ and f. That motivates us to derive an estimator for the coefficients $\mu_j = \langle f, f_j\rangle$. Using the Plancherel isometry (see Theorem A.4), we derive that

$$\begin{aligned} \mu_j = \langle f, f_j\rangle &= \frac{1}{2\pi}\langle f^{\mathrm{ft}}, f_j^{\mathrm{ft}}\rangle = \frac{1}{2\pi} \int f^{\mathrm{ft}}(t)\overline{f_j^{\mathrm{ft}}(t)}\, \mathrm{d}t \\ &= \frac{1}{2\pi} \int h^{\mathrm{ft}}(t) \frac{\overline{f_j^{\mathrm{ft}}(t)}}{g^{\mathrm{ft}}(t)}\, \mathrm{d}t , \end{aligned}$$

where the deconvolution formula (Lemma A.5) has been employed; we recall that $h = f * g$ denotes the density of each Y_j. Therefore, an empirical version for μ_j shall be obtained by replacing $h^{\mathrm{ft}}(t)$ by the empirical characteristic function (2.2). The resulting estimator

$$\hat{\mu}_j = \frac{1}{2\pi n} \sum_{j=1}^{n} \int \exp(\mathrm{i}tY_j) \frac{\overline{f_j^{\mathrm{ft}}(t)}}{g^{\mathrm{ft}}(t)}\, \mathrm{d}t \tag{2.13}$$

is well-define, that is, the integral exists, whenever g^{ft} is nonvanishing and f_j^{ft} is compactly supported. That motivates us to define the density estimator

$$\hat{f}(x) = \sum_{|j| \leq J_n} \hat{\mu}_j f_j(x), \tag{2.14}$$

where the smoothing parameter J_n and the specific orthonormal basis are still to be selected.

In some cases, it is useful to generalize the estimator (2.14) so that the order of the basis functions f_j may be changed. Therefore, we define some finite set $\mathcal{J}_n$ containing only integers; and we specify

$$\hat{f}(x) = \sum_{j \in \mathcal{J}_n} \hat{\mu}_j f_j(x). \tag{2.15}$$

Of course, estimator (2.14) is still included by putting $\mathcal{J}_n = \{-J_n, \ldots, J_n\}$.

Popular orthonormal bases of $L_2(\mathbb{R})$ may be derived from some substructures of wavelets, which have received remarkable attention in numerical analysis, approximation theory as well as statistical science during the last decades. In view of our current deconvolution problem, those wavelets studied by [96] with compactly supported Fourier transforms seem to be particularly appropriate. The wavelet bases consist of the collection

$$\{f_l\}_{l \in \mathbb{N}} = \{\varphi_{m,k} : k \text{ integer}\} \cup \{\psi_{j,k} : k \text{ integer}, j = m, m+1, \ldots\}, \tag{2.16}$$

where $\varphi_{m,k}(x) = 2^{m/2}\varphi(2^m x - k)$ and $\psi_{j,k}(x) = 2^{j/2}\psi(2^j x - k)$, while m is an arbitrary but fixed positive integer. The scaling function φ and the wavelet function ψ are defined via their Fourier transforms; we put

$$\varphi^{\mathrm{ft}}(t) = \big[P([t-\pi, t+\pi])\big]^{1/2}, \; \psi^{\mathrm{ft}}(t) = \exp(-\mathrm{i}t/2)\,\big[P([|t|/2-\pi, |t|-\pi])\big]^{1/2},$$

for some probability measure P, which is concentrated on the interval $[-\pi/3, \pi/3]$. From there, we may derive that φ^{ft} and ψ^{ft} are supported on $[-4\pi/3, 4\pi/3]$ and $[-8\pi/3, -2\pi/3] \cup [2\pi/3, 8\pi/3]$, respectively. Also, we have $\varphi^{\mathrm{ft}}(t) = 1$ for all $|t| \leq 2\pi/3$. That wavelet basis is used in [107].

At this stage, we restrict our consideration to the linear space generated by the $\varphi_{m,k}$, k integer. More precisely, we put $\mathcal{J}_n = \{\varphi_{m_n,k} : |k| \leq K_n\}$ for some integer-valued sequences $(K_n)_n \uparrow \infty$ and $(m_n)_n$. The resulting estimator will be called the linear wavelet estimator. We mention that $\{\varphi_{m_n,k} : k \text{ integer}\}$ is no orthonormal basis of $L_2(\mathbb{R})$ but of its subspace

$$L_2^{\omega_n}(\mathbb{R}) = \{f \in L_2(\mathbb{R}) : f^{\mathrm{ft}} \text{ is supported on } [-\omega_n, \omega_n]\}, \quad \omega_n = 4\pi 2^{m_n}/3.$$

To prove that claim, we first show that all $\varphi^{\mathrm{ft}}_{m_n,k}$ are supported on $[-\omega_n, \omega_n]$. This can be seen using Lemma A.1(e), leading to

$$\begin{aligned}\varphi^{\mathrm{ft}}_{m,k}(t) &= 2^{-m/2}\exp(\mathrm{i}kt/2^m)\varphi^{\mathrm{ft}}(t/2^m),\\ \psi^{\mathrm{ft}}_{j,k}(t) &= 2^{-j/2}\exp(\mathrm{i}kt/2^j)\psi^{\mathrm{ft}}(t/2^j).\end{aligned}$$

Furthermore, we easily derive that any $L_2^{\omega_n}(\mathbb{R})$ is a Hilbert space itself. Linearity follows straightforward, while $L_2(\mathbb{R})$-closedness can be seen as follows: assume some sequence $(f_n)_n \subseteq L_2^{\omega_n}(\mathbb{R})$ tending to some $f \in L_2(\mathbb{R})$ with respect to the $L_2(\mathbb{R})$-norm. Then, by Parseval's identity (Theorem A.4), we obtain

$$\int_{|t|>\omega_n} \left|f^{\mathrm{ft}}(t)\right|^2 \mathrm{d}t \leq 2\int_{|t|>\omega_n} \left|f^{\mathrm{ft}}(t) - f_n^{\mathrm{ft}}(t)\right|^2 \mathrm{d}t + 2\underbrace{\int_{|t|>\omega_n} \left|f_n^{\mathrm{ft}}(t)\right|^2 \mathrm{d}t}_{=0}$$

$$\leq 2\|f^{\mathrm{ft}} - f_n^{\mathrm{ft}}\|_2^2 = 4\pi\|f - f_n\|_2^2 \stackrel{n\to\infty}{\longrightarrow} 0,$$

so that we may conclude $f^{\mathrm{ft}}(t) = 0$ for Lebesgue almost all $|t| > \omega_n$ and, hence, $f = 0$. Also see [65].

Now, let us show the orthonormality of the basis $\{f_j\}_{j\in\mathbb{N}}$. Again, by Parseval's identity, we derive

$$\begin{aligned}
\langle \varphi_{m,k}, \varphi_{m,k'} \rangle &= \frac{1}{2\pi}\int \varphi^{\mathrm{ft}}_{m,k}(t)\overline{\varphi^{\mathrm{ft}}_{m,k'}(t)}\,\mathrm{d}t \\
&= \frac{1}{2^{m+1}\pi}\int \exp\left(\mathrm{i}t(k-k')/2^m\right)\left|\varphi^{\mathrm{ft}}(t/2^m)\right|^2 \mathrm{d}t \\
&= \frac{1}{2\pi}\int \exp\left(\mathrm{i}t(k-k')\right)\left|\varphi^{\mathrm{ft}}(t)\right|^2 \mathrm{d}t \\
&= \frac{1}{2\pi}\int \exp\left(\mathrm{i}t(k-k')\right)P([t-\pi, t+\pi])\,\mathrm{d}t \\
&= \frac{1}{2\pi}\int \exp\left(\mathrm{i}t(k-k')\right)\int \chi_{[-\pi,\pi]}(x-t)\,\mathrm{d}P(x)\,\mathrm{d}t
\end{aligned}$$

$$= \frac{1}{2\pi}\int \exp\left(\mathrm{i}y(k-k')\right)\mathrm{d}P(y)\underbrace{\int_{-\pi}^{\pi} \exp\left(\mathrm{i}t(k-k')\right)\mathrm{d}t}_{=2\pi\delta_{k,k'}}$$

$$= \delta_{k,k'}.$$

Furthermore, completeness of the basis $\{\varphi_{m,k} : k \text{ integer}\}$ shall be shown, that is, any $f \in L_2^{\omega_n}(\mathbb{R})$ shall be approximable in that basis with respect to the $L_2(\mathbb{R})$-norm. As the Fourier transform is a quasi-isomorphism from $L_2(\mathbb{R})$ to itself (see Theorem A.4), that is, the Fourier transform is norm-preventing up to the constant scaling factor $\sqrt{2\pi}$, completeness is equivalent with the fact that the Fourier polynomials $\sqrt{1/(2\pi)}\exp(ij\cdot)$, j integer, are an orthonormal basis of $L_2([-\pi,\pi])$. The latter fact is one of the basic results in Fourier analysis. We show that fact in Theorem A.7.

The MISE – as defined in (2.10) – of the linear wavelet estimator is studied in the following proposition.

Proposition 2.4. *Consider the linear wavelet deconvolution estimator*

$$\hat{f}(x) = \sum_{|k|\leq K_n} \hat{\mu}_k\,\varphi_{m_n,k}(x),$$

where we replace f_j by $\varphi_{m_n,k}$ in (2.13) to define the estimator $\hat{\mu}_k$. Assume that $f \in L_2(\mathbb{R})$ and $g^{\mathrm{ft}}(t) \neq 0$ for all $t \in \mathbb{R}$. Then, we have

$$E\|\hat{f} - f\|_2^2 \leq \frac{4}{3} \cdot \frac{2^{m_n}}{n} \Big(\min_{|t| \leq 4\pi 2^{m_n}/3} \big|g^{\mathrm{ft}}(t)\big|^2 \Big)^{-1} + \sum_{|k| > K_n} \big|\langle f, \varphi_{m_n,k}\rangle\big|^2 + \frac{1}{2\pi} \int_{|t| > 4\pi 2^{m_n}/3} \big|f^{\mathrm{ft}}(t)\big|^2 \, \mathrm{d}t \, .$$

Proof. By Parseval's identity (Theorem A.4), we obtain

$$\begin{aligned} E\|\hat{f} - f\|_2^2 &= E\Big\| \sum_{|k| \leq K_n} \hat{\mu}_k \varphi_{m_n,k} - \sum_k \langle f, \varphi_{m_n,k}\rangle \varphi_{m_n,k} \Big\|_2^2 \\ &\qquad + \frac{1}{2\pi} \int_{|t| > 4\pi 2^{m_n}/3} \big|f^{\mathrm{ft}}(t)\big|^2 \, \mathrm{d}t \\ &= \sum_{|k| \leq K_n} E\big|\hat{\mu}_k - \langle f, \varphi_{m_n,k}\rangle\big|^2 + \sum_{|k| > K_n} \big|\langle f, \varphi_{m_n,k}\rangle\big|^2 \\ &\qquad + \frac{1}{2\pi} \int_{|t| > 4\pi 2^{m_n}/3} \big|f^{\mathrm{ft}}(t)\big|^2 \, \mathrm{d}t \\ &= \sum_{|k| \leq K_n} \operatorname{var} \hat{\mu}_k + \sum_{|k| \leq K_n} \big|E\hat{\mu}_k - \langle f, \varphi_{m_n,k}\rangle\big|^2 \\ &\qquad + \frac{1}{2\pi} \int_{|t| > 4\pi 2^{m_n}/3} \big|f^{\mathrm{ft}}(t)\big|^2 \, \mathrm{d}t + \sum_{|k| > K_n} \big|\langle f, \varphi_{m_n,k}\rangle\big|^2 . \end{aligned}$$

We have used the decomposition

$$f = f^{\omega_n} + f^{\omega_n, r},$$

for $\omega_n = 4\pi 2^{m_n}/3$, where f^{ω_n} is the orthogonal projection of f onto the space $L_2^{\omega_n}(\mathbb{R})$; hence, we have $(f^{\omega_n})^{\mathrm{ft}}(t) = f^{\mathrm{ft}}(t)$ on $t \in [-\omega_n, \omega_n]$. Then, $f^{\omega_n, r}$ denotes the orthogonal complement, whose Fourier transform is supported on $\mathbb{R} \backslash [-\omega_n, \omega_n]$. Also, we have applied that $\{\varphi_{m_n,k} : k \text{ integer}\}$ is an orthonormal basis of $L_2^{\omega_n}(\mathbb{R})$.

Let us consider the expectation and the variance of $\hat{\mu}_k$ by the Plancherel isometry and Parseval's identity and the independence of the data with respect to the variance.

$$\begin{aligned} E\hat{\mu}_k &= \frac{1}{2\pi} \int E \exp(itY_1) \overline{\varphi_{m_n,k}^{\mathrm{ft}}(t)} / g^{\mathrm{ft}}(t) \, \mathrm{d}t \\ &= \frac{1}{2\pi} \int f^{\mathrm{ft}}(t) \overline{\varphi_{m_n,k}^{\mathrm{ft}}(t)} \, \mathrm{d}t \\ &= \langle f, \varphi_{m_n,k}\rangle \, , \end{aligned}$$

$$\begin{aligned}
\operatorname{var} \hat{\mu}_k &\leq \frac{1}{(2\pi)^2 n} E\Big| \int \exp(itY_1)\overline{\varphi^{\mathrm{ft}}_{m_n,k}(t)}/g^{\mathrm{ft}}(t)\, \mathrm{d}t\Big|^2 \\
&= \frac{1}{(2\pi)^2 n} \int \Big| \int \exp(ity)\overline{\varphi^{\mathrm{ft}}_{m_n,k}(t)}/g^{\mathrm{ft}}(t)\, \mathrm{d}t\Big|^2 [f * g](y)\, \mathrm{d}y \\
&= \frac{1}{(2\pi)^2 n} \int \Big| \langle \chi_{[-\omega_n,\omega_n]} \exp(iy\cdot)/g^{\mathrm{ft}}, \varphi^{\mathrm{ft}}_{m_n,k}\rangle\Big|^2 [f * g](y)\, \mathrm{d}y\,.
\end{aligned}$$

Hence, we obtain, with respect to the variance term, that

$$\begin{aligned}
&\sum_{|k|\leq K_n} \operatorname{var} \hat{\mu}_k \\
&\leq \frac{1}{(2\pi)^2 n} \int \sum_{|k|\leq K_n} \Big| \langle \chi_{[-\omega_n,\omega_n]} \exp(iy\cdot)/g^{\mathrm{ft}}, \varphi^{\mathrm{ft}}_{m_n,k}\rangle\Big|^2 [f * g](y)\, \mathrm{d}y \\
&\leq \frac{1}{2\pi n} \iint \big| \chi_{[-\omega_n,\omega_n]}(t) \exp(ity)/g^{\mathrm{ft}}(t)\big|^2 \mathrm{d}t\, [f * g](y)\, \mathrm{d}y \\
&\leq \frac{2^{m_n+2}}{3} \cdot \frac{1}{n} \cdot \Big(\min_{|t|\leq \omega_n} \big|g^{\mathrm{ft}}(t)\big|^2\Big)^{-1}.
\end{aligned}$$

Again, it is essential that $\{\varphi_{m_n,k} : k \text{ integer}\}$ is an orthonormal basis of $L_2^{\omega_n}(\mathbb{R})$. That gives us

$$\sum_{|k|\leq K_n} \big|\langle f, \varphi_{m_n,k}\rangle\big|^2 \leq \|f\|_2^2\,,$$

for all $f \in L_2^{\omega_n}(\mathbb{R})$, which has been used in the above equation. Also, we use the fact that $f * g$ integrates to one. Inserting that result gives us the desired upper bound. ∎

We learn from Proposition 2.4 that the parameter K_n needs no special calibration. We can fix that the larger the K_n the smaller the upper bound of the MISE. The optimal selection $K_n = \infty$ is prohibited by computational practicability only. On the other hand, when the parameter m_n increases one term of the upper bound becomes larger while the other one decreases. Therefore, the selection of m_n, which regularizes the cut-off region in the Fourier domain, must be optimized. Also, 2^{m_n} corresponds to one by the bandwidth b in comparison to the kernel deconvolution estimator (2.7).

The fact that the optimal choice of m_n depends on the smoothness degree of f motivates people to consider the nonlinear wavelet deconvolution estimator, which uses the residual part of the wavelet basis (2.16), in addition. This estimator is mentioned in [107]. It leans on the so-called multiresolution decomposition for wavelets, given by

$$f = \sum_k \langle \varphi_{m,k}, f\rangle\, \varphi_{m,k} + \sum_k \sum_{j=m}^{\infty} \langle \psi_{j,k}, f\rangle\, \psi_{j,k}\,,$$

for all $f \in L_2(\mathbb{R})$. Transferred to our notation and approach, the nonlinear wavelet deconvolution estimator has the shape

$$\hat{f}_{nl}(x) = \sum_{|k| \leq K_n} \hat{\mu}_k \varphi_{m_n,k}(x) + \sum_{j=m_n}^{m_n+r} \Big[\sum_{|k| \leq L_n} \hat{b}_{j,k} \psi_{j,k}(x) \Big] \chi_{(\delta^2_{j,n},\infty)} \Big(\sum_{|k| \leq L_n} |\hat{b}_{j,k}|^2 \Big), \tag{2.17}$$

where

$$\hat{b}_{j,k} = \frac{1}{2\pi n} \sum_{l=1}^{n} \exp(\mathrm{i}tY_l) \overline{\psi_{j,k}(t)} / g^{\mathrm{ft}}(t)\, \mathrm{d}t\,.$$

and the positive parameters $K_n, L_n, \delta_{j,n}, m_n, r$ are still to be chosen. Nevertheless, their selection does not require knowledge of the smoothness degree of f. By the term $\chi_{(\delta^2_{j,n},\infty)}\Big(\sum_{|k|\leq L_n} |\hat{b}_{j,k}|^2\Big)$, the estimation of the coefficients $\hat{b}_{j,k}$ while j runs from m_n to $m_n + r$ cuts off automatically when j gets too large. In that case, the empirical coefficients are put equal to zero. Therefore, the nonlinear part of the estimator may be seen as a testing procedure for the selection of parameter m_n, in the terminology used for the linear estimator. For details, especially with respect to the parameter selection, see the work of [107]. Another strategy to fill the lack of knowledge of the smoothness of f is to employ the linear estimator and choose parameter m_n empirically by penalizing complexity. That approach is studied in [20, 21].

Another approach to wavelet deconvolution is given by [48]. The wavelet estimator is mainly designed to be considered with respect to its MISE as it is constructed based on an orthonormal bases. The consideration of the MSE as in Proposition 2.1 for the kernel estimator is more difficult.

2.2.3 Ridge-Parameter Approach

Obviously, well-definiteness of the kernel method and the wavelet approach requires the assumption that g^{ft} vanishes nowhere in order to avoid division by zero, see (2.7) and (2.13). Therefore, the condition

$$g^{\mathrm{ft}}(t) \neq 0\,, \qquad \forall t \in \mathbb{R}\,, \tag{2.18}$$

has become common in deconvolution topics although there are important densities that do not satisfy (2.18), for instance the uniform density g on the interval $[-1, 1]$ having the Fourier transform $g^{\mathrm{ft}}(t) = (\sin t)/t$. Hence, all integer multiples of π are zeros of g^{ft}.

Instead of multiplying by $K^{\mathrm{ft}}(tb)$ as in kernel estimation, we may regularize the empirical version of $f^{\mathrm{ft}}(t)$ in (2.3) by preventing the denominator from being too close to zero in a direct way. We consider

$$\hat{\Psi}_{X;\mathrm{ridge}}(t) = \frac{1}{n}\sum_{j=1}^{n}\exp(\mathrm{i}tY_j)\frac{g^{\mathrm{ft}}(-t)}{\max\{|g^{\mathrm{ft}}(t)|^2\,,\,\rho_n(t)\}}\,, \tag{2.19}$$

with a continuous and positive-valued so-called ridge function $\rho_n(t)$. Whenever $|g^{\mathrm{ft}}(t)|^2 \geq \rho_n(t)$, we have coincidence of $\hat{\Psi}_{X;\mathrm{ridge}}(t)$ and (2.3). Otherwise, if $|g^{\mathrm{ft}}(t)|$ gets too small, the ridge-regularization becomes efficient. In any case we fix $|g^{\mathrm{ft}}(t)|/\rho_n(t)$ as an upper bound on $|\hat{\Psi}_{X;\mathrm{ridge}}(t)|$ so that square-integrability of (2.19) is guaranteed by appropriate selection of $\rho_n(t)$ and additional kernel regularization can be omitted. When $|g^{\mathrm{ft}}|$ decays very slowly (such as $|t|^{-\nu}$ with $\nu \in (0,1)$), it is useful to introduce a third parameter $r > 0$ in order to regularize, see [58] for details. The ridge function shall increase in $|t|$ at a sufficiently high order (as $|t| \to \infty$) but decrease in n at a sufficiently low order (as $n \to \infty$). That motivates the parametric approach to $\rho_n(t)$,

$$\rho_n(t) = |t|^{\zeta} n^{-\eta}, \tag{2.20}$$

with two nonnegative ridge parameters ζ and η. Regardless of g and its possible zeros, the condition $\zeta > 1/2$ suffices to ensure well-definiteness of the final ridge-parameter deconvolution density estimator

$$\hat{f}(x) = \frac{1}{2\pi}\int \exp(-\mathrm{i}tx)\hat{\Psi}_{X;\mathrm{ridge}}(t)\,\mathrm{d}t\,, \tag{2.21}$$

even if (2.18) is violated by allowing for some isolated zeros of g^{ft}. The ridge-parameter estimator has been introduced in [58] with a slightly different incorporation of the ridge parameters and another smoothing parameter $r > 0$ as indicated earlier. The following proposition gives us an upper bound on the MISE (see (2.10)) of estimator (2.21).

Proposition 2.5. *Select the ridge function according to (2.20). Defining the set*

$$G_n = \{t \in \mathbb{R} : |g^{\mathrm{ft}}(t)|^2 < \rho_n(t)\}\,,$$

the MISE of the estimator (2.21) satisfies

$$MISE(\hat{f}, f) \leq \frac{1}{2\pi\, n}\int_{G_n}\frac{|g^{\mathrm{ft}}(t)|^2}{\rho_n^2(t)}\,\mathrm{d}t + \frac{1}{2\pi\, n}\int_{\mathbb{R}\setminus G_n}|g^{\mathrm{ft}}(t)|^{-2}\,\mathrm{d}t + \frac{1}{2\pi}\int_{G_n}\left|f^{\mathrm{ft}}(t)\right|^2\,\mathrm{d}t\,.$$

Proof. As in the proof of Proposition 2.2, we use Fubini's theorem and Parseval's identity and the variance-bias decomposition of $E\left|\hat{\Psi}_{X;\mathrm{ridge}}(t) - f^{\mathrm{ft}}(t)\right|^2$ to obtain

$$\begin{aligned}
\mathrm{MISE}(\hat f, f) &= \frac{1}{2\pi}\int E\big|\hat\Psi_{X;\mathrm{ridge}}(t) - f^{\mathrm{ft}}(t)\big|^2 \,\mathrm{d}t \\
&= \frac{1}{2\pi\, n}\int \frac{|g^{\mathrm{ft}}(t)|^2}{\max\{|g^{\mathrm{ft}}(t)|^4\,,\,\rho_n^2(t)\}}\,\mathrm{var}\,\exp(itY_j)\,\mathrm{d}t \\
&\qquad + \frac{1}{2\pi}\int\Big|\frac{|g^{\mathrm{ft}}(t)|^2}{\max\{|g^{\mathrm{ft}}(t)|^2\,,\,\rho_n(t)\}} - 1\Big|^2 \big|f^{\mathrm{ft}}(t)\big|^2\,\mathrm{d}t \\
&= \frac{1}{2\pi\, n}\int_{G_n} \frac{|g^{\mathrm{ft}}(t)|^2}{\rho_n^2(t)}\big[1 - |g^{\mathrm{ft}}(t) f^{\mathrm{ft}}(t)|^2\big]\,\mathrm{d}t \\
&\qquad + \frac{1}{2\pi\, n}\int_{\mathbb{R}\setminus G_n} |g^{\mathrm{ft}}(t)|^{-2}\big[1 - |g^{\mathrm{ft}}(t) f^{\mathrm{ft}}(t)|^2\big]\,\mathrm{d}t \\
&\qquad + \tfrac{1}{2\pi}\int_{G_n}\Big|\underbrace{\frac{|g^{\mathrm{ft}}(t)|^2}{\rho_n(t)}}_{\in[0,1]} - 1\Big|^2 \big|f^{\mathrm{ft}}(t)\big|^2 \mathrm{d}t\,.
\end{aligned}$$

Then, $|g^{\mathrm{ft}}(t) f^{\mathrm{ft}}(t)| \le 1$, for all $t \in \mathbb{R}$, leads to the desired upper bound. ■

Finally, we mention that, in the case of uniform g, an approach apart from Fourier techniques is given in the paper of [54]. There, the authors concentrate on estimating the distribution rather than the density function.

2.3 General Consistency

To evaluate the quality of an estimator we may consider the estimation error, that is, the distance between the density estimator $\hat f$ and the true density to be estimated, also referred to as the target density f. The meaning of distance can be made precise in several reasonable ways; unlike in parametric statistics, where the goal is to estimate a real-valued parameter or a finite-dimensional vector with real-valued components and, hence, the Euclidean metric can be used as the standard interpretation of a distance. Therefore, we keep our consideration as general as possible by defining a semi-metric space (F, d), that is, the mapping $d : F \times F \to [0,\infty)$ satisfies all criteria of a metric except positive definiteness, which is replaced by positive semi-definiteness. This means that $d(f, \tilde f) = 0$ does not necessarily imply $f = \tilde f$. Sometimes one is not interested in the density function f as a whole but only in some parts or characteristics, for example, in $f(x)$ at a specific point $x \in \mathbb{R}$. The latter situation is included in our framework by choosing the pointwise semi-metric $\mathrm{d}(f, \tilde f) = \big|f(x) - \tilde f(x)\big|$. As we want to measure the estimation error, we must assume that both the target density f and the estimator $\hat f$ lie in F almost surely. To increase generality we may consider the estimation error weighted by kth power with $k \ge 1$. Large k implies that huge estimation error is more severely penalized than small error. Of course, the smaller the estimation error the better the estimator.

A basic asymptotic criterion of quality for an estimator is consistency. An estimator $\hat{f}$ is called consistent if $\mathrm{d}^k(\hat{f}, f)$ converges to zero as the sample size n tends to infinity for any fixed f (independent of n), which is admitted to be the true density. Therefore, consistency actually refers to an estimation sequence $(\hat{f}_n)_n$ depending on the number of observations rather than to a single estimator. Roughly speaking, the estimation error shall approximately vanish for huge n. According to the different choices of the semi-metric d and the different definitions of convergence of a sequence consisting of random variables in probability theory, different precise definitions of consistency are possible: if $\mathrm{d}^k(\hat{f}, f)$ converges to zero almost surely (a.s.), the estimator $\hat{f}$ is said to be *strongly consistent*; if $\mathrm{d}^k(\hat{f}, f)$ converges to zero in probability, the estimator $\hat{f}$ satisfies *weak consistency*; the case where the so-called risk, that is, the expectation of the weighted estimation error, $E\,\mathrm{d}^k(\hat{f}, f)$, tends to zero is referred to as *mean consistency*. In the previous sections, we have already defined two special kinds of risks, namely the MSE and the MISE (see (2.8) and (2.10), respectively). For the MSE, we put d equal to the pointwise semi-metric in some $x \in \mathbb{R}$; while, for the MISE, we put d equal to the metric induced by the $\|\cdot\|_2$-norm. In both cases, we set $k = 2$. The choice $k = 2$ makes the mathematical investigation of the risk easier, since we are able to use some Hilbert space properties and common bias-variance decomposition. We learn from probability theory that strong consistency implies weak consistency. Also, weak consistency follows from mean consistency by Markov's inequality as

$$P\big[\mathrm{d}^k(\hat{f}, f) \geq \varepsilon\big] \leq \varepsilon^{-1}\, E\,\mathrm{d}^k(\hat{f}, f)\,.$$

However, neither strong nor mean consistency imply each other without any further conditions.

As a necessary condition for consistency, identifiability is required. In the density deconvolution context, identifiability means that

$$f * g = \tilde{f} * g \Longrightarrow \mathrm{d}(f, \tilde{f}) = 0 \tag{2.22}$$

for all $f, \tilde{f} \in F$, which are admitted to be the true density. That can be seen as follows: assume that there exists a weakly consistent estimator $\hat{f}$ for f with respect to some semi-metric d and two densities $f, \tilde{f}$ with $f * g = \tilde{f} * g$. Following from weak consistency, this estimator $\hat{f} = \hat{f}_n(\cdot; Y_1, \ldots, Y_n)$ satisfies, for any $\varepsilon > 0$,

$$\begin{aligned}
0 \overset{n\to\infty}{\longleftarrow}\; & P_{f*g}\big[\mathrm{d}^k(\hat{f}_n(\cdot; Y_1, \ldots, Y_n), f) \geq \varepsilon\big] + P_{\tilde{f}*g}\big[\mathrm{d}^k(\hat{f}_n(\cdot; Y_1, \ldots, Y_n), \tilde{f}) \geq \varepsilon\big] \\
& \geq P_{f*g}\big[\mathrm{d}(\hat{f}_n(\cdot; Y_1, \ldots, Y_n), f) \geq \varepsilon^{1/k} \vee \mathrm{d}(\hat{f}_n(\cdot; Y_1, \ldots, Y_n), \tilde{f}) \geq \varepsilon^{1/k}\big] \\
& \geq P_{f*g}\big[\mathrm{d}(\hat{f}_n(\cdot; Y_1, \ldots, Y_n), f) + \mathrm{d}(\hat{f}_n(\cdot; Y_1, \ldots, Y_n), \tilde{f}) \geq 2\varepsilon^{1/k}\big] \\
& \geq P_{f*g}\big[\mathrm{d}(f, \tilde{f}) \geq 2\varepsilon^{1/k}\big] = \chi_{[2\varepsilon^{1/k}, \infty)}\big(\mathrm{d}(f, \tilde{f})\big) \geq 0\,,
\end{aligned}$$

where P_h denotes the probability measure under the assumption that any Y_j possesses the density h so that the measures P_{f*g} and $P_{\tilde{f}*g}$ are identical;

further, the triangle inequality for the semi-metric d has been used. The term in the last line of the above inequality does not depend on n, from what follows that $\mathrm{d}(f,\tilde{f}) < 2\varepsilon^{1/k}$ for any $\varepsilon > 0$ and, hence, $\mathrm{d}(f,\tilde{f}) = 0$ so that (2.22) is verified.

Apart from the squared pointwise semi-metric and the $L_2(\mathbb{R})$-metric leading to the MSE and the MISE, respectively, another popular metric used for d is the $L_1(\mathbb{R})$-metric,

$$\mathrm{d}(f,\tilde{f}) = \int \big|f(x) - \tilde{f}(x)\big|\,\mathrm{d}x.$$

As an advantage of that metric, we point out that any density may be considered. Recall that densities are defined as all $L_1(\mathbb{R})$-functions having 1 as their $L_1(\mathbb{R})$-norm, denoted by $\|\cdot\|_1$, and taking nonnegative values almost everywhere in the Lebesgue sense. Then, $\mathrm{d}(f,\tilde{f}) = 0$ implies coincidence of f and $\tilde{f}$ almost everywhere so that both densities induce the same probability measure. For a comprehensive overview of $L_1(\mathbb{R})$-consistency in standard density estimation with incorrupted data, see the book of [35]. First we check (2.22) for the $L_1(\mathbb{R})$-metric. We utilize the result from probability theory that any probability measure and, hence, its density function, whenever existing, are uniquely determined by its characteristic function. Therefore, for d equal to the $L_1(\mathbb{R})$-metric, (2.22) is equivalent with

$$f^{\mathrm{ft}}(t)\cdot g^{\mathrm{ft}}(t) = \tilde{f}^{\mathrm{ft}}(t)\cdot g^{\mathrm{ft}}(t),\ \text{for all } t\in\mathbb{R} \Longrightarrow f^{\mathrm{ft}}(t) = \tilde{f}^{\mathrm{ft}}(t),\ \text{for all } t\in\mathbb{R}\,. \tag{2.23}$$

Introducing the set $N_g = \big\{t\in\mathbb{R} : g^{\mathrm{ft}}(t) = 0\big\}$, we derive that (2.23) holds if and only if N_g does not contain any open, nonempty interval as a subset. That is due to the following consideration: Simple division by $g^{\mathrm{ft}}(t)$ makes $f^{\mathrm{ft}}(t)$ identifiable for any $t \notin N_g$. The above condition implies that $\mathbb{R}\backslash N_g$ lies dense in $\mathbb{R}$, that is, the closure of $\mathbb{R}\backslash N_g$ equals $\mathbb{R}$. As the function f^{ft} is continuous due to Lemma A.1(d), it is uniquely determined by its restriction to the dense domain $\mathbb{R}\backslash N_g$.

On the other hand, if N_g contains some open, nonvoid interval (a,b), we construct the densities $f(x) = \frac{b-a}{2} f_0\big((b-a)x/2\big)$ with the density

$$f_0(x) = \frac{1-\cos(x)}{\pi x^2} \tag{2.24}$$

having the triangle-shaped Fourier transform $f^{\mathrm{ft}}(t) = (1-|t|)_+$. That claim can easily be verified by calculating the inverse Fourier transform of f^{ft} explicitly by integration by parts and using Theorem A.2. Furthermore, we specify

$$\tilde{f}(x) = f(x)\cdot\big[1+\cos\big((a+b)t/2\big)\big],$$

whose Fourier transform is equal to

$$\tilde{f}^{\mathrm{ft}}(t) = f^{\mathrm{ft}}(t) + \frac{1}{2} f^{\mathrm{ft}}(t-(a+b)/2) + \frac{1}{2} f^{\mathrm{ft}}(t+(a+b)/2),$$

where Lemma A.1 has been applied from the appendix chapter. To check that $\tilde{f}$ is a density, we derive nonnegativity and integrability by its definition and the fact that $|\cos x| \leq 1$ for all $x \in \mathbb{R}$. Since f^{ft} is supported on $[-(b-a)/2, (b-a)/2]$, we have coincidence of f^{ft} and $\tilde{f}^{\mathrm{ft}}$ on $\mathbb{R}\backslash([-b,-a] \cup [a,b])$ and, hence, $\tilde{f}^{\mathrm{ft}}(0) = f^{\mathrm{ft}}(0) = 1$ so that $\tilde{f}$ integrates to one. In particular, zero is not contained in $[a,b]$ because $g^{\mathrm{ft}}(0) = 1$ and g^{ft} is continuous. As $g^{\mathrm{ft}}(-t) = \overline{g^{\mathrm{ft}}(t)} = 0$ for all $t \in N_g$, it follows that $f^{\mathrm{ft}}(t)g^{\mathrm{ft}}(t) = \tilde{f}^{\mathrm{ft}}(t)g^{\mathrm{ft}}(t)$ for all t while $f^{\mathrm{ft}}(t) \neq \tilde{f}^{\mathrm{ft}}(t)$ on $t \in (a,b)$. Therefore, (2.22) is violated. The latter implication has been shown in [34] for some different types of densities. Thus, we have proved that there exists no consistent estimator of f in the additive measurement error model (2.1) whenever some open, nonvoid set is subset of N_g. Furthermore, the negation of that latter condition is equivalent with identifiability with respect to the $L_1(\mathbb{R})$-metric.

Those results are no longer valid when f is assumed to be compactly supported, see [50] in the circular deconvolution model when estimating the density of some angle and [92] for general compactly supported models. The idea behind those approaches is the fact that the Fourier transform f^{ft} is analytic (i.e., pointwise representable by its Taylor series on the whole real line) if f has compact support. That result can be used to define some continuation of the deconvolution estimator to N_g by Bessel functions or Legendre polynomials in the Fourier domain.

The question whether identifiability is sufficient for the existence of a consistent estimator is difficult to address. Meister [88] shows that a mean consistent estimator exists as long as (2.22) is valid; nevertheless, a less conventional weaker metric is applied compared to the $L_1(\mathbb{R})$-distance. Devroye [34] proved mean consistency with respect to the $L_1(\mathbb{R})$-metric under the slightly stronger condition that N_g has Lebesgue measure zero. In spite of the existence of some densities where N_g does not contain any open, nonvoid set but has positive Lebesgue measure (see [88]), those densities may be seen as rather pathological and not important for practical applications. Therefore, we restrict our consideration to error densities where the Lebesgue measure of N_g is equal to zero. On the other hand, we derive strong consistency, which has rarely been studied in deconvolution topics.

First, we face the problem that neither f nor g may be assumed to be in $L_2(\mathbb{R})$. Therefore, we manipulate the data by some additional independent contamination. More concretely, we generate some data $\varepsilon'_{1,n}, \ldots, \varepsilon'_{n,n}$, independently of the given observations in (2.1), with some density $\delta_n^{-1}\varphi(\cdot/\delta_n)$, $\delta_n > 0$, where we may insert any fixed density $\varphi \in L_2(\mathbb{R})$. Then, we add some further noise to the data; leading to the dataset $Y'_{1,n}, \ldots, Y'_{n,n}$, where

$$Y_{j,n} = Y_j + \varepsilon'_{j,n} = X_j + \varepsilon'_{j,n} + \varepsilon_j \,.$$

Then, the $Y'_{j,n}$ have the density function $f * g * \big(\delta_n^{-1}\varphi(\delta_n^{-1}\cdot)\big)$ where

$$\|f * g * \big(\delta_n^{-1}\varphi(\delta_n^{-1}\cdot)\big)\|_2^2 = \int \Big| \int \delta_n^{-1}\varphi(\delta_n^{-1}(x-y))\,\big[f*g\big](y)\,\mathrm{d}y\Big|^2 \mathrm{d}x$$
$$\leq \int \int \delta_n^{-2}\varphi^2(\delta_n^{-1}(x-y))\,\big[f*g\big](y)\,\mathrm{d}y\,\mathrm{d}x$$
$$= \int \delta_n^{-2}\varphi^2(\delta_n^{-1}x)\,\mathrm{d}x \cdot \|f*g\|_1 = \delta_n^{-1}\|\varphi\|_2^2\,,$$

so that $f * g * \big(\delta_n^{-1}\varphi(\delta_n^{-1}\cdot)\big) \in L_2(\mathbb{R})$. We have used the standard inequality $E|Z|^2 \geq |EZ|^2$ for any random variable Z in the above equation. Apparently, the characteristic function of $Y'_{j,n}$ is equal to $f^{\mathrm{ft}}g^{\mathrm{ft}}\varphi^{\mathrm{ft}}(\delta_n\cdot)$ so that we shall approximate the density of the observation Y_j by putting $\delta_n \to 0$. We define the estimator $\hat{f}_{X,2}$ by

$$\hat{f}_{X,2}(x;Y_1,\ldots,Y_n) = \hat{f}(x;Y'_{1,n},\ldots,Y'_{n,n}),$$

where $\hat{f}$ denotes the ridge-parameter estimator (2.21). We choose this type of deconvolution estimator because it is able to handle the case where g^{ft} has some zeros, and we are still focussed on the most general case in this section. Furthermore, $\hat{f}_{X,2}$ may be well-defined as an $L_2(\mathbb{R})$-function; however, we are not guaranteed that it lies in $L_1(\mathbb{R})$. Therefore, it is truncated, leading to the final estimator

$$\hat{f}_{X,3} = \hat{f}_{X,2} \cdot \chi_{[-R_n,R_n]}, \tag{2.25}$$

with some positive-valued parameter sequence $(R_n)_n$. As both $\hat{f}_X$ and $\chi_{[-R_n,R_n]}$ are included in $L_2(\mathbb{R})$ we may fix that $\hat{f}_{X,3} \in L_1(\mathbb{R})$.

We give the following theorem.

Theorem 2.6. (strong consistency) *Consider the additive measurement error model (2.1) and the randomized deconvolution density estimator (2.25). Assume that g^{ft} vanishes almost nowhere in the Lebesgue sense. Choose $\varphi(x) = 1/[\pi(1+x^2)]$ so that $\varphi^{\mathrm{ft}}(t) = \exp(-|t|)$ in the construction of estimator (2.25). Then, the estimator (2.25) satisfies*

$$\|\hat{f}_{X,3} - f\|_1 \overset{n\to\infty}{\longrightarrow} 0 \quad a.s.,$$

for any density f under the parameter selection

(i) $\eta > 0\,,\ \zeta > 1/2\,,$
(ii) $\delta_n \downarrow 0\,,$
(iii) $\delta_n^{-5}\int \chi_{G_n}(t)(1+|t|)^{-2}dt \overset{n\to\infty}{\longrightarrow} 0\,,$
(iv) $\delta_n^{-1}n^{2\eta-1} = O\big(n^{-\xi}\big)\,,$ *for some* $\xi > 0\,,$
(v) $R_n = \delta_n^{-1}\,.$

Therein, G_n is defined in Proposition 2.5. Note that $\chi_{G_n}(t) = 1$ implies $|t|^\zeta n^{-\eta} \geq |g^{\mathrm{ft}}(t)|^2$. By assumption, we have $|g^{\mathrm{ft}}(t)| > 0$ for almost all $t \in \mathbb{R}$. Therefore, we have the pointwise limit

$$\lim_{n\to\infty} \chi_{G_n}(t) = 0$$

for almost all t. By dominated convergence, we obtain that

$$\int \chi_{G_n}(t)(1+|t|)^{-2}\,\mathrm{d}t \stackrel{n\to\infty}{\longrightarrow} 0\,.$$

Also, note that this integral is known as it only depends on the known error density g and the ridge parameters η and ζ. Therefore, one should start the parameter selection with η and ζ so that (i) is satisfied. Then, select δ_n so that (ii), (iii), and (iv) are fulfilled. We mention that $(\delta_n)_n$ shall converge to zero in view of (ii); while (iii) and (iv) restrict the speed of that convergence. Then, choose $(R_n)_n$ according to (v). We mention that one might be able to relax the conditions for choosing the parameters. Nevertheless, we focus on the fact that strong consistency is attainable under general assumptions on f rather than minimal conditions for the parameter choice.

Proof. We derive that

$$\begin{aligned}
\|\hat{f}_{X,3} - f\|_1 \leq & \int\limits_{|x|>R_n} f(x)\,\mathrm{d}x + \int\limits_{|x|\leq R_n} \big|\hat{f}_{X,2}(x) - \big(f * [\delta_n^{-1}\varphi(\delta_n^{-1}\cdot)]\big)(x)\big|\,\mathrm{d}x \\
& \qquad + \big\|f - \big(f * [\delta_n^{-1}\varphi(\delta_n^{-1}\cdot)]\big)\big\|_1 \\
\leq & \int_{|x|>R_n} f(x)\,\mathrm{d}x + (2R_n)^{1/2} \cdot \big\|\hat{f}_{X,2} - \big(f * [\delta_n^{-1}\varphi(\delta_n^{-1}\cdot)]\big)\big\|_2 \\
& \qquad + \big\|f - \big(f * [\delta_n^{-1}\varphi(\delta_n^{-1}\cdot)]\big)\big\|_1\,,
\end{aligned} \tag{2.26}$$

by the triangle inequality with respect to $\|\cdot\|_1$ and the Cauchy–Schwarz inequality. As $(R_n)_n \uparrow \infty$ by (v), we may fix that the first (deterministic) term

$$\int_{|x|>R_n} f(x)\,\mathrm{d}x \to 0$$

for any density f by dominated convergence. Note that any deterministic sequence converging to zero in the sense of elementary analysis may be viewed as a sequence of degenerated random variables, which converges to zero almost surely. In that spirit, we may also treat the third term. There we use the fact that the set of all continuous functions lies dense in $L_1(\mathbb{R})$, that is, for any $\varepsilon > 0$ and $f \in L_1(\mathbb{R})$, there is some continuous $\tilde{f} \in L_1(\mathbb{R})$ with $\|\tilde{f} - f\|_1 < \varepsilon$. To prove that result, we use Lemma A.3 from the appendix, implying that the continuous and bounded functions in $L_1(\mathbb{R})$ are dense in $L_2(\mathbb{R})$ with respect to $\|\cdot\|_2$. For any $\varepsilon > 0$, we first pick some $S > 0$ so that $\int_{|x|>S} f(x)\,dx \leq \varepsilon/3$. Then, we fix some continuous and bounded $\overline{f} \in L_1(\mathbb{R})$ with $\|f - \overline{f}\|_2 \leq \varepsilon/(3\sqrt{2S})$ and, then, some $R > 0$ satisfying

$$\int_{S<|x|<S+R} \big(|f| + |\overline{f}|\big)\,dx \leq \varepsilon/3\,.$$

Using the triangle inequality in $L_1(\mathbb{R})$ and the Cauchy–Schwarz inequality, we derive that $\|f - \tilde{f}\|_1 \leq \varepsilon$ when putting $\tilde{f} = \overline{f} \cdot q$ where the function $q(x)$ is equal to 1 on $x \in [-S, S]$, equal to 0 on $x \notin [-R-S, R+S]$, and equal to the linear connection on $|x| \in [R, R+S]$ so that q is continuous on the whole real line. Obviously, $\tilde{f}$ is continuous and integrable on the whole real line. Equipped with that result, we obtain that

$$\begin{aligned}\big\|f - \big(f * [\delta_n^{-1}\varphi(\delta_n^{-1}\cdot)]\big)\big\|_1 &\leq \|f - \tilde{f}\|_1 + \|(f - \tilde{f}) * [\delta_n^{-1}\varphi(\delta_n^{-1}\cdot)]\|_1 \\ &\quad + \|\tilde{f} - \big(\tilde{f} * [\delta_n^{-1}\varphi(\delta_n^{-1}\cdot)]\big)\|_1 \\ &\leq 2\|f - \tilde{f}\|_1 + \iint \big|\tilde{f}(x - \delta_n y) - \tilde{f}(x)\big|\,dx\,|\varphi(y)|\,dy.\end{aligned}$$

As shown earlier, we may select some continuous $\tilde{f} \in L_1(\mathbb{R})$ so that the first term above becomes smaller than any positive number. Therefore, we have to prove that the second term tends to zero as $n \to \infty$. The inner integral in the second term is bounded above by $2\|\tilde{f}\|_1$. Hence, we may fix $2\|\tilde{f}\|_1\,|\varphi(\cdot)|$ as an integrable upper bound on the outer integral so that the convergence of the third term in (2.26) to zero follows by dominated convergence if we can show that the inner integral converges to zero pointwise for any $y \in \mathbb{R}$. For that purpose, we consider

$$\begin{aligned}\int \big|\tilde{f}(x - \delta_n y) - \tilde{f}(x)\big|\,dx &\leq \int_{|x|\leq S} \big|\tilde{f}(x - \delta_n y) - \tilde{f}(x)\big|\,dx \\ &\quad + 2\int_{|x|>S-\sup_m \delta_m|y|} \big|\tilde{f}(x)\big|\,dx\,,\end{aligned}$$

for any $S > 0$. Again, assume some arbitrary $\varepsilon > 0$. First we choose S large enough so that the latter addend is bounded above by $\varepsilon/2$. Then, using the continuity of $\tilde{f}$ and (ii), the first addend tends to zero by dominated convergence and, hence, has $\varepsilon/2$ as an upper bound for n sufficiently large. Also, note that continuity of $\tilde{f}$ implies boundedness of this function on $[-S, S]$. Thus, the third term in (2.26) converges to zero (almost surely).

It remains to be shown that the second term in (2.26) converges to zero almost surely. Applying Parseval's identity (Theorem A.4) and bias-variance decomposition, we obtain that

$$R_n \cdot \|\hat f_{X,2} - \big(f * [\delta_n^{-1}\varphi(\delta_n^{-1}\cdot)]\big)\|_2^2 = \frac{R_n}{2\pi}\,\|\hat f_{X,2}^{\mathrm{ft}} - f^{\mathrm{ft}}\varphi^{\mathrm{ft}}(\delta_n\cdot)\|_2^2$$
$$\le \frac{R_n}{\pi}\int \Big|\frac{g^{\mathrm{ft}}(-t)}{\max\{\rho_n(t)\,,\,|g^{\mathrm{ft}}(t)|^2\}}\Big|^2 \Big|\frac{1}{n}\sum_{j=1}^n \big(\exp(\mathrm{i}tY'_{j,n}) - E\,\exp(\mathrm{i}tY'_{j,n})\big)\Big|^2\,\mathrm{d}t$$
$$+\ \frac{R_n}{\pi}\int \big|f^{\mathrm{ft}}(t)\varphi^{\mathrm{ft}}(\delta_n t)\big|^2\Big|\frac{|g^{\mathrm{ft}}(t)|^2}{\max\{\rho_n(t)\,,\,|g^{\mathrm{ft}}(t)|^2\}} - 1\Big|^2\,\mathrm{d}t\,. \tag{2.27}$$

We realize that the second (deterministic) term above has the upper bound

$$O(R_n)\cdot \int_{G_n} \big|\varphi^{\mathrm{ft}}(\delta_n t)\big|^2\Big|\frac{|g^{\mathrm{ft}}(t)|^2}{\rho_n(t)} - 1\Big|^2\,\mathrm{d}t \le O(R_n)\cdot\int_{G_n}\big|\varphi^{\mathrm{ft}}(\delta_n t)\big|^2\,\mathrm{d}t$$
$$\le O(R_n)\cdot\int_{G_n\cap[-\delta_n^{-2},\delta_n^{-2}]}\big|\varphi^{\mathrm{ft}}(\delta_n t)\big|^2\,\mathrm{d}t + O(R_n)\cdot\int_{G_n\backslash[-\delta_n^{-2},\delta_n^{-2}]}\big|\varphi^{\mathrm{ft}}(\delta_n t)\big|^2\,\mathrm{d}t$$
$$\le O(R_n)\cdot\int_{G_n\cap[-\delta_n^{-2},\delta_n^{-2}]}\big|\varphi^{\mathrm{ft}}(\delta_n t)\big|^2\,\mathrm{d}t + O\big(R_n/\delta_n\big)\cdot\int_{\mathbb{R}\backslash[-\delta_n^{-1},\delta_n^{-1}]}\big|\varphi^{\mathrm{ft}}(t)\big|^2\,\mathrm{d}t$$
$$\le O\big(\delta_n^{-1}\big)\cdot\int_{G_n\cap[-\delta_n^{-2},\delta_n^{-2}]}\big|\varphi^{\mathrm{ft}}(\delta_n t)\big|^2\,\mathrm{d}t + O\big(\delta_n^{-2}\exp(-2\delta_n^{-1})\big)\,,$$

where condition (v) and the specific selection of φ, according to the theorem, have been inserted in the last step. Following from (ii), the latter addend tends to zero. The first addend is bounded above by

$$O\big(\delta_n^{-1}\big)\cdot\int_{G_n\cap[-\delta_n^{-2},\delta_n^{-2}]}(1+|t|)^2(1+|t|)^{-2}\big|\varphi^{\mathrm{ft}}(\delta_n t)\big|^2\,\mathrm{d}t$$
$$\le O\big(\delta_n^{-1}\big)\underbrace{\Big(\sup_{t\in[-\delta_n^{-2},\delta_n^{-2}]}(1+|t|)^2\big|\varphi^{\mathrm{ft}}(\delta_n t)\big|^2\Big)}_{=O(\delta_n^{-4})}\cdot\int\chi_{G_n}(t)(1+|t|)^{-2}\,\mathrm{d}t\,.$$

Then, we have shown that the second term in (2.27) converges to zero (almost surely) as a direct consequence of (iii).

Focussing on the first term in (2.27), we introduce the notation

$$\lambda_n(t) = \Big|\frac{g^{\mathrm{ft}}(-t)}{\max\{\rho_n(t), |g^{\mathrm{ft}}(t)|^2\}}\Big|^2,$$

and $\psi_j(t) = \exp(\mathrm{i}tY'_{j,n}) - E\,\exp(itY'_{j,n})$ so that this term satisfies

$$T_{1,n} = O(\delta_n^{-1})\cdot\frac{1}{n^2}\sum_{j_1,j_2=1}^n\int\lambda_n(t)\psi_{j_1}(t)\psi_{j_2}(-t)\,\mathrm{d}t\,.$$

For any $\varepsilon > 0$, let us consider the probability

$$P\big[|T_{1,n}|\ge\varepsilon\big] = P\big[|T_{1,n}|^s\ge\varepsilon^s\big] \le \varepsilon^{-s}\,E\big|T_{1,n}\big|^s\,, \tag{2.28}$$

by Markov's inequality, for any integer $s > 1$. Then, we study this expectation.

$$E\big|T_{1,n}\big|^s = O\big(\delta_n^{-s} n^{-2s}\big) \cdot \sum_{j_1,j_2=1}^{n} \cdots \sum_{j_{2s-1},j_{2s}=1}^{n} \int \cdots \int \prod_{k=1}^{s} \lambda_n(t_k) \cdot E \prod_{k=1}^{s} \psi_{j_{2k-1}}(t_k) \psi_{j_{2k}}(-t_k)\, \mathrm{d}t_1 \cdots \mathrm{d}t_s\,.$$

Whenever the set $\{j_1, \ldots, j_{2s}\}$ contains more than s different elements there exists at least one j_l so that $j_{l'} \neq j_l$ for all $l' \in \{1, \ldots, 2s\} \backslash \{l\}$. In this case, the random variable ψ_{j_l} is independent of the σ-algebra generated by the $\psi_{j'_l}$, $l' \neq l$, so that we have

$$E \prod_{k=1}^{s} \psi_{j_{2k-1}}(t_k) \psi_{j_{2k}}(-t_k) = \underbrace{E \psi_{j_l}(t_{\lceil l/2 \rceil})}_{=0} \cdot E \prod_{k=1 \wedge 2k-1 \neq l}^{s} \psi_{j_{2k-1}}(t_k) \cdot \prod_{k=1 \wedge 2k \neq l}^{s} \psi_{j_{2k}}(-t_k) = 0\,.$$

In the countercase, that is, $\#\{j_1, \ldots, j_{2s}\} \leq s$, we apply the rough upper bound

$$\Big| E \prod_{k=1}^{s} \psi_{j_{2k-1}}(t_k) \psi_{j_{2k}}(-t_k) \Big| \leq E \prod_{k=1}^{s} \big| \psi_{j_{2k-1}}(t_k) \psi_{j_{2k}}(-t_k) \big| \leq 4^s\,,$$

as $E|\psi_j(t)| \leq E\big|\exp(itY'_{j,n})\big| + \big|E \exp(itY'_{j,n})\big| \leq 2$. Therefore, we conclude that

$$E\big|T_{1,n}\big|^s \leq O\big(\delta_n^{-s} n^{-2s}\big) \sum_{\underline{j} \in J_n} \int \cdots \int \prod_{k=1}^{s} |\lambda_n(t_k)|\, \mathrm{d}t_1 \cdots \mathrm{d}t_s\,,$$

where the set J_n collects all vectors $(j_1, \ldots, j_{2s})$ satisfying $\#\{j_1, \ldots, j_{2s}\} \leq s$. By elementary results from combinatorics, we derive that $\#J_n \leq O\big(n^s\big)$. This leads to

$$E\big|T_{1,n}\big|^s \leq O\big(\delta_n^{-s} n^{-s}\big) \Big(\int |\lambda_n(t)|\, \mathrm{d}t \Big)^s\,.$$

We consider that

$$\int |\lambda_n(t)|\, \mathrm{d}t \leq \int \min\{|g^{\mathrm{ft}}(t)|^{-2}, n^{2\eta}|t|^{-2\zeta}\}\, \mathrm{d}t = O(1) + O\big(n^{2\eta}\big)\,,$$

where we have used $\zeta > 1/2$ and the existence of some $w > 0$ so that $\inf_{[-w,w]} |g^{\mathrm{ft}}(t)| > 0$. The latter fact holds true since $g^{\mathrm{ft}}(0) = 1$ and g^{ft} is continuous on the whole real line (see Lemma A.1, appendix). Summarizing, we obtain that

$$E\big|T_{1,n}\big|^s \leq O\big(\big[\delta_n^{-1} n^{2\eta-1}\big]^s\big) .$$

The condition (iv) gives us

$$E\big|T_{1,n}\big|^s \leq O(n^{-2}),$$

when selecting $s > 2/\xi$, where the choice of s has still been open so far. Then, by (2.28), we have for any $\varepsilon > 0$,

$$\sum_{n=1}^{\infty} P\big[|T_{1,n}| > \varepsilon\big] \leq O\big(\varepsilon^{-s}\big) \sum_{n=1}^{\infty} n^{-2} < \infty .$$

It follows from there by the Borel–Cantelli Lemma from probability theory that $(T_{1,n})_n$ converges to zero almost surely as $n \to \infty$. Therefore, the upper bound in (2.27) tends to zero almost surely, what finally completes the proof. ■

Thus, we have established the most basic asymptotic property of an estimator, namely consistency under quite general conditions. In the following, we focus on more detailed asymptotic characteristics under stronger assumptions.

2.4 Optimal Convergence Rates

In the previous section, we studied consistency, that is, the fact that the distance between the estimator and the true density converges to zero in some stochastic sense. To evaluate the quality of the estimator more precisely, we are not only interested in sole consistency but also in how fast this convergence takes place; therefore, we study the convergence rates, which shall be as fast as possible, of course. Furthermore, we consider uniform consistency rather than its individual version as in the previous section. This means that we consider the convergence rates of

$$\sup_{f \in \mathcal{F}} E \, \mathrm{d}^k(\hat{f}, f) ,$$

where $\mathcal{F}$ is some density class consisting of all densities which are admitted to be the true one. Deterministic a-priori knowledge about the target density can be included in the definition of $\mathcal{F}$. In parametric estimation problems, $\mathcal{F}$ can be described by finitely many real-valued parameters, for instance, the family of the normal densities $\mathcal{F} = \big\{N(\mu, \sigma^2) \, : \, \mu \in \mathbb{R}, \sigma > 0\big\}$. On the other hand, in nonparametric problems, $\mathcal{F}$ does not have such an easy shape. When considering uniform consistency, we must appreciate that $f \in \mathcal{F}$ may change in n, which tends to infinity. Hence, individual consistency, that is, convergence for some fixed $f \in \mathcal{F}$ independently of n, does not necessarily imply uniform consistency.

2.4.1 Smoothness Classes/Types of Error Densities

We focus on the MSE and the MISE again, see (2.8) and (2.10) for their definition; hence, we choose $k = 2$ and d equal to the pointwise semi-metric at some specific $x \in \mathbb{R}$ and the $L_2(\mathbb{R})$-distance, respectively. Then, the convergence rates $O(n^{-1})$ are typical for parametric estimation problems. Only in few models, one may obtain faster rates as in compactly supported models; for example, when estimating the uniform density on $[0, \theta]$ with unknown $\theta > 0$, one can achieve the rate $O(n^{-2})$. However, in nonparametric problems of curve estimation, one cannot derive any rate if nothing else is assumed than the fact that f is a density (also see [35]). All that holds true for the error-free case. Therefore, in the case of corrupted data, the situation may become even worse. On the other hand, under the assumption of smoothness conditions on f, which prevent the density function from oscillating too intensively, we are able to establish convergence rates.

The probably most common smoothness conditions are of Hölder-type. First we may assume that f is β-fold continuously differentiable in a neighborhood of x and the βth derivative of f is locally bounded around some $x \in \mathbb{R}$, that is,

$$\left|f^{(\beta)}(y)\right| \leq C, \text{ for all } y \in [x - \delta, x + \delta],$$

where $C, \delta > 0$. Then, β is called the smoothness degree of f. To extend the definition of the smoothness degree to noninteger $\beta > 0$, we may assume the Hölder condition

$$\left|f^{(\lfloor\beta\rfloor)}(y) - f^{(\lfloor\beta\rfloor)}(\tilde{y})\right| \leq C\,|y - \tilde{y}|^{\beta - \lfloor\beta\rfloor}, \text{ for all } y, \tilde{y} \in [x - \delta, x + \delta]. \quad (2.29)$$

Then we can define the Hölder class $\mathcal{F} = \mathcal{F}_{\beta,C,\delta;x}$ consisting of all densities satisfying (2.29) and $|f(x)| \leq C$ for all $x \in \mathbb{R}$. We show that uniform local bounds are also attainable for the lower order derivatives of f when assuming $f \in \mathcal{F}_{\beta,C,\delta;x}$. Those conditions could also be included in the definition of the Hölder classes; however, for the sake of generality, we prove that they follow from the current definition.

Lemma 2.7. *(a) Assume that a function $f : \mathbb{R} \to \mathbb{R}$ satisfies condition (2.29) and is bounded above by C_0 on its restriction to the interval $[x-\delta, x+\delta]$. Then, we have*

$$\left|f^{(j)}(y)\right| \leq C_j\,, \textit{ for all } y \in [x - \delta, x + \delta]\,, j = 0, \ldots, \lfloor\beta\rfloor,$$

where the C_j depend on C_0, C, β, δ only.

Proof. First, split $I = [x - \delta, x + \delta]$ into $3^{\lfloor\beta\rfloor}$ disjoint intervals I_k, which have the length $2 \times 3^{-\lfloor\beta\rfloor}\delta$ and cover I completely. By assumption, f has the upper bound C_0 on $I_1, I_3, I_5, \ldots, I_{3\lfloor\beta\rfloor}$. It follows from elementary analysis that there exist some $x_1, x_2, \ldots$ contained in $I_1 \cup I_2 \cup I_3, I_7 \cup I_8 \cup I_9, \ldots, I_{3\text{-}\lfloor\beta\rfloor-2} \cup I_{3\text{-}\lfloor\beta\rfloor-1} \cup I_{3\text{-}\lfloor\beta\rfloor}$, respectively, so that $|f'(x_k)| \leq D_1$ for some

D_1 only depending on C_0, δ, β. As the distances between those unified intervals are also bounded away from zero, we can fix the existence of some $x_1' \in I_1 \cup \cdots \cup I_9, x_2' \in I_{19} \cup \cdots \cup I_{27}, \ldots$, with $|f''(x_k')| \leq D_2$ for some D_2, accordingly; etc.. Hence, by induction, we may derive that, for any $j = 0, \ldots, \lfloor \beta \rfloor$, we have some fixed D_j and some $y_j \in I$ so that $|f^{(j)}(y_j)| \leq D_j$. Now assume a sequence $(f_n)_n$ of functions satisfying the conditions imposed on f in the lemma and some sequence $(z_n)_n \subseteq I$ so that $|f_n^{(j)}(z_n)|$ diverges to infinity as $n \to \infty$ and some fixed $j \in \{0, \ldots, \lfloor \beta \rfloor - 1\}$. Considering the existence of some $y_n \in I$ so that $|f_n^{(j)}(y_n)| \leq D_j$, we conclude that there exist some $w_n \in I$ so that $|f_n^{(j+1)}(w_n)|$ tends to infinity, such that the same situation occurs for the stage $j+1$ as for the jth derivative. The boundedness of I has been used in this step. Again, by induction, we obtain that situation for $j = \lfloor \beta \rfloor$ so that we receive contradiction to the Hölder condition (2.29). Therefore, we conclude uniform boundedness of all the first jth order derivatives for any function satisfying the conditions imposed on f. ■

We give the remark that the condition of boundedness of f on the interval $[x - \delta, x + \delta]$ can be removed if f is a density function. However, as we need boundedness of f on the whole real line anyway, we will not study this relaxation.

When considering the MISE rather than the MSE, Hölder constraints seem inappropriate as smoothness is imposed locally around some specific x. Instead, we assume the existence of a uniform upper bound on the $L_2(\mathbb{R})$-norms of the first β derivatives of f. This is equivalent with (A.8), see the appendix for more details on that. Therefore, we assume the so-called Sobolev condition

$$\int \left|f^{\mathrm{ft}}(t)\right|^2 |t|^{2\beta}\, \mathrm{d}t \leq C, \tag{2.30}$$

which is easily extendable to noninteger smoothness degree β. All densities satisfying (2.30) are collected in the Sobolev class $\mathcal{F} = \mathcal{F}_{\beta,C;L_2}$. We notice that (2.30) implies

$$\begin{aligned} \|f\|_2^2 &= \frac{1}{2\pi} \int \left|f^{\mathrm{ft}}(t)\right|^2 \mathrm{d}t \\ &= \frac{1}{2\pi} \int_{-1}^{1} \underbrace{\left|f^{\mathrm{ft}}(t)\right|^2}_{\leq 1} \mathrm{d}t + \frac{1}{2\pi} \int_{|t|>1} \left|f^{\mathrm{ft}}(t)\right|^2 |t|^{2\beta} \underbrace{|t|^{-2\beta}}_{\leq 1} \mathrm{d}t \\ &\leq \frac{2 + C}{2\pi} \end{aligned}$$

by Parseval's identity (see Theorem A.4) and the fact that f is a density (see appendix (A.1)). Therefore, the set $\mathcal{F} = \mathcal{F}_{\beta,C;L_2}$ is included in $L_2(\mathbb{R})$ and uniformly bounded with respect to the $\|\cdot\|_2$-norm.

Now we turn to the conditions that are required with respect to the error density g. As, in the Fourier domain, division by g^{ft} occurs, we may expect better results for larger $|g^{\mathrm{ft}}|$. As already seen in Sect. 2.3, the zero set of g^{ft} must be restricted to establish consistency. To derive convergence rates we need stronger assumptions that classify the tail behavior of $|g^{\mathrm{ft}}|$. Two major types of error densities are intensively studied in deconvolution literature: First, the so-called ordinary smooth error densities whose Fourier transform satisfy

$$C_1(1+|t|)^{-\alpha} \leq \left|g^{\mathrm{ft}}(t)\right| \leq C_2(1+|t|)^{-\alpha}, \quad \text{for all } t \in \mathbb{R}, \tag{2.31}$$

for some $C_2 > C_1 > 0$ and $\alpha > 0$. Those densities are characterized by the fact that their Fourier transforms decay in some finite power. Densities whose Fourier transforms have exponential tails are called supersmooth; they are defined by

$$C_1 \exp(-d_1|t|^\gamma) \leq \left|g^{\mathrm{ft}}(t)\right| \leq C_2 \exp(-d_2|t|^\gamma), \quad \text{for all } t \in \mathbb{R}, \tag{2.32}$$

for some $C_2 > C_1 > 0$, $0 < d_2 < d_1$, $\gamma > 0$. This is a slight modification of the terminology that was introduced by [43]. For both types, the Fourier transform decrease about monotonously, that is, their upper and lower bound coincide up to some constant factors or scaling parameters. As some examples for supersmooth error densities, we give the normal densities $f_N(x) = (2\pi\sigma^2)^{-1/2} \exp\left(-|x-\mu|^2/[2\sigma^2]\right)$ with $f_N^{\mathrm{ft}}(t) = \exp(i\mu t) \cdot \exp\left(-\sigma^2|t|^2/2\right)$, which are very popular in measurement error models. In general, it is believed that measurement error is due to many slight impacts so that using normal distributions seems obvious due to the central limit theorem from probability theory. They are included in (2.32) for $\gamma = 2$. As an example for $\gamma = 1$, we mention the Cauchy density $f_{\mathrm{C}}(x) = 1/\left[\pi(1+x^2)\right]$, with $f_{\mathrm{C}}^{\mathrm{ft}}(t) = \exp(-|t|)$. A famous example for ordinary smooth error densities is the Laplace density (also known as double-exponential density), defined by $f_{\mathrm{L}}(x) = (1/2)\cdot\exp(-|x|)$, with $f_{\mathrm{L}}^{\mathrm{ft}}(t) = 1/(1+t^2)$. So we have $\alpha = 2$. Considering the m-fold self-convolution of f_{L}, we obtain ordinary smooth densities with the degree $2m$, since the corresponding Fourier transform is equal to $\left[f_{\mathrm{L}}^{\mathrm{ft}}\right]^m$.

Furthermore, we mention that only the lower bound on $|g^{\mathrm{ft}}|$ in (2.31) and (2.32) will be required to derive an upper bound on the convergence rates, while only the upper bound on $|g^{\mathrm{ft}}|$ will be needed to prove a theoretical lower bound on those rates. However, one does usually not split those definitions, as rate optimality can only be derived if both bounds are imposed.

Note that (2.31) and (2.32) imply that g^{ft} does not vanish. Nevertheless, there are still important error densities left whose structure is classified neither by (2.31) nor by (2.32), such as uniform densities. A discussion on those nonstandard types will be given in Sect. 2.4.3.

2.4.2 Mean Squared Error: Upper Bounds

Equipped with the assumptions and definitions of the previous subsection, we are able to derive the convergence rates of the mean squared error uniformly over the Hölder class $\mathcal{F}_{\beta,C,\delta;x}$ (see (2.29)), that is, the speed of convergence of the term

$$\sup_{f\in\mathcal{F}_{\beta,C,\delta;x}} E\big|\hat{f}(x)-f(x)\big|^2,$$

as the sample size n tends to infinity. For the MSE-asymptotics, that is, the pointwise case, the deconvolution kernel estimator (2.7) seems appropriate, unlike wavelet estimators, which are designed to handle the MISE-asymptotics, that is, the $L_2(\mathbb{R})$-risk.

We use a kernel function $K\in L_1(\mathbb{R})\cap L_2(\mathbb{R})$ whose Fourier transform is supported on $[-1,1]$. We derive by Proposition 2.1 that

$$\begin{aligned}
\sup_{f\in\mathcal{F}_{\beta,C,\delta;x}} E\big|\hat{f}(x)-f(x)\big|^2 &\le \sup_{f\in\mathcal{F}_{\beta,C,\delta;x}} \operatorname{var}\hat{f}(x) \\
&\qquad + \sup_{f\in\mathcal{F}_{\beta,C,\delta;x}} \big|E\hat{f}(x)-f(x)\big|^2 \\
&\le O(1/n)\cdot\Big(\min_{|t|\le 1/b}|g^{\mathrm{ft}}(t)|\Big)^{-2}\cdot\int |K^{\mathrm{ft}}(tb)|^2\,\mathrm{d}t \\
&\qquad + \sup_{f\in\mathcal{F}_{\beta,C,\delta;x}} \big|\big[K_b*f\big](x)-f(x)\big|^2 \\
&\le O\big(1/(bn)\big)\cdot\Big(\min_{|t|\le 1/b}|g^{\mathrm{ft}}(t)|\Big)^{-2}\cdot\int |K^{\mathrm{ft}}(s)|^2\,\mathrm{d}s \\
&\qquad + \sup_{f\in\mathcal{F}_{\beta,C,\delta;x}} \big|\big[K_b*f\big](x)-f(x)\big|^2,
\end{aligned} \tag{2.33}$$

as all $f\in\mathcal{F}_{\beta,C,\delta;x}$ are bounded by C following from the definition of that density class. This result may be combined with

$$\|f*g\|_\infty \le \int\Big(\sup_{x\in\mathbb{R}}\big|f(x-y)\big|\Big)g(y)\,\mathrm{d}y = \|f\|_\infty,$$

as g integrates to one. We have substituted the integration variable by $s=bt$ in the first term of (2.33). Then, assuming that $K\in L_2(\mathbb{R})$, we may apply Parseval's identity (see Theorem A.4) so that we derive

$$O\big(1/[bn]\big)\cdot\Big(\min_{|t|\le 1/b}|g^{\mathrm{ft}}(t)|\Big)^{-2}, \tag{2.34}$$

as an upper bound on the variance term.

The second addend in (2.33), which represents the bias term and which is denoted by B_n in the sequel, needs more intensive consideration. Indeed, it corresponds to the bias term in standard density estimation in the error-free case. Nevertheless, the selection of the kernel function is not easy. By simple substitution, the bias term is equal to

$$\begin{aligned}
B_n &= \sup_{f\in\mathcal{F}_{\beta,C,\delta;x}} \Big|\int K(z)\big[f(x-zb)-f(x)\big]\,\mathrm{d}z\Big|^2 \\
&\le \sup_{f\in\mathcal{F}_{\beta,C,\delta;x}} \Big|\int_{|z|\le\delta/b} K(z)\big[f(x-zb)-f(x)\big]\,\mathrm{d}z\Big|^2 \\
&\qquad + \sup_{f\in\mathcal{F}_{\beta,C,\delta;x}} \Big|\int_{|z|>\delta/b} K(z)\big[f(x-zb)-f(x)\big]\,\mathrm{d}z\Big|^2,
\end{aligned} \tag{2.35}$$

when arranging $\int K(z)\,\mathrm{d}z = 1$. Those two addends above are denoted by $B_{n,1}$ and $B_{n,2}$, respectively. With respect to the term $B_{n,1}$, we may use that f is $\lfloor\beta\rfloor$-fold differentiable on $[x-\delta,x+\delta]$. Hence, we may employ the Taylor expansion of f with the degree $\lfloor\beta\rfloor-1$, leading to

$$\begin{aligned}
B_{n,1} &= \sup_{f\in\mathcal{F}_{\beta,C,\delta;x}} \Big|\int_{|z|\le\delta/b} K(z)\cdot\Big(\sum_{j=1}^{\lfloor\beta\rfloor-1}\frac{1}{j!}f^{(j)}(x)(-zb)^j \\
&\qquad + \frac{1}{\lfloor\beta\rfloor!}f^{(\lfloor\beta\rfloor)}(\xi_{b,z})(-zb)^{\lfloor\beta\rfloor}\Big)\,\mathrm{d}z\Big|^2 \\
&= \sup_{f\in\mathcal{F}_{\beta,C,\delta;x}} \Big|\int_{|z|\le\delta/b} K(z)\cdot\Big(\sum_{j=1}^{\lfloor\beta\rfloor}\frac{1}{j!}f^{(j)}(x)(-zb)^j \\
&\qquad + \frac{1}{\lfloor\beta\rfloor!}\big(f^{(\lfloor\beta\rfloor)}(\xi_{b,z})-f^{(\lfloor\beta\rfloor)}(x)\big)(-zb)^{\lfloor\beta\rfloor}\Big)\,\mathrm{d}z\Big|^2 \\
&\le 2\sup_{f\in\mathcal{F}_{\beta,C,\delta;x}} \Big|\sum_{j=1}^{\lfloor\beta\rfloor}(-1)^j\int_{|z|\le\delta/b} K(z)z^j\,\mathrm{d}z\cdot\frac{b^j}{j!}f^{(j)}(x)\Big|^2 \\
&\qquad + \frac{2C^2}{\lfloor\beta\rfloor!^2}b^{2\lfloor\beta\rfloor}\Big(\int|K(z)||z|^\beta\,\mathrm{d}z\Big)^2\cdot b^{2\beta-2\lfloor\beta\rfloor},
\end{aligned} \tag{2.36}$$

where $\xi_{z,b}$ denotes some real number between x and $x-zb$, which is specified no further. It follows from the Lagrange representation of the residual term of the Taylor expansion. To derive the bound on the second addend in (2.36), we have used the Hölder condition (2.29); then, that term is bounded above by $O\big(b^{2\beta}\big)$ whenever $\int|K(z)||z|^\beta\,\mathrm{d}z<\infty$, which can be arranged by appropriate choice of the kernel function. With respect to the first term in (2.36), we assume validity of the following system of equation,

$$\int K(z)z^j\,\mathrm{d}z = 0, \text{ for all } j=1,\ldots,\lfloor\beta\rfloor.$$

Verifying those conditions is subject to find a suitable kernel function. Then, the first term in (2.36) is equal to

$$
\begin{aligned}
& 2 \sup_{f\in\mathcal{F}_{\beta,C,\delta;x}} \Big|\sum_{j=1}^{\lfloor\beta\rfloor}(-1)^j \int_{|z|>\delta/b} K(z)z^j \,\mathrm{d}z \cdot \frac{b^j}{j!} f^{(j)}(x)\Big|^2 \\
& \leq O(1) \sup_{f\in\mathcal{F}_{\beta,C,\delta;x}} \sum_{j=1}^{\lfloor\beta\rfloor} b^{2j} \Big|\int_{|z|>\delta/b} |K(z)||z|^j \,\mathrm{d}z \cdot f^{(j)}(x)\Big|^2 \\
& \leq O(1) \sum_{j=1}^{\lfloor\beta\rfloor} b^{2j} \cdot \Big(\int |K(z)||z|^\beta \underbrace{|z|^{j-\beta}}_{\leq(\delta/b)^{j-\beta}} \,\mathrm{d}z\Big)^2 \\
& = O\big(b^{2\beta}\big),
\end{aligned}
$$

where we have used Lemma 2.7 to impose a uniform upper bound on the $|f^{(j)}(x)|$, $j = 0, \ldots, \lfloor\beta\rfloor$. Summarizing, we have shown $B_{n,1} = O\big(b^{2\beta}\big)$.

Concerning the term $B_{n,2}$, we use $\|f\|_\infty \leq C$ so that

$$
\begin{aligned}
B_{n,2} & \leq (2C)^2 \cdot \Big(\int_{|z|>\delta/b} |K(z)| \,\mathrm{d}z\Big)^2 \leq O\big(b^{2\beta}\big) \cdot \Big(\int_{|z|>\delta/b} |K(z)||z|^\beta \,\mathrm{d}z\Big)^2 \\
& = O\big(b^{2\beta}\big),
\end{aligned}
$$

where we use the condition $\int |K(z)||z|^\beta \,\mathrm{d}z < \infty$, which has already been assumed. Therefore, the bias term B_n as a total is bounded above by $O\big(b^{2\beta}\big)$. Combining that result with (2.34), we obtain

$$
\sup_{f\in\mathcal{F}_{\beta,C,\delta;x}} E\big|\hat{f}(x) - f(x)\big|^2 \leq O\big(1/[bn]\big) \cdot \big(\min_{|t|\leq 1/b} |g^{\mathrm{ft}}(t)|\big)^{-2} + O\big(b^{2\beta}\big). \quad (2.37)
$$

However, we must not forget about the conditions on the kernel function that we have assumed. They can be summarized as follows:

$$
\begin{aligned}
& K \in L_1(\mathbb{R}) \cap L_2(\mathbb{R}),\ K^{\mathrm{ft}} \text{ supported on } [-1,1], \\
& \int K(z)z^j \,\mathrm{d}z = \delta_{j,0}, \text{ for all } j = 0, \ldots, \lfloor\beta\rfloor, \\
& \int |K(z)||z|^\beta \,\mathrm{d}z < \infty, \qquad (2.38)
\end{aligned}
$$

where $\delta_{j,k}$ is equal to 1 if $j = k$; and zero otherwise. We call such a kernel a *β-order kernel.* It remains to be shown that such a β-order kernel exists for any $\beta > 0$. Therefore, we consider a function L, which is supported on $[-1, 1]$, equal to one on $[-1/2, 1/2]$ and $(\lfloor\beta\rfloor + 3)$-fold continuously differentiable on the whole real line. Such a function L can be constructed by polynomial fitting in the intervals $[-1, -1/2]$ and $[1/2, 1]$. Therefore, as $L \in L_2(\mathbb{R})$, there exists some $K \in L_2(\mathbb{R})$ so that $K^{\mathrm{ft}} = L$ and $K(z) = (2\pi)^{-1} \int \exp(-itz)L(t) \,\mathrm{d}t$, where the integral must be viewed as an $L_2(\mathbb{R})$-limit (see Theorem A.4). Since we also have $L \in L_1(\mathbb{R})$, the integral may also be interpreted as a common Lebesgue integral on the real line. As L, along with all of its derivatives with the degree $\leq \lfloor\beta\rfloor + 3$, are equal to zero outside the interval $[-1, 1]$, we may establish

$$|K(z)| \leq \frac{1}{2\pi} |z|^{-j} \int \left|L^{(j)}(t)\right| \mathrm{d}t$$

for $|z| \geq 1$ and $j = 0, \ldots, \lfloor\beta\rfloor + 3$ by integration by parts. Furthermore, we have $\int_{-1}^{1} |K(x)||x|^k \, \mathrm{d}x \leq 2\|K\|_2 < \infty$, for all integer $k \geq 0$ by the Cauchy–Schwarz inequality, so that we receive $K \in L_1(\mathbb{R})$ and

$$\int |K(z)||z|^\beta \, \mathrm{d}z < \infty .$$

Further, observe that

$$\delta_{k,0} = \frac{\mathrm{d}^k}{\mathrm{d}t^k} K^{\mathrm{ft}}(t)\Big|_{t=0} = \int \left[\frac{\mathrm{d}^k}{\mathrm{d}t^k} \exp(\mathrm{i}tx)\right]_{t=0} K(x) \, \mathrm{d}x = \mathrm{i}^k \int K(x) x^k \, \mathrm{d}x \,,$$

so that (2.38) has been verified for that specific kernel K. Apparently, any γ-order kernel with $\gamma > \beta$ is also a β-order kernel. Therefore, no precise knowledge of β is required to construct the kernel K, but only some number which is larger than β. There are also superkernels whose order is infinite. For the construction of those kernels, we would have to assume that L is differentiable infinitely often rather than $(\lfloor\beta\rfloor + 3)$-fold differentiability. Their construction is theoretically possible but becomes more difficult. On the other hand, those superkernels are useful when considering the MISE, so their investigation will be deferred to Sect. 2.4.3.

Sometimes, density functions are suggested for kernels. In the error-free case, they have the advantage that the outcome density estimator itself is a density function. Such kernels are quite convenient for the purpose of establishing consistency (see [35]). However, we learn from (2.38) that no density function can be used as a kernel with the order ≥ 2 since, otherwise, $\int K(z) z^2 \, \mathrm{d}z = 0$ implies $K \equiv 0$ so that K does not integrate to one. Therefore, to derive higher order kernels, we need to allow for those K that take some negative values as those constructed earlier.

For the practical choice of the kernel functions, we also mention the paper of [27] where the performance of different kernels is studied by intensive numerical simulations.

Now, as the existence of β-order kernels has been proved, the formula (2.37) is evident. As usual in kernel smoothing, the variance term increases when $b > 0$ becomes smaller, while the bias term decreases. Hence, for optimal selection of b, it must be calibrated appropriately. Let us consider (2.37) for ordinary smooth and supersmooth error densities, (2.31) and (2.32), as introduced in the previous subsection. Quite obviously, we have

$$\sup_{f \in \mathcal{F}_{\beta,C,\delta;x}} E\left|\hat{f}(x) - f(x)\right|^2 \leq O\left(b^{-1-2\alpha}/n\right) + O\left(b^{2\beta}\right),$$

$$\sup_{f \in \mathcal{F}_{\beta,C,\delta;x}} E\left|\hat{f}(x) - f(x)\right|^2 \leq O\left[\exp\left(2d_1 b^{-\gamma}\right)/(bn)\right] + O\left(b^{2\beta}\right)$$

for ordinary smooth and supersmooth error densities g, respectively. Apparently, for ordinary smooth g, the optimal convergence rates are achieved when both variance and bias term converge to zero with the same rate. That occurs when choosing $b \asymp n^{-1/(2\beta+2\alpha+1)}$, that is, $bn^{1/(2\beta+2\alpha+1)} \in (c_0, c_1)$ for all integer n and some $c_1 > c_0 > 0$. Inserting that selection of the bandwidth gives us the asymptotic upper bound $O\big(n^{-2\beta/(2\beta+2\alpha+1)}\big)$ on the MSE. With respect to the supersmooth case, we propose to select the bandwidth $b = c_b^{-1/\gamma} \cdot (\log n)^{-1/\gamma}$ with $c_b \in (0, 1/(2d_1))$. Then, the variance term is as $O\big((\log n)^{1/\gamma} n^{2d_1 c_b - 1}\big)$ so that it converges to zero faster than some algebraic rate, that is, $n^{-\xi}$ with some $\xi > 0$. Unfortunately, the bias term converges to zero only with the logarithmic rate $O\big((\log n)^{-2\beta/\gamma}\big)$, which dominates the variance term. Therefore, we have shown the following theorem.

Theorem 2.8. *Assume that $\beta > 0$, choose a β-order kernel according to (2.38). All densities f, which are bounded by C and satisfy (2.29), are collected in the class $\mathcal{F}_{\beta,c,\delta;x}$. Then,*
(a) for ordinary smooth error densities g satisfying (2.31), we select the bandwidth $b \asymp n^{-1/(2\beta+2\alpha+1)}$ in order to derive that

$$\sup_{f \in \mathcal{F}_{\beta,C,\delta;x}} E\big|\hat{f}(x) - f(x)\big|^2 = O\big(n^{-2\beta/(2\beta+2\alpha+1)}\big) .$$

(b) for supersmooth error densities g satisfying (2.32), we select the bandwidth $b = c_b^{-1/\gamma} \cdot (\log n)^{-1/\gamma}$ with $c_b \in (0, 1/(2d_1))$ in order to derive that

$$\sup_{f \in \mathcal{F}_{\beta,C,\delta;x}} E\big|\hat{f}(x) - f(x)\big|^2 = O\big((\log n)^{-2\beta/\gamma}\big) .$$

In comparison to the convergence rates established in nonparametric curve estimation in the error-free case (they are as $n^{-2\beta/(2\beta+1)}$, e.g., [119]), we realize that the contamination of the data causes deterioration of the rates both for ordinary smooth and supersmooth error densities. In the ordinary smooth case, we see that the error-free rate is included when putting $\alpha = 0$. Note that the case of no contamination can be included by putting $\varepsilon_j = 0$ a.s. for all $j = 1, \ldots, n$. In this case, g corresponds to the Dirac-functional (called δ-function in theoretical physics), that is, g is a degenerated density and the corresponding distribution function G is equal to $G = \chi_{[0,\infty)}$. Then, $g^{\mathrm{ft}} \equiv 1$ so that we have $\alpha = 0$ in (2.31), indeed. While in the ordinary smooth case, the convergence rates still have an algebraic shape, they are very slow (logarithmic) in the supersmooth case. This is a big problem since the supersmooth normal density occurs as the error density rather frequently. There are some ways to improve the rates by changing the statistical experiment as indicated in the following subsection.

Those convergence rates have first been investigated in [43].

Somehow as a pay-back for that slow rates under supersmooth contamination, we realize that the bandwidth choice that leads to the above rates does not require knowledge of the smoothness degree β of the target density f. Then the bandwidth selection may be called self-adaptive. However, for ordinary smooth g, the parameter β occurs in the selection rule for b. Assuming knowledge of β seems unrealistic in many practical applications. However, when considering the MSE, there does obviously not exist any method of omitting that assumption without losing some speed of convergence. That was even shown in the error-free case (see [81, 82, 86]). Nevertheless, when switching to the MISE, there will be some adaptive methods of bandwidth selection as studied in Sect. 2.5.

Finally, we mention that one could also consider the convergence rates with respect to the uniform metric, that is, $\sup_{f\in\mathcal{F}} E\big\|\hat{f}-f\big\|_\infty^2$. This, however, is a very strong metric. While one is able to keep the uniform rates on the bias term when changing the validity of the smoothness conditions from local to global – more precisely, one shall assume that $f \in \mathcal{F}_{\beta,C,\delta;x}$ for any $x \in \mathbb{R}$ – the treatment of the variance term is more difficult. As in the error-free case, the application of some Brownian bridges (e.g., [76]) is required, which causes some logarithmic loss in the convergence rates. General consistency with respect to $\|\cdot\|_\infty$ is considered in [85].

2.4.3 Mean Integrated Squared Error: Upper Bounds

Now, we are interested in the convergence rates that the deconvolution kernel estimator (2.7) achieves when f lies in some Sobolev class $\mathcal{F}_{\beta,C;L_2}$ as defined in (2.30). We recall the definition of the MISE (see (2.10)) and Proposition 2.2 giving us an equivalent representation of the MISE. As in the previous subsection, we choose a kernel function $K \in L_2(\mathbb{R})$ so that K^{ft} is supported on the interval $[-1,1]$ and bounded by 1. However, with respect to the MISE, we do not assume that $K \in L_1(\mathbb{R})$. Then, by Proposition 2.2, we derive that

$$\begin{aligned}\sup_{f\in\mathcal{F}_{\beta,C;L_2}} \mathrm{MISE}(\hat{f},f) &= O\big(n^{-1}b^{-1}\big(\min_{|t|\le 1/b}\big|g^{\mathrm{ft}}(t)\big|\big)^{-2}\big)\\ &\quad + \sup_{f\in\mathcal{F}_{\beta,C;L_2}} \frac{1}{2\pi}\int \big|K^{\mathrm{ft}}(tb)-1\big|^2|f^{\mathrm{ft}}(t)|^2\,dt.\end{aligned} \tag{2.39}$$

The latter addend in (2.39) reflects the bias term. By (2.30), we have

$$\begin{aligned}&\sup_{f\in\mathcal{F}_{\beta,C;L_2}} \frac{1}{2\pi}\int \big|K^{\mathrm{ft}}(tb)-1\big|^2|f^{\mathrm{ft}}(t)|^2\,dt\\ &= \sup_{f\in\mathcal{F}_{\beta,C;L_2}} \frac{1}{2\pi}\int \big|K^{\mathrm{ft}}(tb)-1\big|^2|f^{\mathrm{ft}}(t)|^2|t|^{2\beta}|t|^{-2\beta}\,dt\\ &\le \Big(\sup_{t\neq 0}\big|K^{\mathrm{ft}}(tb)-1\big|^2|t|^{-2\beta}\Big)\cdot \sup_{f\in\mathcal{F}_{\beta,C;L_2}} \frac{1}{2\pi}\int |f^{\mathrm{ft}}(t)|^2|t|^{2\beta}\,dt\end{aligned}$$

$$\leq \frac{C}{2\pi} \sup_{s \neq 0} \left|K^{\mathrm{ft}}(s) - 1\right|^2 |s|^{-2\beta} \cdot b^{2\beta}$$
$$= O\big(b^{2\beta}\big),$$

if we select K so that $|K^{\mathrm{ft}}(t) - 1| = O(|t|^\beta)$ as $|t| \to 0$. Summarizing, we have assumed with respect to the kernel function K that

$$K \in L_2(\mathbb{R}), \, K^{\mathrm{ft}} \text{ supported on } [-1,1], \|K^{\mathrm{ft}}\|_\infty \leq 1,$$
$$\sup_{t \neq 0} |t|^{-\beta} |K^{\mathrm{ft}}(t) - 1| < \infty. \tag{2.40}$$

Critically, one may argue that continuity of K^{ft} is not guaranteed as long as we do not assume $K \in L_1(\mathbb{R})$. However, the validity of the conditions imposed on K^{ft} in (2.40) may be restricted to all $t \in \mathbb{R}$ except a Lebesgue zero set. We refer to kernels satisfying the totality of the conditions (2.40) as *Sobolev-β-order kernels*. Their definition can be compared to the β-order kernels in (2.38).

The condition (2.40) says that $K^{\mathrm{ft}}(t)$ shall be very close to one for small $|t|$, while its support shall still be included in $[-1,1]$. Hence, the kernel K whose Fourier transform is equal to $\chi_{[-1,1]}$ seems to be convenient. Let us derive an explicit formula for that kernel by Fourier inversion (see Theorem A.4).

$$K(x) = \frac{1}{2\pi} \int \exp(-\mathrm{i}tx)\chi_{[-1,1]}(t)\, \mathrm{d}t = \frac{1}{2\pi} \frac{\exp(-\mathrm{i}x) - \exp(\mathrm{i}x)}{-\mathrm{i}x} = \frac{\sin x}{\pi x}. \tag{2.41}$$

That kernel is referred to as the *sinc-kernel*. Indeed, we realize that it satisfies the conditions (2.40) for all $\beta > 0$. Therefore, the sinc-kernel is also called an ∞-order kernel or a superkernel.

Hence, when using a Sobolev-β-order kernel, we obtain that the MISE is uniformly bounded above as follows.

$$\sup_{f \in \mathcal{F}_{\beta,C;L_2}} \mathrm{MISE}(\hat{f}, f) = O\big(n^{-1} b^{-1} \max_{|t| \leq 1/b} \left|g^{\mathrm{ft}}(t)\right|^{-2}\big) + O\big(b^{2\beta}\big). \tag{2.42}$$

In fact, this bound corresponds to the bound on the MSE in (2.37) when replacing our Sobolev conditions by Hölder conditions with the same smoothness degree. Therefore, the conditions of ordinary smooth and supersmooth error densities as well as the bandwidth selection can be adopted from Theorem 2.8 so that we can give the analogous result for the MISE.

Theorem 2.9. *Assume that $\beta > 0$; choose a Sobolev-β-order kernel according to (2.40). All densities satisfying (2.30) are collected in the density class $\mathcal{F}_{\beta,C;L_2}$. Then,*
(a) for ordinary smooth error densities g satisfying (2.31), we select the bandwidth $b \asymp n^{-1/(2\beta+2\alpha+1)}$ to derive that

$$\sup_{f \in \mathcal{F}_{\beta,C;L_2}} E\|\hat{f} - f\|_2^2 = O\big(n^{-2\beta/(2\beta+2\alpha+1)}\big).$$

(b) for supersmooth error densities g satisfying (2.32), we select the bandwidth $b = c_b^{-1/\gamma} \cdot (\log n)^{-1/\gamma}$ with $c_b \in (0, 1/(2d_1))$ to derive that

$$\sup_{f \in \mathcal{F}_{\beta,C;L_2}} E\|\hat{f} - f\|_2^2 = O\big((\log n)^{-2\beta/\gamma}\big).$$

Again, the bandwidth selection is adaptive, that is, independent of the smoothness degree of the target density f, only in the supersmooth case. As mentioned in the previous subsection, there are methods for adaptive bandwidth selection in the ordinary smooth case as well, which will be discussed in Sect. 2.5. As an attractive feature, we may use the sinc-kernel (2.41), which satisfies the requirements of the kernel function as stipulated in Theorem 2.9 without knowing the smoothness degree β. Then, adaption is only a matter of the bandwidth selection. Nevertheless, the use of the sinc-kernel is sometimes seen critically as numerical simulations indicate that the density estimator shows rather oscillatory behavior, due to Gibbs phenomena. Further, some papers dealing with the error-free case (e.g., [42, 115]) indicate that different kernels shall be used when optimizing the asymptotic constants and not only the convergence rates. For an intensive study on the performance of kernels in density deconvolution based on numerical simulations, we refer to [27]. In that paper, the authors recommend to use the kernel K having the Fourier transform $K^{\mathrm{ft}}(t) = (1 - t^2)^3 \chi_{[-1,1]}(t)$, although its Sobolev order is just 2 (finite).

The convergence rates of the MISE have been studied in [46], where different types of smoothness classes for f are used. Also, in that paper, the integral contained in the MISE is not considered on the whole real line but on a compact interval.

In the case of supersmooth contamination, there are some ways of changing the model assumptions so that the convergence rates become algebraic (i.e., $n^{-\xi}$ for some $\xi > 0$). The first method assumes a low noise level. Unlike in the standard model (2.1) where the error density g is always seen as fixed, we alter the observation scheme to

$$Y_j = X_j + \sigma_n \varepsilon_j, \quad \text{for all } j = 1, \ldots, n, \tag{2.43}$$

where the X_j, ε_j are as in (2.1), but the scaling parameter (also referred to as the noise level) σ_n may now depend on the sample size n. We assume that $(\sigma_n)_n \downarrow 0$. Therefore, model (2.43) can only be employed if the contamination effect is low compared to the variance of the random variables X_j. Let us consider the model for centered standard normal contamination (i.e., $E\varepsilon_j = 0$, $\operatorname{var} \varepsilon_j = 1$). Apparently, the error density is equal to $g = N(0, \sigma_n^2)$, that is, the normal density with mean zero and the variance σ_n^2. If σ_n was bounded away from zero, then we would achieve convergence rates not faster than

$(\log n)^{-\beta/\gamma}$, according to Theorem 2.9. On the other hand, when $(\sigma_n)_n \downarrow 0$, we are able to improve the speed of convergence. When using the sinc-kernel, the deconvolution kernel estimator has the upper bound

$$\sup_{f\in\mathcal{F}_{\beta,C;L_2}} \mathrm{MISE}(\hat{f},f) = O\big[n^{-1}b^{-1}\max_{|t|\le 1/b}\exp\big(\sigma_n^2 t^2\big)\big] + O\big(b^{2\beta}\big),$$

according to (2.42). Hence, if $\sigma_n = O\big(n^{-1/(2\beta+1)}\big)$, then we may choose the bandwidth $b \asymp n^{-1/(2\beta+1)}$ so that the convergence rates

$$\sup_{f\in\mathcal{F}_{\beta,C;L_2}} \mathrm{MISE}(\hat{f},f) = O\big(n^{-2\beta/(2\beta+1)}\big)$$

occur; they are not only algebraic but they are even identical with those rates derived in density estimation in the error-free case (e.g., [119]). Therefore, if $(\sigma_n)_n$ tends to zero sufficiently fast, then there is no loss of the speed of convergence by data contamination at all. The model of decreasing σ_n has been studied in [45]; it is also used for low-order approximations of the deconvolution estimator (see [15]).

Another method of establishing non-logarithmic rates assumes more restrictive smoothness conditions on f. Instead of the Sobolev conditions $f \in \mathcal{F}_{\beta,C;L_2}$ as used in Theorem 2.9, we may permit only densities with exponential Fourier tails to lie in our density class. Hence, we may define the density class $\mathcal{F}_{\delta,d,C;\exp}$ containing those densities, which satisfy

$$\int \big|f^{\mathrm{ft}}(t)\big|^2 \exp(d|t|^\delta)\,dt \le C.$$

Of course, that condition is much stronger than $f \in \mathcal{F}_{\beta,C;L_2}$, since it rules out those densities, which are differentiable only finitely many often. However, supersmooth densities (see (2.32)) with $\gamma > \delta$ are still included in $\mathcal{F}_{\delta,d,C;\exp}$. One can show that $\mathcal{F}_{\delta,d,C;\exp}$ still contains a comprehensive nonparametric class of densities. When using the sinc-kernel for the deconvolution kernel estimator, the rates of the bias term change while the variance is not affected. Revisiting (2.39), we derive with respect to the bias that

$$\begin{aligned}
&\sup_{f\in\mathcal{F}_{\delta,d,C;\exp}} \frac{1}{2\pi}\int \big|K^{\mathrm{ft}}(tb)-1\big|^2 |f^{\mathrm{ft}}(t)|^2\,dt \\
&= \sup_{f\in\mathcal{F}_{\delta,d,C;\exp}} \frac{1}{2\pi}\int_{|t|>1/b} \exp(-\mathrm{d}|t|^\delta)\,\exp(\mathrm{d}|t|^\delta)|f^{\mathrm{ft}}(t)|^2\,dt \\
&\le \frac{C}{2\pi}\exp(-\mathrm{d}b^{-\delta}).
\end{aligned}$$

Therefore, we derive, as the uniform upper bound of the MISE,

$$\sup_{f\in\mathcal{F}_{\delta,d,C;\exp}} \mathrm{MISE}(\hat{f},f) = O\big[n^{-1}b^{-1}\max_{|t|\le 1/b}\big|g^{\mathrm{ft}}(t)\big|^{-2}\big] + O\big(\exp(-\mathrm{d}b^{-\delta})\big).$$

If g is ordinary smooth (see (2.31)), then our upper bound is equal to

$$O\big(n^{-1}b^{-1-2\alpha}\big) + O\big(\exp(-db^{-\delta})\big).$$

Selecting the bandwidth $b = c_b^{-1/\delta}\big(\log n\big)^{-1/\delta}$ with $c_b > 1/d$ leads to the convergence rates $O\big(n^{-1}\big(\log n\big)^{(1+2\alpha)/\delta}\big)$. Therefore, we have improved the convergence rates compared to Theorem 2.9(a). Returning to the supersmooth case, we suggest to select the bandwidth $b = b_n$ so that

$$n^{-1}b_n^{-1}\exp(2d_1 b_n^{-\gamma}) = \exp(-db_n^{-\delta}).$$

The left side of the above equation decreases montonously from $+\infty$ to 0 as b_n runs from 0 to ∞, while the right side increases from 0 to 1. As both sides are continuous in b_n there must be a unique solution of this equation. That choice of b_n leads to optimal rates in the supersmooth case. Although the equation is explicitly solvable in rare cases only, we recognize that algebraic rates are available whenever $\gamma < \delta$, for example, when putting $b = c_b^{-1/\gamma} \cdot (\log n)^{-1/\gamma}$ with $c_b \in (0, 1/(2d_1))$. For those classes of supersmooth target densities, see the papers of [9, 10, 20, 107] for instance.

When g is a uniform density then g^{ft} has some isolated zeros and shows oscillatory behavior. Therefore, it is not contained in the framework of the error density classes (2.31) and (2.32). The uniform density g on $[-1, 1]$, for example, has the Fourier transform $g^{\mathrm{ft}}(t) = (\sin t)/t$. Such nonstandard error densities, which may be called Fourier-oscillating, are studied in [58] and [94]. They can be defined by the condition

$$C_1|t|^{-\nu}|\sin(\lambda t)|^{\mu} \le |g^{\mathrm{ft}}(t)| \le C_2|t|^{-\nu}|\sin(\lambda t)|^{\mu}$$

for all $|t| > T > 0$, while $|g^{\mathrm{ft}}|$ is bounded away from zero on $(-T, T)$. We have $C_2 \ge C_1 > 0$ and $\lambda > 0$. The parameter $\nu > 0$ describes the tail behavior of the Fourier transform as for ordinary smooth and supersmooth g. The new parameter $\mu > 0$ reflects the order of the periodic zeros. μ-fold self-convolutions of uniform densities are included as well as their convolution with an ordinary smooth density ($\nu > \mu$).

In [58], it is shown that the convergence rates from Theorem 2.9(a) can be kept when putting $\alpha = \nu$, in rare cases only. First, the usual Sobolev class $\mathcal{F}_{\beta,C;L_2}$ is changed into a uniform upper bound on f^{ft}, given by

$$|f^{\mathrm{ft}}(t)| \le C|t|^{-\beta-1/2},$$

for some constant $C > 0$ and all $t \in \mathbb{R}$. But even under those more restrictive smoothness assumptions, the convergence rates are equal to

$$\max\big\{n^{-1/(2\mu)}, n^{-2\beta/(2\beta+2\nu+1)}\big\},$$

where, in the coincidence case, of both terms, some further logarithmic loss occurs. The authors establish those rates by using the ridge-parameter estimator as in (2.21). Unfortunately, stronger smoothness constraints cannot improve the upper bound $n^{-1/(2\mu)}$. The authors prove optimality of those convergence rates with respect to any estimator.

When we return to the standard Sobolev conditions on f, and assume additional smoothness conditions on f^{ft} locally around the isolated zeros of g^{ft}, which can be justified by moment restrictions, then the convergence rates can be improved by a specific estimator introduced in [94]. It is based on local polynomial approximation in the Fourier domain. Intuitively, $f^{\mathrm{ft}}(t)$ in small neighborhoods of the zeros is approximated by some empirical information of f^{ft} acquired outside those sets. Meister [94] establishes rate optimality in the given setting, where completely new and rather complicated rates occur. When considering compactly supported functions f, so that all moments of f exist and increase geometrically, it is even possible to keep the classical deconvolution rates $n^{-2\beta/(2\beta+2\nu+1)}$ up to some logarithmic loss, again.

Hall and Meister [58] also consider error densities g, which are the convolution of uniform densities and supersmooth densities. In the situation where the rates are slow (logarithmic) anyway as in Theorem 2.9(b), no further loss of the rates occurs, even under standard Sobolev conditions without any additional moment restrictions.

2.4.4 Asymptotic Normality

In this subsection, we focus on a more precise characterization of the asymptotic behavior of the deconvolution kernel estimator (2.7). To motivate the following result, we focus on the difference between estimator (2.7) and the target density f. It may be decomposed into some bias term and a centered random variable. We have

$$\hat{f}(x) - f(x) = \underbrace{\hat{f}(x) - E\hat{f}(x)}_{=R(x)} + \underbrace{E\hat{f}(x) - f(x)}_{=Q(x)}.$$

The bias term $Q(x)$ is purely deterministic. The first part $R(x)$ represents the random deviation between the estimator and its expectation. We are interested in the distribution of $R(x)$ in order to quantify that random deviation. Considering large sample sizes again, that is, $n \to \infty$, $Q(x)$ shall tend to zero as a deterministic sequence; and $R(x)$ should also converge to the degenerated random variable $R \equiv 0$ a.s. in distribution. That follows from consistency. Scaling $R(x)$ with one by the standard deviation of $\hat{f}(x)$, we receive the standardized random variable

$$S(x) = \frac{\hat{f}(x) - E\hat{f}(x)}{\left(\operatorname{var}\hat{f}(x)\right)^{1/2}},$$

whose asymptotic distribution shall be studied. The estimator $\hat{f}(x)$ is said to be *asymptotically normal* if $S(x)$ tends to the standard normal distribution

$N(0,1)$ in distribution as $n \to \infty$. We recall the definition of convergence in distribution: a sequence $(S_n)_n$ of random variables is said to converge in distribution to some random variable S if the distribution functions $F_n(x)$ of the S_n converge to that of S – denoted by $F(x)$ – pointwise for any $x \in \mathbb{R}$ where F is continuous.

To give some motivation for this criterion, we may consider the basic problem of estimating the expectation of a distribution function based on direct i.i.d. observation $A_1, \ldots, A_n$ from this distribution. Assuming that the second moment of A_1 exists, the average of the data is an asymptotically normal estimator, following from the central limit theorem. Hence, asymptotic normality in terms of density deconvolution can be seen as an extension of that famous probabilistic theorem to a particular nonparametric problem.

In the following consideration, we restrict to the real part of estimator (2.7) as already suggested below its definition. We realize that estimator (2.7) may be written as the average of random variables, which are measurable in the σ-algebra generated by just one observation. We have

$$\hat{f}(x) = \frac{1}{n}\sum_{j=1}^{n} A_{j,n}(b),$$

where

$$A_{j,n}(b) = \frac{1}{2\pi}\,\mathrm{Re}\int \exp(-itx)K^{\mathrm{ft}}(tb)\ \exp(itY_j)/g^{\mathrm{ft}}(t)\ dt.$$

Therefore, $\hat{f}$ is a so-called linear estimator. Further, we may fix that

$$\operatorname{var}\hat{f}(x) = \frac{1}{n^2}\sum_{j=1}^{n}\operatorname{var} A_{j,n}(b) = \frac{1}{n}\operatorname{var} A_{1,n}(b).$$

As a consequence of the Portmanteau theorem from probability theory, convergence in distribution of some $(S_n)_n$ to S is equivalent with pointwise convergence of the characteristic functions of S_n to that of S. That will be our basic strategy to derive asymptotic normality. We obtain that

$$\begin{aligned}
E\,\exp\big(itS(x)\big) &= E\,\exp\Big(itn^{1/2}\big(\operatorname{var} A_{1,n}(b)\big)^{-1/2}\frac{1}{n}\sum_{j=1}^{n}\big(A_{j,n}(b)-EA_{j,n}(b)\big)\Big)\\
&= E\prod_{j=1}^{n}\exp\Big[itn^{-1/2}\big(\operatorname{var} A_{1,n}(b)\big)^{-1/2}\big(A_{j,n}(b)-EA_{j,n}(b)\big)\Big]\\
&= \Big(E\,\exp\Big[itn^{-1/2}\big(\operatorname{var} A_{1,n}(b)\big)^{-1/2}\big(A_{1,n}(b)-EA_{1,n}(b)\big)\Big]\Big)^{n}\\
&= \Big(E\sum_{j=0}^{\infty}\frac{1}{j!}(it)^{j}n^{-j/2}\big(\operatorname{var} A_{1,n}(b)\big)^{-j/2}\big(A_{1,n}(b)-EA_{1,n}(b)\big)^{j}\Big)^{n}\\
&= \Big(1 + itn^{-1/2}\big(\operatorname{var} A_{1,n}(b)\big)^{-1/2}\underbrace{E\big(A_{1,n}(b)-EA_{1,n}(b)\big)}_{=0}
\end{aligned}$$

$$- \frac{1}{2} t^2 n^{-1} \underbrace{\big(\mathrm{var}\, A_{1,n}(b)\big)^{-1} E\big(A_{1,n}(b) - EA_{1,n}(b)\big)^2}_{=1}$$

$$+ \sum_{j=3}^{\infty} \frac{1}{j!} (\mathrm{i}t)^j n^{-j/2} \big(\mathrm{var}\, A_{1,n}(b)\big)^{-j/2} E\big(A_{1,n}(b) - EA_{1,n}(b)\big)^j \Big)^n,$$

where the independence of the $Y_1, \ldots, Y_n$ and the power series of exp is utilized. From analysis we learn that, whenever

$$\Big| \sum_{j=3}^{\infty} \frac{1}{j!} (\mathrm{i}t)^j n^{-j/2} \big(\mathrm{var}\, A_{1,n}(b)\big)^{-j/2} E\big(A_{1,n}(b) - EA_{1,n}(b)\big)^j \Big| = o(1/n), \quad (2.44)$$

then we have

$$E \, \exp\big(\mathrm{i}tS(x)\big) \stackrel{n\to\infty}{\longrightarrow} \exp(-t^2/2)\,, \quad \text{for all}\, t \in \mathbb{R},$$

where we recognize the limit function on the right side as the Fourier transform of the standard normal density. Hence, (2.44) is all that remains to be verified to establish asymptotic normality. Therefore, we consider

$$\begin{aligned} & E\big|A_{1,n}(b) - EA_{1,n}(b)\big|^j \\ & \le \Big(\frac{1}{2\pi}\Big)^j \int \Big| \int \exp\big(-\mathrm{i}t(x-y)\big) K^{\mathrm{ft}}(tb)/g^{\mathrm{ft}}(t)\, \mathrm{d}t\Big|^j \big[f * g\big](y)\, \mathrm{d}y \\ & \le \|f * g\|_\infty \cdot \Big(\frac{1}{2\pi}\Big)^j \int \Big| \int \exp(-\mathrm{i}ty) K^{\mathrm{ft}}(tb)/g^{\mathrm{ft}}(t)\, \mathrm{d}t\Big|^j \, \mathrm{d}y \\ & \le O(1) \cdot \Big(\int \big|K^{\mathrm{ft}}(tb)/g^{\mathrm{ft}}(t)\big|^{j/(j-1)}\, \mathrm{d}t\Big)^{j-1} \\ & \le d^j b^{1-j-\alpha j} \end{aligned}$$

under the assumption of ordinary smooth g, see (2.31), with some appropriate constant $d > 0$. Also, we assume that $f * g$ is bounded and K^{ft} is supported on $[-1, 1]$, as in Theorem 2.8, and bounded above by 1. We have used the Hausdorff–Young inequality saying that $\|f^{\mathrm{ft}}/\sqrt{2\pi}\|_j \le \|f\|_{j/(j-1)}$ for $j > 1$, where $\|\cdot\|_j$ denotes the $L_j(\mathbb{R})$-norm (see appendix, (A.5)). Also, some lower bound on the variance of $A_{1,n}(b)$ is required. We obtain that

$$\mathrm{var}\, A_{1,n}(b) \ge \mathrm{const} \cdot b^{-1-2\alpha}\,.$$

Otherwise, we would have an efficiently smaller upper bound on the MSE compared to Proposition 2.1 so that faster uniform convergence rates could be established in comparison to Theorem 2.8. This, however, contradicts Theorem 2.13. Hence, we refer to the following section for details. Roughly, we notice that the lower bound proposed for the variance corresponds to the upper bound derived above for $j = 2$. Then, we conclude that

$$\Big|\sum_{j=3}^{\infty}\frac{1}{j!}(\mathrm{it})^j n^{-j/2}\big(\mathrm{var}\,A_{1,n}(b)\big)^{-j/2}E\big(A_{1,n}(b)-EA_{1,n}(b)\big)^j\Big|$$

$$\leq \sum_{j=3}^{\infty}\frac{1}{j!}|d't|^j n^{-j/2}b^{j/2+j\alpha}b^{1-j-\alpha j} \leq b\cdot\sum_{j=3}^{\infty}\frac{1}{j!}|d't/\sqrt{nb}|^j$$

$$\leq n^{-3/2}b^{-1/2}\cdot\sum_{j=0}^{\infty}\frac{1}{(j+3)!}|d't/\sqrt{nb}|^j$$

$$\leq n^{-1}O\big(n^{-1/2}b^{-1/2}\big)\cdot\exp\big(|d't/\sqrt{nb}|\big)\,,$$

with some constant $d' > 0$, which needs no further specification. We realize that (2.44) is satisfied whenever $n^{-1}b^{-1} \stackrel{n\to\infty}{\longrightarrow} 0$. For any ordinary smooth g with $\alpha > 0$, the selection of b which gives us the convergence rates established in Theorem 2.8(a) fulfills the above requirement. Therefore, we may derive the following proposition.

Proposition 2.10. *Assume the additive measurement error model (2.1); the error density g is supposed to satisfy (2.31). Further, assume a kernel function K, where K^{ft} is supported on $[-1,1]$ and bounded by 1; and $\|f*g\|_\infty < \infty$. Choose the bandwidth b so that $nb \stackrel{n\to\infty}{\longrightarrow} \infty$. Then, with respect to the deconvolution kernel density estimator $\hat{f}(x)$ at some $x\in\mathbb{R}$, the standardized error*

$$\frac{\hat{f}(x)-E\hat{f}(x)}{\big(var\,\hat{f}(x)\big)^{1/2}}$$

converges to a standard normally distributed random variable in distribution as $n\to\infty$.

Therefore, for large n, one may use the approximation

$$\hat{f}(x) \approx E\hat{f}(x) + \Delta,$$

where Δ is a normally distributed random variable with mean 0 and the variance equal to that of $\hat{f}(x)$. Such approximations can be applied to derive asymptotic confidence intervals; when changing the pointwise semi-metric to the uniform metric $\sup_{x\in I}|\hat{f}(x)-E\hat{f}(x)|$, one is able to construct confidence bands (see, e.g., [2]). Also, asymptotic normality is applicable to testing procedures to calculate the error of hypothesis approximately (e.g., [7, 67]).

The first approach to asymptotic normality in density deconvolution is given by [44]. Under supersmooth contamination as in (2.32), we do not have asymptotic normality, in general. For papers on that, see, for example, [123, 124] and [68], where, in the latter note, the squared integrated error $\|\hat{f}-f\|_2^2$ has been studied rather than the pointwise error. Holzmann and Boysen [68] show that the asymptotic distribution of the integrated squared error turns to χ_1^2 when the smoothness parameter γ is too large.

2.4.5 Mean Squared Error: Lower Bounds

In the Sect. 2.4.2 and 2.4.3, we derive upper bounds on the convergence rates achieved by the deconvolution kernel estimator. The current and the following subsection are dedicated to the question whether those rates can be improved with respect to any density estimator based on the given data. As an arbitrary density estimator of f, which uses the empirical information contained in the observations $Y_1, \ldots, Y_n$ drawn from (2.1), we may consider any mapping

$$\hat{f} : \mathbb{R} \times \mathbb{R}^n \to \mathbb{C},$$

where we insert the vector $(x; Y_1, \ldots, Y_n)$ as the argument of $\hat{f}$. Recalling the general framework of Sect. 2.3, we assume that $\hat{f}(\cdot; Y_1, \ldots, Y_n)$ as well as the true density f lie in some semi-metric space (F, d) almost surely. Further, we assume that $f \in \mathcal{F}$ for some density class $\mathcal{F} \subseteq F$. As some examples for $\mathcal{F}$, we can consider the Hölder and Sobolev classes as introduced in the previous two subsections. However, we keep our conditions general at this stage.

To give a lower bound on the convergence rates with respect to an arbitrary estimator, we need some mathematical prerequisites. As an important tool used to derive some information distance, we introduce the Hellinger distance

$$H(h, \tilde{h}) = \Big(\int \big(\sqrt{h(x)} - \sqrt{\tilde{h}(x)}\big)^2 \, \mathrm{d}x \Big)^{1/2}, \tag{2.45}$$

between some densities $h, \tilde{h}$. In fact, the Hellinger distance is well defined for all densities h and $\tilde{h}$ because $\sqrt{h}$ is contained in $L_2(\mathbb{R})$ for any density h, which is nonnegative and a member of $L_1(\mathbb{R})$ by definition. Using rather elementary calculus, we obtain

$$\begin{aligned} H^2(h, \tilde{h}) &= \int h(x) \, \mathrm{d}x - 2 \int \sqrt{h(x)\tilde{h}(x)} \, \mathrm{d}x + \int \tilde{h}(x) \, \mathrm{d}x \\ &= 2\Big(1 - \int \sqrt{h(x)}\sqrt{\tilde{h}(x)} \, \mathrm{d}x\Big), \end{aligned}$$

as h and $\tilde{h}$ integrate to one. Therefore, we have the following equation.

$$\int \sqrt{h(x)}\sqrt{\tilde{h}(x)} \, \mathrm{d}x = 1 - \frac{1}{2} H^2(h, \tilde{h}) \tag{2.46}$$

for all densities $h, \tilde{h}$. Furthermore, we derive that

$$\begin{aligned} H^2(h, \tilde{h}) &= \int \big(\sqrt{h(x)} - \sqrt{\tilde{h}(x)}\big)^2 \, \mathrm{d}x = \int \Big(\frac{h(x) - \tilde{h}(x)}{\sqrt{h(x)} + \underbrace{\sqrt{\tilde{h}(x)}}_{\geq 0}} \Big)^2 \, \mathrm{d}x \\ &\leq \int \big(h(x) - \tilde{h}(x)\big)^2 h^{-1}(x) \, \mathrm{d}x = \chi^2(h, \tilde{h}). \end{aligned} \tag{2.47}$$

That latter kind of distance between the densities h and $\tilde{h}$ is called the χ^2-distance. Unlike the Hellinger distance, it is not well defined for all densities; for example, h must be assumed to be nonvanishing, etc. However, we recognize from its definition (in particular, from the nonnegativity of the function to be integrated) that the χ^2-distance may be put equal to $+\infty$ whenever the integral does not exist. Then, the inequality in (2.47) formally holds true for all densities h and $\tilde{h}$. Therefore, we learn that the squared Hellinger distance between two densities is always smaller or equal to the χ^2-distance.

As another important result, we give the following lemma.

Lemma 2.11. (LeCam's inequality) *For all multivariate (d-dimensional, $d \geq 1$) densities $h, \tilde{h}$, we have*

$$\frac{1}{2}\Big(\int \sqrt{h(x)}\sqrt{\tilde{h}(x)}\,\mathrm{d}x\Big)^2 \leq \int \min\{h(x), \tilde{h}(x)\}\,\mathrm{d}x.$$

Proof. Let us consider

$$\begin{aligned}\int \sqrt{h(x)\tilde{h}(x)}\,\mathrm{d}x &= \int \sqrt{\min\{h(x),\tilde{h}(x)\}}\sqrt{\max\{h(x),\tilde{h}(x)\}}\,\mathrm{d}x \\ &\leq \Big(\int \min\{h(x),\tilde{h}(x)\}\,\mathrm{d}x\Big)^{1/2}\Big(\int \underbrace{\max\{h(x),\tilde{h}(x)\}}_{\leq h(x)+\tilde{h}(x)}\,\mathrm{d}x\Big)^{1/2} \\ &\leq \sqrt{2}\Big(\int \min\{h(x),\tilde{h}(x)\}\,\mathrm{d}x\Big)^{1/2},\end{aligned}$$

where the Cauchy–Schwarz inequality has been used with respect to the inner product in $L_2(\mathbb{R})$. Then, the lemma follows by elementary calculation. ■

For more intensive consideration of those distances between densities, see the book of [33]. For our purpose, the tools derived above are satisfactory.

To give a survey on the general structure of the proofs of lower bounds, we need two competing sequences $(h_n)_n$ and $(\tilde{h}_n)_n$ consisting of density functions h_n and $\tilde{h}_n$. More concretely, we imagine that h_n and $\tilde{h}_n$ compete for being the true density of the data. Based on the given observations, we shall be able to distinguish between the setting when either h_n or $\tilde{h}_n$ is the observation density. Nevertheless, there may be some densities for which the decision between h_n and $\tilde{h}_n$ is very hard. That difficulty will be described by a specific distance between h_n and $\tilde{h}_n$, referring to the empirical information contained in the dataset drawn from density h_n and $\tilde{h}_n$, respectively. We see that the χ^2-distance reflects this distinction very well in deconvolution problems. In the density deconvolution model, we apparently have $h_n = f_n * g$ and $\tilde{h}_n = \tilde{f}_n * g$, where $(f_n)_n$ and $(\tilde{f}_n)_n$ are also density sequences in the class $\mathcal{F}$ consisting of all densities admitted to be the true f. Then, the distance between $f_n(x)$ and $\tilde{f}_n(x)$ will give us a lower bound on the convergence rates when considering the MSE.

Those readers who are familiar with the lower bound proofs in the error-free case might wonder why the χ^2-distance is preferred to the Hellinger distance. The reason is that, in deconvolution problems, the Fourier transforms of h_n and $\tilde{h}_n$ must be involved. For that purpose, the best strategy is to estimate the Hellinger distance by the χ^2-distance and, then, by the $L_2(\mathbb{R})$-distance so that Parseval's identity (Theorem A.4) may be employed. Besides, there is obviously no other easy way to represent the Hellinger distance by the Fourier transforms of h_n^{ft} and $\tilde{h}_n^{\mathrm{ft}}$.

Our considerations are formalized in the following proposition.

Proposition 2.12. *Assume two density sequences $(f_n)_n$ and $(\tilde{f}_n)_n$ in some density class $\mathcal{F}$, and a semi-metric space (F, d) along with some $k \geq 1$, where $\mathcal{F} \subseteq F$. Then, with respect to an arbitrary estimator sequence $\hat{f}_n$ of f based on the i.i.d. data $Y_1, \ldots, Y_n$ having the density $h = f * g$, $f \in \mathcal{F}$, which satisfies $\hat{f}_n \in F$ almost surely*

$$\sup_{f \in \mathcal{F}} E\, d^k(\hat{f}_n, f) \geq const \cdot d^k(f_n, \tilde{f}_n)\,,$$

for n sufficiently large, if we have

$$\chi^2(h_n, \tilde{h}_n) = O(1/n), \tag{2.48}$$

where $h_n = f_n * g$ and $\tilde{h}_n = \tilde{f}_n * g$.

Proof. As $f_n, \tilde{f}_n \in \mathcal{F}$ for all n and the data $Y_1, \ldots, Y_n$ are i.i.d., the notation E_{f_n} and $E_{\tilde{f}_n}$ indicates that any Y_j occurring in the estimator has the density h_n and $\tilde{h}_n$, respectively, and we have

$$\begin{aligned}
&\sup_{f \in \mathcal{F}} E\, d^k(\hat{f}_n, f) \geq \frac{1}{2}\Big(E_{f_n}\, d^k(\hat{f}_n, f_n) + E_{\tilde{f}_n}\, d^k(\hat{f}_n, \tilde{f}_n)\Big) \\
&= \frac{1}{2}\Big(\int \cdots \int d^k\big(\hat{f}_n(\cdot; y_1, \ldots, y_n), f_n\big) \prod_{j=1}^n h_n(y_j)\, \mathrm{d}y_1 \cdots \mathrm{d}y_n \\
&\qquad + \int \cdots \int d^k\big(\hat{f}_n(\cdot; y_1, \ldots, y_n), \tilde{f}_n\big) \prod_{j=1}^n \tilde{h}_n(y_j)\, \mathrm{d}y_1 \cdots dy_n\Big) \\
&\geq \frac{1}{2}\Big(\int \cdots \int \big[d^k\big(\hat{f}_n(\cdot; y_1, \ldots, y_n), f_n\big) + d^k\big(\hat{f}_n(\cdot; y_1, \ldots, y_n), \tilde{f}_n\big)\big] \\
&\qquad \cdot \min\Big\{\prod_{j=1}^n h_n(y_j), \prod_{j=1}^n \tilde{h}_n(y_j)\Big\}\, \mathrm{d}y_1 \cdots \mathrm{d}y_n\Big) \\
&\geq 2^{-k}\Big(\int \cdots \int \big[d\big(\hat{f}_n(\cdot; y_1, \ldots, y_n), f_n\big) + d\big(\hat{f}_n(\cdot; y_1, \ldots, y_n), \tilde{f}_n\big)\big]^k \\
&\qquad \cdot \min\Big\{\prod_{j=1}^n h_n(y_j)\,, \prod_{j=1}^n \tilde{h}_n(y_j)\Big\}\, \mathrm{d}y_1 \cdots \mathrm{d}y_n\Big) \\
&\geq 2^{-k}\, d^k(f_n, \tilde{f}_n) \int \cdots \int \min\Big\{\prod_{j=1}^n h_n(y_j)\,, \prod_{j=1}^n \tilde{h}_n(y_j)\Big\}\, \mathrm{d}y_1 \cdots \mathrm{d}y_n,
\end{aligned} \tag{2.49}$$

where we have used the elementary inequality $a^k + b^k \geq 2^{1-k} \cdot (a+b)^k$ for all $a, b > 0$, $k \geq 1$ and, then, the triangle inequality with respect to the semi-metric d. Apparently, the proof is completed when we show that

$$\int \cdots \int \min\Big\{\prod_{j=1}^n h_n(y_j), \prod_{j=1}^n \tilde{h}_n(y_j)\Big\} \, dy_1 \cdots dy_n \geq \text{const} > 0 \quad (2.50)$$

holds true for all integers $n > N$ and some fixed integer N. By Lemma 2.11 – applied to the multivariate densities $\prod_{j=1}^n h_n(\cdot_j)$ and $\prod_{j=1}^n \tilde{h}_n(\cdot_j)$ – we have

$$\begin{aligned}
&\int \cdots \int \min\Big\{\prod_{j=1}^n h_n(y_j), \prod_{j=1}^n \tilde{h}_n(y_j)\Big\} \, dy_1 \cdots dy_n \\
&\geq \frac{1}{2}\Big(\int \cdots \int \Big(\prod_{j=1}^n h_n(y_j)\Big)^{1/2} \Big(\prod_{j=1}^n \tilde{h}_n(y_j)\Big)^{1/2} dy_1 \cdots dy_n\Big)^2 \\
&= \frac{1}{2}\Big(\int \sqrt{h_n(y)}\sqrt{\tilde{h}_n(y)} \, dy\Big)^{2n}.
\end{aligned}$$

Using (2.46) and (2.47), we have

$$\begin{aligned}
\frac{1}{2}\Big(\int \sqrt{h_n(y)}\sqrt{\tilde{h}_n(y)} \, dy\Big)^{2n} &= \frac{1}{2}\Big(1 - \frac{1}{2}H^2(h_n, \tilde{h}_n)\Big)^{2n} \\
&\geq \frac{1}{2}\Big(1 - \frac{1}{2}\chi^2(h_n, \tilde{h}_n)\Big)^{2n} \\
&\geq \frac{1}{2}\Big(1 - c/n\Big)^{2n}
\end{aligned}$$

for n sufficiently large and some constant $c > 0$, where condition (2.48) has been employed in the last step. From analysis, we learn that the last term in the above equation converges to $\exp(-2c)/2 > 0$ as $n \to \infty$; hence we may conclude that, for n large enough, the latter term is bounded away from zero so that the validity of (2.50) is evident now. That completes the proof of the proposition. ■

Proposition 2.12 provides the general framework for deriving lower bounds on the convergence rates with respect to the MSE. Its application is as follows: We have to construct two density sequences $(f_n)_n$ and $(\tilde{f}_n)_n$ where we have to check $f_n, \tilde{f}_n \in \mathcal{F}$ for all n. Then, we have to verify (2.48), that is, the χ^2-distance of $f_n * g$ and $\tilde{f}_n * g$ must tend to zero with the rate $1/n$. Then, we conclude that there exists no estimator that achieves convergence rates faster than $d^k(f_n, \tilde{f}_n)$. We put $d(f, \tilde{f}) = |f(x) - \tilde{f}(x)|$ and $k = 2$ as we want to study the MSE. However, the appropriate selection of those sequences is not an easy task as they have to capture the whole difficulty of the specific estimation problem. In the deconvolution setting, the distance between the convolutions of these densities with g shall converge to zero very fast, while convergence of the term $|f_n(x) - \tilde{f}_n(x)|^2$ shall be rather slow. Such sequences are useful to show lower bounds.

Now we turn to our specific Hölder class $\mathcal{F} = \mathcal{F}_{\beta,C,\delta;x}$, as defined in (2.29), and focus on the specific rates derived in Theorem 2.8. For the sake of simplicity, we restrict our consideration to the case $x = 0$, although the results are extendable to any other $x \in \mathbb{R}$. To construct the sequences $(f_n)_n$ and $(\tilde{f}_n)_n$ from Proposition 2.12 explicitly, we consider the density

$$f_0(x) = \frac{1 - \cos x}{\pi x^2},$$

from (2.24) again, with its triangle-shaped Fourier transform $f_0^{\mathrm{ft}}(t) = (1 - |t|)_+$. Also, we introduce the supersmooth Cauchy density

$$f_1(x) = 1/\big[\pi(1 + x^2)\big], \tag{2.51}$$

where $f_1^{\mathrm{ft}}(t) = \exp(-|t|)$. For both (2.24) and (2.51), the Fourier transform is elementarily calculable by Fourier inversion (see Theorem A.2). Now we define the density sequences from Proposition 2.12 by

$$\begin{aligned} f_n(x) &= f_1(x), \\ \tilde{f}_n(x) &= f_1(x) + b_n \cos(3x/a_n) f_0(x/a_n), \end{aligned} \tag{2.52}$$

where $(a_n)_n$ and $(b_n)_n$ denote some positive-valued sequences tending to zero; they will be specified later. First, we will check that $\tilde{f}_n$ is a density function. We realize that $f_1(x) \geq f_1(x/a_n) \geq c \cdot f_0(x/a_n) \geq 0$ for all x and n large enough and some appropriate constant $c > 0$ as f_1 is an even function and monotonously decreasing on $[0, \infty)$. Then, we conclude that

$$\tilde{f}_n(x) \geq f_1(x) \cdot \big[1 - b_n/c\big] \geq \frac{1}{2} f_1(x) > 0$$

for all n sufficiently large as $(b_n)_n \downarrow 0$ and $|\cos x| \leq 1$ for all real numbers x. In fact, we are only interested in large n as we focus on the convergence rates. Then, nonnegativity of $\tilde{f}_n$ follows. As $\tilde{f}_n$ is a linear combination of finitely many $L_1(\mathbb{R})$-functions, we may also fix $\tilde{f}_n \in L_1(\mathbb{R})$. Let us now consider the Fourier transforms of f_n and $\tilde{f}_n$. We have

$$\begin{aligned} f_n^{\mathrm{ft}}(t) &= f_1^{\mathrm{ft}}(t), \\ \tilde{f}_n^{\mathrm{ft}}(t) &= f_1^{\mathrm{ft}}(t) + \frac{a_n b_n}{2} f_0^{\mathrm{ft}}\big(a_n t + 3\big) + \frac{a_n b_n}{2} f_0^{\mathrm{ft}}\big(a_n t - 3\big), \end{aligned}$$

where we have used Lemma A.1(a), (e), and (f) in the appendix to calculate those Fourier transforms. Since f_0^{ft} is supported on $[-1, 1]$, we have $\int \tilde{f}_n(x)\,\mathrm{d}x = \tilde{f}_n^{\mathrm{ft}}(0) = 1$, so that we are guaranteed that both f_n and $\tilde{f}_n$ are density functions for n sufficiently large.

Next, our goal is to verify that $f_n, \tilde{f}_n \in \mathcal{F}_{\beta,C,\delta;x=0}$ for n sufficiently large, at least. With respect to f_n, we notice that we are dealing with a constant sequence, that is, independent of n. As f_n is differentiable infinitely often and bounded on the whole real line, we may arrange the constant C sufficiently large so that $f_n \in \mathcal{F}_{\beta,C,\delta;0}$ because, for any $|y|, |\tilde{y}| \leq \delta$, we have

$$\begin{aligned} \left|f_n^{(\lfloor\beta\rfloor)}(\tilde{y}) - f_n^{(\lfloor\beta\rfloor)}(y)\right| &\leq \min\left\{2 \sup_{|z|\leq\delta} \left|f_n^{(\lfloor\beta\rfloor)}(z)\right|, |\tilde{y} - y| \cdot \sup_{|z|\leq\delta} \left|f_n^{(\lfloor\beta\rfloor+1)}(z)\right|\right\} \\ &\leq O(1)\min\{1, |\tilde{y} - y|\} \leq O(1) \cdot |\tilde{y} - y|^{\beta-\lfloor\beta\rfloor}. \end{aligned} \tag{2.53}$$

The only property needed is the fact that continuous functions are bounded on compact intervals.

Concerning $\tilde{f}_n$, we easily derive boundedness on the whole real line as f_1 and f_0 are bounded functions, too, and $(b_n)_n \downarrow 0$. Now we calculate the $\lfloor\beta\rfloor$th derivative of $\tilde{f}_n$. We obtain that

$$\tilde{f}_n^{(\lfloor\beta\rfloor)}(y) = f_1^{(\lfloor\beta\rfloor)}(y) + b_n a_n^{-\lfloor\beta\rfloor} \varphi^{(\lfloor\beta\rfloor)}(x/a_n),$$

where $\varphi(x) = \cos(3x) f_0(x)$. We may fix that any derivative of φ is bounded on the whole of $\mathbb{R}$. This is a technical result that can be shown by elementary analysis. Also, we can show that $\varphi^{(\lfloor\beta\rfloor)}$ satisfies the following Hölder condition as in (2.53).

$$\left|\varphi^{(\lfloor\beta\rfloor)}(\tilde{y}) - \varphi^{(\lfloor\beta\rfloor)}(y)\right| \leq O(1) \cdot |\tilde{y} - y|^{\beta-\lfloor\beta\rfloor},$$

for all $y \in \mathbb{R}$. Then, we may conclude that, for any $|y| \leq \delta$, $|\tilde{y}| \leq \delta$,

$$\begin{aligned} \left|\tilde{f}_n^{(\lfloor\beta\rfloor)}(\tilde{y}) - \tilde{f}_n^{(\lfloor\beta\rfloor)}(y)\right| &\leq \left|f_1^{(\lfloor\beta\rfloor)}(\tilde{y}) - f_1^{(\lfloor\beta\rfloor)}(y)\right| \\ &\qquad + O(1) \cdot b_n a_n^{-\lfloor\beta\rfloor} \, |y|^{\beta-\lfloor\beta\rfloor} a_n^{\lfloor\beta\rfloor-\beta} \\ &\leq \left[O(1) + O\left(b_n a_n^{-\beta}\right)\right] \cdot |\tilde{y} - y|^{\beta-\lfloor\beta\rfloor}. \end{aligned}$$

Finally, we can verify the membership of both f_n and $\tilde{f}_n$ in $\mathcal{F}_{\beta,C,\delta;x}$ with $C > 0$ sufficiently large under the selection

$$b_n = \text{const} \cdot a_n^{\beta}. \tag{2.54}$$

Condition (2.48) remains to be shown. Let us consider the χ^2-distance between h_n and $\tilde{h}_n$. First, we obtain that

$$\begin{aligned} h_n(x) &= \int f_n(x-y) g(y)\, dy \geq \int f_1(x-y) g(y)\, dy \\ &\geq \frac{1}{\pi} \int \frac{1}{1 + 2x^2 + 2y^2} g(y)\, dy \geq \frac{1}{\pi} \int_{|y|\leq\eta} \frac{1}{1 + 2x^2 + 2y^2} g(y)\, dy \\ &\geq \frac{1}{\pi} \frac{1}{1 + 2x^2 + 2\eta^2} \int_{|y|\leq\eta} g(y)\, dy \geq C_\eta \cdot (1 + x^2)^{-1} \int_{|y|\leq\eta} g(y)\, dy, \end{aligned} \tag{2.55}$$

for any $\eta > 0$ and some corresponding constant $C_\eta > 0$. We have used the nonnegativity of the density function g. As g integrates to one, we may select

$\eta > 0$ sufficiently large (but independently of n) so that $\int_{|y|\leq\eta} g(y)\,\mathrm{d}y \geq 1/2$. Then, we obtain that

$$h_n(x) \geq \text{const}\cdot(1+x^2)^{-1}, \text{ for all } x \in \mathbb{R}.$$

Using the linearity of the convolution operation and the above equation, we derive that

$$\begin{aligned}\chi^2(h_n,\tilde{h}_n) &= \int h_n^{-1}(x)\,\big|\big[g*(f_n-\tilde{f}_n)\big](x)\big|^2\,\mathrm{d}x\\ &\leq O(1)\cdot\int(1+x^2)\big|\big[g*(f_n-\tilde{f}_n)\big](x)\big|^2\,\mathrm{d}x \qquad (2.56)\end{aligned}$$

As the convolution of any $L_2(\mathbb{R})$-function with some density lies in $L_2(\mathbb{R})$, we have $\Delta = g*\big(f_n-\tilde{f}_n\big) \in L_2(\mathbb{R})$. We use Fourier inversion in $L_2(\mathbb{R})$, Parseval's identity (see Theorem A.4) and Lemma A.6 from the appendix on Fourier-analytic tools to establish

$$\int\big|x\Delta(x)\big|^2\,\mathrm{d}x = \frac{1}{4\pi^2}\int\big|x\big(\Delta^{\mathrm{ft}}\big)^{\mathrm{ft}}(-x)\big|^2\,\mathrm{d}x = \frac{1}{2\pi}\int\big|(\Delta^{\mathrm{ft}})'(t)\big|^2\,\mathrm{d}x, \tag{2.57}$$

where $\big(\Delta^{\mathrm{ft}}\big)'$ is to be understood as a weak derivative (again, see appendix on Fourier analysis). We have

$$\begin{aligned}\Delta^{\mathrm{ft}}(t) = g^{\mathrm{ft}}(t)\big(f_n^{\mathrm{ft}}(t)-\tilde{f}_n^{\mathrm{ft}}(t)\big) &= g^{\mathrm{ft}}(t)\frac{a_nb_n}{2}f_0^{\mathrm{ft}}(a_nt+3)\\ &\quad + g^{\mathrm{ft}}(t)\frac{a_nb_n}{2}f_0^{\mathrm{ft}}(a_nt-3).\end{aligned}$$

Hence,

$$\begin{aligned}(\Delta^{\mathrm{ft}})'(t) &= (g^{\mathrm{ft}})'(t)\frac{a_nb_n}{2}f_0^{\mathrm{ft}}(a_nt+3) + (g^{\mathrm{ft}})'(t)\frac{a_nb_n}{2}f_0^{\mathrm{ft}}(a_nt-3)\\ &\quad + g^{\mathrm{ft}}(t)\frac{a_n^2b_n}{2}(f_0^{\mathrm{ft}})'(a_nt+3) + g^{\mathrm{ft}}(t)\frac{a_n^2b_n}{2}(f_0^{\mathrm{ft}})'(a_nt-3).\end{aligned}$$

Using those results along with Parseval's identity, we obtain that (2.56) is bounded above by

$$\begin{aligned}&O(a_n^2b_n^2)\cdot\int\big|g^{\mathrm{ft}}(t)f_0^{\mathrm{ft}}(a_nt+3)\big|^2\,\mathrm{d}t + O(a_n^2b_n^2)\cdot\int\big|g^{\mathrm{ft}}(t)f_0^{\mathrm{ft}}(a_nt-3)\big|^2\,\mathrm{d}t\\ &+ O(a_n^2b_n^2)\cdot\int\big|(g^{\mathrm{ft}})'(t)f_0^{\mathrm{ft}}(a_nt+3)\big|^2\,\mathrm{d}t\\ &\qquad\qquad + O(a_n^2b_n^2)\cdot\int\big|(g^{\mathrm{ft}})'(t)f_0^{\mathrm{ft}}(a_nt-3)\big|^2\,\mathrm{d}t\\ &+ O(a_n^4b_n^2)\cdot\int\big|g^{\mathrm{ft}}(t)(f_0^{\mathrm{ft}})'(a_nt+3)\big|^2\,\mathrm{d}t\\ &\qquad\qquad + O(a_n^4b_n^2)\cdot\int\big|g^{\mathrm{ft}}(t)(f_0^{\mathrm{ft}})'(a_nt-3)\big|^2\,\mathrm{d}t.\end{aligned}$$

Therefore, we assume that g^{ft} is continuously differentiable on the real line, and that $(g^{\mathrm{ft}})'$ has $\mathrm{const} \cdot |t|^{-\alpha}$ or $\mathrm{const} \cdot \exp\big(- d'|t|^{\gamma}\big)$ for some $d' > 0$ as an upper bound if g satisfies (2.31) or (2.32), respectively. Although $f_0^{\mathrm{ft}}(t) = (1-|t|)_+$ is not differentiable at $t \in \{-1,0,1\}$, it is a standard example for a weakly differentiable $L_2(\mathbb{R})$-function. Its weak derivative is equal to $(f_0^{\mathrm{ft}})'(t) = \chi_{(-1,0)}(t) - \chi_{(0,1)}(t)$. Therefore, both $f_0^{\mathrm{ft}}(a_n \cdot +3)$ and $(f_0^{\mathrm{ft}})'(a_n \cdot +3)$ are bounded by 1 and supported on $[-4/a_n, -2/a_n] \cup [2/a_n, 4/a_n]$ so that (2.56) is bounded above by

$$O\Big(a_n^2 b_n^2 \int_{2/a_n \le |t| \le 4/a_n} \big(\big|g^{\mathrm{ft}}(t)\big|^2 + \big|(g^{\mathrm{ft}})'(t)\big|^2\big)\, \mathrm{d}t\Big),$$

as $(a_n)_n \downarrow 0$ has already been assumed. Inserting the upper bounds on g^{ft} and $(g^{\mathrm{ft}})'$ as assumed above leads to the following upper bounds on (2.56),

$$O\big(a_n^{1+2\alpha} b_n^2\big) \qquad \text{and} \qquad O\big(a_n \exp(-D' a_n^{-\gamma}) b_n^2\big),$$

for some constant $D' > 0$ as the upper bounds in the ordinary smooth and the supersmooth case, respectively. We recall that the selection (2.54) satisfies the smoothness constraints. Finally, we derive that (2.56) is bounded above by

$$O\big(a_n^{1+2\alpha+2\beta}\big) \qquad \text{and} \qquad O\big(a_n^{1+2\beta} \exp(-D' a_n^{-\gamma})\big),$$

respectively. Consequently, the condition (2.48) in Proposition 2.12 is satisfied under the selection $a_n = \mathrm{const} \cdot n^{-1/(1+2\beta+2\alpha)}$ and $a_n = \big(c_a \cdot \log n\big)^{-1/\gamma}$, respectively, with some constant $c_a > 1/D'$. Now, we are in the position to apply Proposition 2.12. It follows from there that

$$\begin{aligned}\sup_{f \in \mathcal{F}_{\beta,C,\delta;0}} E\big|\hat f(0) - f(0)\big|^2 \ge \mathrm{const} \cdot \big|f_n(0) - \tilde f_n(0)\big|^2 &\ge \mathrm{const} \cdot b_n^2\, f_0^2(0) \\ &\ge \mathrm{const} \cdot b_n^2.\end{aligned}$$

Combining (2.54) with the selection of $(a_n)_n$ as above, we have shown the following theorem.

Theorem 2.13. *Assume that $\beta > 1/2$, $C, \delta > 0$ sufficiently large; and consider the density class $\mathcal{F}_{\beta,C,\delta;x}$.*
(a) Assume that the error density g satisfies (2.31) and, in addition, $\big|(g^{\mathrm{ft}})'(t)\big| \le \mathrm{const} \cdot |t|^{-\alpha}$. Then, for an arbitrary estimator $\hat f$ of f, based on the i.i.d. data $Y_1, \ldots, Y_n$ from model (2.1), we have

$$\sup_{f \in \mathcal{F}_{\beta,C,\delta;x}} E\big|\hat f(x) - f(x)\big|^2 \ge \mathrm{const} \cdot n^{-2\beta/(2\beta+2\alpha+1)},$$

for n sufficiently large.
(b) Assume that the error density g satisfies (2.32) and, in addition, $\big|(g^{\mathrm{ft}})'(t)\big| \le \mathrm{const} \cdot \exp(-D'|t|^{\gamma})$. Then, for an arbitrary estimator $\hat f$ of f, based on the i.i.d. data $Y_1, \ldots, Y_n$ from model (2.1), we have

$$\sup_{f \in \mathcal{F}_{\beta,C,\delta;x}} E\big|\hat{f}(x) - f(x)\big|^2 \geq const \cdot \big(\log n\big)^{-2\beta/\gamma}.$$

for n sufficiently large.

We notice that the lower bounds on the convergence rates established by Theorem 2.13 with respect to an arbitrary density estimator correspond to the upper bounds derived in Theorem 2.8 for the deconvolution kernel estimator (2.7). Therefore, we may conclude that the deconvolution kernel density estimator attains optimal convergence rates under common smoothness constraints on the target density f for ordinary smooth and supersmooth error densities g, which satisfy some slight additional technical conditions with respect to the derivative of their Fourier transforms. The optimality of those rates was first proved by [43] where slightly different assumptions and techniques are used.

Rate-optimality has become an important criterion to evaluate the quality of an estimation procedure in nonparametric statistics in general. The investigation of optimal rates of convergence and, in some problems, the even more precise study of optimal asymptotic constants are often referred to as minimax theory. However, nowadays, it is sometimes criticized that research is focused too heavily on the asymptotics as, in practice, we only have finitely many data. In econometric applications, the sample sizes are usually larger than in biometrics where asymptotic results may be less valuable. Nevertheless, one should consider that asymptotics have not been developed to make nonparametric statistics mathematically more attractive; but rather due to the fact that so far one has not been able to prove that some estimator minimizes the MSE or the MISE for some fixed finite sample size for the overwhelming majority of problems in nonparametric function estimation. The finite sample size performance is frequently studied by numerical simulations. However, then, we can only simulate an estimator for finitely many target densities f, but certainly not for a whole nonparametric class of densities. Therefore, we are not guaranteed that the specific true density for real data is included in the simulations. Hence, both asymptotics and numerical simulations are just two approaches to reality, which all have some disadvantages. Hence, both asymptotics and numerical simulations should be considered to evaluate the quality and practical performance of an estimator.

2.4.6 Mean Integrated Squared Error: Lower Bounds

In the spirit of the previous subsection, we show that the convergence rates derived for the MISE in Theorem 2.9 are optimal with respect to any estimator of f, too. However, the competing density sequences derived in (2.52) to establish lower bounds for the MSE are not useful to obtain sufficiently large lower bounds on the rates of the MISE.

Instead of two competing density sequences, we consider a randomized class of densities with respect to the MISE, which are all admitted to be the true f. Furthermore, f shall be contained in the Sobolev class $\mathcal{F}_{\beta,C;L_2}$ as defined in (2.30). We introduce the parametric set $\mathcal{F}'_n$ of all the densities

$$f_\theta(x) = \frac{1}{2}\big(f_0(x) + f_1(x)\big) + b_n \sum_{j=K_n}^{2K_n} \theta_j \cos(2jx) f_0(x),$$

where the parameter sequence $(b_n) \downarrow 0$ and the integer-valued sequence $(K_n) \uparrow \infty$ are still to be chosen. The densities f_0 and f_1 are as in the proof of Theorem 2.13 and defined in (2.24) and (2.51). We write $\theta = \big(\theta_{K_n}, \ldots, \theta_{2K_n}\big)$, where each component satisfies $\theta_j \in \{0,1\}$. Then, we can show that any $f \in \mathcal{F}'_n$ is a density under suitable selection of $(b_n)_n$. As f_0 and f_1 are density functions and, hence, integrable on the real line, we notice that all f_θ are contained in $L_1(\mathbb{R})$. Furthermore, as $f_1(x) \geq 0$ and $|\cos x| \leq 1$ for all $x \in \mathbb{R}$, we have

$$f_\theta(x) \geq \frac{1}{2} f_1(x) + f_0(x) \cdot \Big(\frac{1}{2} - b_n K_n\Big),$$

where we stipulate that $b_n \leq 1/(2K_n)$ so that nonnegativity of any f_θ and, even stronger, the inequality

$$f(x) \geq \frac{1}{2} f_1(x), \tag{2.58}$$

for all $x \in \mathbb{R}$ and all $f \in \mathcal{F}'_n$ can be ensured. It remains to be shown that f_θ integrates to one. Therefore, we consider the Fourier transform

$$f_\theta^{\mathrm{ft}}(t) = \frac{1}{2}\big(f_0^{\mathrm{ft}}(t) + f_1^{\mathrm{ft}}(t)\big) + \frac{1}{2} b_n \sum_{j=K_n}^{2K_n} \theta_j f_0^{\mathrm{ft}}(t-2j) + \frac{1}{2} b_n \sum_{j=K_n}^{2K_n} \theta_j f_0^{\mathrm{ft}}(t+2j), \tag{2.59}$$

where we have used Lemma A.1(a) and (f) from the appendix. As derived in the previous section, we have $f_0^{\mathrm{ft}}(t) = (1-|t|)_+$ so that $f_0^{\mathrm{ft}}(\pm 2j) = 0$ for all $j \in \{K_n, \ldots, 2K_n\}$. Since f_0 and f_1 are densities we have $f_0^{\mathrm{ft}}(0) = f_1^{\mathrm{ft}}(0) = 1$, implying that

$$\int f_\theta(x)\,\mathrm{d}x = f_\theta^{\mathrm{ft}}(0) = 1.$$

Finally, we conclude that all $f \in \mathcal{F}'_n$ are density functions.

Furthermore, all the f_θ shall be included in $\mathcal{F}_{\beta,C;L_2}$, that is, we have to show $\mathcal{F}'_n \subseteq \mathcal{F}_{\beta,C;L_2}$. We consider

$$\begin{aligned}\int \big|f_\theta^{\mathrm{ft}}(t)\big|^2 |t|^{2\beta}\,\mathrm{d}t \leq \frac{1}{2} \int \big|f_0^{\mathrm{ft}}(t) + f_1^{\mathrm{ft}}(t)\big|^2 |t|^{2\beta}\,\mathrm{d}t \\ + 2 \sum_{K_n \leq |j| \leq 2K_n} \int_{2j-1}^{2j+1} b_n^2 \theta_j \big|f_0^{\mathrm{ft}}(t+2j)\big|^2 |t|^{2\beta}\,\mathrm{d}t\end{aligned}$$

$$\leq \text{const} + O\big(b_n^2\big) \sum_{j=K_n}^{2K_n} j^{2\beta}$$
$$\leq \text{const} + O\big(b_n^2 K_n^{2\beta+1}\big),$$

where we have used that f_1^{ft} has exponential tails and f_0^{ft} is supported on $[-1,1]$; further, as the $|f_0^{\text{ft}}(t+2j)|$ are bounded by 1 and supported on $[-2j-1,-2j+1]$, their supports are almost disjoint (i.e., disjoint up to a Lebesgue zero set) so that the integral can be split into the sum of those over the intervals $[-2j-1,-2j+1]$. Hence, we notice that, for C sufficiently large, we can fix $\mathcal{F}_n' \subseteq \mathcal{F}_{\beta,C;L_2}$ under the selection

$$b_n = \text{const} \cdot K_n^{-\beta-1/2}. \tag{2.60}$$

We assume that $\beta > 1/2$ so that the condition $b_n \leq 1/(2K_n)$ from above is not violated.

We introduce some i.i.d. Bernoulli random variables $\hat\theta_{K_n},\ldots,\hat\theta_{2K_n}$ where $P[\hat\theta_{K_n} = 1] = 1/2$. We summarize them as the vector-valued random variable $\hat\theta = (\hat\theta_{K_n},\ldots,\hat\theta_{2K_n})$. Then, we choose the strategy of inserting $\hat\theta$ as the vector θ in the densities f_θ from $\mathcal{F}_n'$. We realize that $f_{\hat\theta}$ is a functional (i.e., function-valued) random variable taking its values in the set $\mathcal{F}_n'$. Using $\mathcal{F}_n' \subseteq \mathcal{F}_{\beta,C;L_2}$, we have

$$\sup_{f\in\mathcal{F}_{\beta,C;L_2}} E\|\hat f - f\|_2^2 \geq \sup\nolimits_{f\in\mathcal{F}_n'} E\|\hat f - f\|_2^2 \geq E_\theta\, E\big(\|\hat f - f_{\hat\theta}\|_2^2 \mid \hat\theta\big),$$

where E_θ denotes the expectation with respect to the random vector $\hat\theta$ and $E(X|Y)$ denotes the conditional expectation of X given Y as usual. Utilizing Parseval's identity (Theorem A.4) and Fubini's theorem, we obtain that the above inequality is continued by

$$= \frac{1}{2\pi} E_\theta\, E\left(\int \left|\hat f^{\text{ft}}(t) - f_{\hat\theta}^{\text{ft}}(t)\right|^2 \mathrm{d}t \mid \hat\theta\right)$$
$$\geq \frac{1}{2\pi}\sum_{j=K_n}^{2K_n} E_\theta\, E\left(\int_{2j-1}^{2j+1} \left|\hat f^{\text{ft}}(t) - f_{\hat\theta}^{\text{ft}}(t)\right|^2 \mathrm{d}t \mid \hat\theta\right)$$
$$= \frac{1}{2\pi}\sum_{j=K_n}^{2K_n} E_\theta\, E\left(\int_{2j-1}^{2j+1} \left|\hat f^{\text{ft}}(t) - f_1^{\text{ft}}(t)/2 - b_n\hat\theta_j f_0^{\text{ft}}(t-2j)\right|^2 \mathrm{d}t \mid \hat\theta\right)$$
$$= \frac{1}{2\pi}\sum_{j=K_n}^{2K_n} E_\theta\, \frac{1}{2}\left(E\left(\int_{2j-1}^{2j+1} \left|\hat f^{\text{ft}}(t) - f_1^{\text{ft}}(t)/2 - b_n f_0^{\text{ft}}(t-2j)\right|^2 \mathrm{d}t \middle| \hat\theta_{j,0}\right)\right.$$
$$\left. + E\left(\int_{2j-1}^{2j+1} \left|\hat f^{\text{ft}}(t) - f_1^{\text{ft}}(t)/2\right|^2 \mathrm{d}t \middle| \hat\theta_{j,1}\right)\right),$$

where, in the latter step, we have calculated the expectation with respect to $\hat\theta_j$. Furthermore, we use the notation $\hat\theta_{j,b} = (\hat\theta_{K_n},\ldots,\hat\theta_{j-1}, b, \hat\theta_{j+1},\ldots,\hat\theta_{2K_n})$

for $b \in \{0,1\}$. Then, inserting the integral representation of the expectation, the above term is bounded below by

$$\frac{1}{2\pi}\sum_{j=K_n}^{2K_n} E_\theta \frac{1}{2}\Big(\int\cdots\int\int_{2j-1}^{2j+1} \big|\hat f^{\rm ft}(t;y_1,\ldots,y_n) - f_1^{\rm ft}(t)/2 - b_n f_0^{\rm ft}(t-2j)\big|^2\,dt \cdot \prod_{k=1}^n h_{\hat\theta_{j,0}}(y_k)\,dy_1\cdots dy_n$$

$$+ \int\cdots\int\int_{2j-1}^{2j+1} \big|\hat f^{\rm ft}(t;y_1,\ldots,y_n) - f_1^{\rm ft}(t)/2\big|^2\,dt \cdot \prod_{k=1}^n h_{\hat\theta_{j,1}}(y_k)\,dy_1\cdots dy_n\Big)$$

$$\geq \frac{1}{8\pi} b_n^2 \sum_{j=K_n}^{2K_n} E_\theta \int\cdots\int\int_{2j-1}^{2j+1} \big|f_0^{\rm ft}(t-2j)\big|^2\,dt \cdot \min\Big\{\prod_{k=1}^n h_{\hat\theta_{j,0}}(y_k), \prod_{k=1}^n h_{\hat\theta_{j,1}}(y_k)\Big\}\,dy_1\cdots dy_n$$

$$\geq \text{const}\cdot b_n^2 K_n \min_{j\in\{K_n,\ldots,2K_n\}} E_\theta \int\cdots\int \min\Big\{\prod_{k=1}^n h_{\hat\theta_{j,0}}(y_k), \prod_{k=1}^n h_{\hat\theta_{j,1}}(y_k)\Big\}\,dy_1\cdots dy_n,$$

where $h_{\hat\theta_{j,b}} = f_{\hat\theta_{j,b}} * g$ for $b \in \{0,1\}$. Hence, we have shown that $\sup_{f\in\mathcal{F}_{\beta,C;L_2}} E\|\hat f - f\|_2^2$ is bounded below by $\text{const}\cdot b_n^2 K_n$ if

$$\min_{j\in\{K_n,\ldots,2K_n\}} \min_{\theta\in\{0,1\}^{K_n+1}} \int\cdots\int \min\Big\{\prod_{k=1}^n h_{\theta_{j,0}}(y_k), \prod_{k=1}^n h_{\theta_{j,1}}(y_k)\Big\}dy_1\cdots dy_n \tag{2.61}$$

is bounded away from zero as $n\to\infty$. We can proceed in a similar way as in the proof of Proposition 2.12. Using Lemma 2.11 combined with (2.46) and (2.47), we realize that (2.61) is bounded below by

$$\min_{j\in\{K_n,\ldots,2K_n\}} \min_{\theta\in\{0,1\}^{K_n+1}} \Big(1 - \frac{1}{2}\chi^2(h_{\theta_{j,0}}, h_{\theta_{j,1}})\Big)^{2n}.$$

Therefore, it remains to be shown that

$$\max_{j\in\{K_n,\ldots,2K_n\}} \max_{\theta\in\{0,1\}^{K_n+1}} \chi^2(h_{\theta_{j,0}}, h_{\theta_{j,1}}) \leq c/n, \tag{2.62}$$

where the constant $c > 0$ must not depend on the vector θ nor on j. Following from (2.58), we have

$$h_{\theta_{j,b}}(x) \geq \int f_{\theta_{j,b}}(x-y)g(y)\,dy \geq \frac{1}{2}[f_1 * g](x),$$

for all $x \in \mathbb{R}$ and $b \in \{0,1\}$. Hence, we obtain that

$$\chi^2(h_{\theta_{j,0}}, h_{\theta_{j,1}}) \le 2 \int [f_1 * g]^{-1}(x) \left|h_{\theta_{j,0}}(x) - h_{\theta_{j,1}}(x)\right|^2 dx$$
$$\le 2b_n^2 \int [f_1 * g]^{-1}(x) \left|g * \big(\cos(2j\cdot)f_0\big)\right|^2 dx.$$

It follows as in (2.55) that

$$\max_{\theta \in \{0,1\}^{K_n+1}} \chi^2(h_{\theta_{j,0}}, h_{\theta_{j,1}}) \le O\big(b_n^2\big) \cdot \int (1+x^2) \left|\left[g * \big(\cos(2j\cdot)f_0\big)(x)\right]\right|^2 dx.$$

As in the proof of Theorem 2.13, we apply the Fourier-analytic result (2.57) along with the concepts of weak derivatives (see the appendix on Fourier analysis) to obtain that

$$\begin{aligned}
&\chi^2(h_{\theta_{j,0}}, h_{\theta_{j,1}}) \\
&\le O\big(b_n^2\big) \cdot \int \Big(\left|g^{ft}(t) f_0^{ft}(t-2j)\right|^2 + \left|\left[g^{ft}(\cdot) f_0^{ft}(\cdot - 2j)\right]'(t)\right|^2\Big) dt \\
&\qquad + O\big(b_n^2\big) \cdot \int \Big(\left|g^{ft}(t) f_0^{ft}(t+2j)\right|^2 + \left|\left[g^{ft}(\cdot) f_0^{ft}(\cdot + 2j)\right]'(t)\right|^2\Big) dt \\
&\le O\big(b_n^2\big) \cdot \int \Big(\left|g^{ft}(t) f_0^{ft}(t-2j)\right|^2 + \left|g^{ft}(t) f_0^{ft}(t+2j)\right|^2\Big) dt \\
&\qquad + O\big(b_n^2\big) \cdot \int \Big(\left|(g^{ft})'(t) f_0^{ft}(t-2j)\right|^2 + \left|(g^{ft})'(t) f_0^{ft}(t+2j)\right|^2\Big) dt \\
&\qquad + O\big(b_n^2\big) \cdot \int \Big(\left|g^{ft}(t) (f_0^{ft})'(t-2j)\right|^2 + \left|g^{ft}(t) (f_0^{ft})'(t+2j)\right|^2\Big) dt.
\end{aligned}$$

We assume the conditions imposed on $\big(g^{ft}\big)'$ in Theorem 2.13 for ordinary smooth and supersmooth errors, again. Both f_0^{ft} and its weak derivative are bounded by 1 and supported on $[-1,1]$ (also see the previous subsection), hence $f_0^{ft}(\cdot \pm 2j)$ and its derivative is supported on $[\pm 2j-1, \pm 2j+1]$. Therefore, as $|j| \ge K_n$, we may conclude that

$$\max_{j \in \{K_n,\ldots,2K_n\}} \max_{\theta \in \{0,1\}^{K_n+1}} \chi^2(h_{\theta_{j,0}}, h_{\theta_{j,1}}) = O\big(b_n^2 K_n^{-2\alpha}\big) = O\big(K_n^{-2\beta-2\alpha-1}\big)$$

in the ordinary smooth case, while we have

$$\begin{aligned}
\max_{j \in \{K_n,\ldots,2K_n\}} \max_{\theta \in \{0,1\}^{K_n+1}} \chi^2(h_{\theta_{j,0}}, h_{\theta_{j,1}}) &= O\big(b_n^2 \exp(-D' K_n^\gamma)\big) \\
&= O\big(K_n^{-2\beta-1} \exp(-D' K_n^\gamma)\big)
\end{aligned}$$

for some constant $D' > 0$, where the $O\big(\cdots\big)$-terms do not implicitly contain $|j| = K_n, \ldots, 2K_n$. We have inserted (2.60). Hence, (2.62) is satisfied by the selection

$$K_n = \text{const} \cdot n^{1/(2\beta+2\alpha+1)}$$

and

$$K_n = c_b^{1/\gamma} \cdot \big(\log n\big)^{1/\gamma}, \quad c_b > 1/D',$$

respectively. Therefore, condition (2.61) is verified so that $b_n^2 K_n \asymp K_n^{-2\beta}$ applies as a lower bound in both cases. Inserting the definitions of $(K_n)_n$ as above proves us the following theorem.

Theorem 2.14. *Assume that $\beta > 1/2$, $C > 0$ sufficiently large, and consider the density class $\mathcal{F}_{\beta,C;L_2}$.*
(a) Assume that the error density g satisfies (2.31) and, in addition, $\big|(g^{\mathrm{ft}})'(t)\big| \leq const \cdot |t|^{-\alpha}$. Then, for an arbitrary estimator $\hat{f}$ of f, based on the i.i.d. data $Y_1, \ldots, Y_n$ from model (2.1), we have

$$\sup_{f\in\mathcal{F}_{\beta,C;L_2}} E\|\hat{f} - f\|_2^2 \geq const \cdot n^{-2\beta/(2\beta+2\alpha+1)}$$

for n sufficiently large.
(b) Assume that the error density g satisfies (2.32) and, in addition, $\big|(g^{\mathrm{ft}})'(t)\big| \leq const \cdot \exp(-D'|t|^\gamma)$. Then, for an arbitrary estimator $\hat{f}$ of f, based on the i.i.d. data $Y_1, \ldots, Y_n$ from model (2.1), we have

$$\sup_{f\in\mathcal{F}_{\beta,C;L_2}} E\|\hat{f} - f\|^2 \geq const \cdot \big(\log n\big)^{-2\beta/\gamma}$$

for n sufficiently large.

Comparing the rates in Theorems 2.9 and 2.14, we realize that we have shown rate-optimality for both ordinary and supersmooth error densities with respect to the MISE, too. The same technical conditions on $(g^{\mathrm{ft}})'$ are required as in Theorem 2.13.

2.5 Adaptive Bandwidth Selection

Nonparametric procedures of smoothing usually contain some parameter whose optimal selection is not obvious. In the deconvolution kernel estimator (2.7), the bandwidth b needs to be chosen; application of the linear wavelet estimator (see Proposition 2.4) requires selection of the band-limiting parameter m_n; the ridge-parameter estimator (2.21) also contains two parameters η and ζ. By those smoothing parameters, one regularizes the complexity of the density estimator so that we have a certain balance between the deterministic error (bias), which represents the penalty for estimating a smoothed version of the true density, and the stochastic error (variance), which is due to the random deviation of the inserted data.

The goal of optimizing the convergence rates gives us some guideline to select the smoothing parameters. In the case of the deconvolution kernel estimator, Theorems 2.8 and 2.9 say that the bandwidth must be chosen so that it converges to zero with some typical rate. Only in special cases of supersmooth contamination, referring to part (b) of the Theorems 2.8 and 2.9, one is able to select the bandwidth without any knowledge of the smoothness degree β of the target density f. Note that the other parameters occurring in the selection guideline do not cause any trouble because they refer to the error density g, which is assumed to be known anyway. However, the parameter β is unknown in many practical applications. Furthermore, although those bandwidth choices lead to optimal rates, they do not guarantee good performance of the estimator for finite sample sizes as the selection of some constant factor is still open. All these motivates us to consider alternative ways of selecting the bandwidth.

In terms of the deconvolution kernel estimator and the ridge-parameter approach, the basic idea is to consider the parameters as random variables, which depend on the empirical data. In the kernel estimator, b is no longer a deterministic sequence in the sample size n, but a function depending on the empirical observations

$$b = \hat{b}(Y_1, \ldots, Y_n) .$$

Note that the same dataset $Y_1, \ldots, Y_n$ is used twice: first, to select the bandwidth and, then, to estimate the density f by the deconvolution kernel estimator where the bandwidth $\hat{b}$ has been inserted. Therefore, we have to be careful as, for example, Proposition 2.2 does no longer hold true; the stochastic dependence between the data $Y_1, \ldots, Y_n$ and the bandwidth $\hat{b}$ has to be respected. This will cause some technical difficulties.

In the ideal case, the asymptotic quality of the estimator does not suffer from the plug-in of the fully data-driven bandwidth $\hat{b}$ instead of the theoretically optimal bandwidth b, which depends on some unknown parameters. This criterion is denoted by *adaptivity* of some bandwidth selector. Nevertheless, the term adaptivity can be made precise in several ways. Constructing an adaptive estimator is very difficult when considering the MSE as in Theorem 2.8. Indeed, in the error-free case, it can be shown that the optimal convergence rates cannot be kept if the smoothness parameter β is unknown (see [81, 82, 86]). Usually, the deterioration of the rates is restricted to some logarithmic loss when choosing an appropriate data-driven procedure. However, when considering the MISE as in Theorem 2.9, the non-knowledge of the smoothness degree of f does not cause any deterioration of the optimal convergence rates, and, in some cases, even the asymptotic constants can be kept. For the deconvolution wavelet estimator, we are able to keep the convergence rates in some cases by using the nonlinear wavelet estimator, please see (2.17). This result is shown in [107]. Also, [20, 21] suggest a method of penalizing complexity for the linear wavelet estimator, which achieves adaptivity in the sense that the convergence rates are kept in many situations. This penalty leads to empirical calibration of the parameter m_n. Even more data-driven procedures for bandwidth selection are proposed for the deconvolution kernel estimator. Cross validation methods have first been proposed in [37]. A rigorous theoretical proof of adaptivity for cross validation is given in [64]. Cross validation may also be applied to select the ridge parameter ζ in estimator

(2.21), see [58]. We will focus on that method in the next subsection. Delaigle and Gijbels [23, 24, 25] suggest plug-in and bootstrapping methods in order to choose h suitably, and investigate the theoretical and numerical properties.

The basic idea of all those procedures is the phenomenon that the ISE (integrated squared error: $\|\hat{f}-f\|_2^2$) or/and its expectation – the MISE – can be decomposed so that some parts are fully accessible by the data; others do not depend on the smoothing parameter to be chosen, and the residual terms are estimable with a faster rate compared to f in the squared $L_2(\mathbb{R})$-metric. For instance, better rates can be established for functionals such as $\int \hat{f}(x)f(x)\,\mathrm{d}x$ or $\int f^2(x)\,\mathrm{d}x$. Indeed, the reason for that is not an improvement of the power of the sample size n in the convergence rates – as one might think, but the power of the smoothness parameter changes compared to the variance term of the MISE. As the smoothness parameter typically tends to either zero or infinity as $n \to \infty$, this fact has some effects on the best achievable convergence rates. Hence, one may define a data-driven quantity, which mimics the MISE or the ISE with sufficient accuracy. This will be illustrated in the following section.

2.5.1 Cross Validation

Cross validation (CV) is one of the most popular data-driven procedures of bandwidth selection for kernel estimators, in general. In the error-free case, that is, standard kernel density estimation, CV has been established in the papers of [55, 57, 120] among many others; where a comprehensive investigation of adaptivity is given in the third work. As mentioned earlier, CV is considered in deconvolution kernel estimation in the papers of, for example, [37, 64].

First, we explain the underlying idea. Consider that the ISE can be decomposed by the binomial theorem in $L_2(\mathbb{R})$ as follows.

$$\mathrm{ISE}(b) \;=\; \|\hat{f}_b - f\|_2^2 \;=\; \|\hat{f}_b\|_2^2 \;-\; 2\,\mathrm{Re}\,\langle \hat{f}_b, f\rangle \;+\; \|f\|_2^2\,,$$

where $\hat{f}_b$ denotes the deconvolution kernel estimator (2.7) with the bandwidth b and we write $\langle\cdot,\cdot\rangle$ for the $L_2(\mathbb{R})$-inner product. Of course, our goal is to choose b appropriately so that the ISE is minimized; hence, we will change the notation from $\hat{f}$ to $\hat{f}_b$ in the sequel. The first addend in the above equation, namely $\|\hat{f}_b\|_2^2$, is known; the third term is unknown but independent of b. Therefore, for the purpose of minimizing the ISE over $b > 0$, it can just be removed. Nevertheless, the second term contains both f and b in its construction; hence, it needs to be estimated. By Plancherel's isometry (Theorem A.4), we derive that

$$\begin{aligned} I(b) = \langle \hat{f}_b, f\rangle &= \frac{1}{2\pi}\int \hat{f}_b^{\mathrm{ft}}(t)\overline{f^{\mathrm{ft}}(t)}\,\mathrm{d}t \\ &= \frac{1}{2\pi}\int K^{\mathrm{ft}}(tb)\frac{1}{n}\sum_{j=1}^{n}\exp(itY_j)f^{\mathrm{ft}}(-t)/g^{\mathrm{ft}}(t)\,\mathrm{d}t. \end{aligned}$$

We obtain an empirically accessible version of $I(b)$ by replacing $f^{\mathrm{ft}}(-t)$ by $\frac{1}{n-1}\sum_{k=1 \wedge k \neq j}^{n} \exp(-\mathrm{i}tY_k)/g^{\mathrm{ft}}(-t)$ in each addend $j = 1, \ldots, n$ in the above sum. We have used a leave-one-out technique to avoid stochastic dependence between $\exp(\mathrm{i}tY_j)$ and $\exp(-\mathrm{i}tY_k)$. Then, we define an empirical version of $I(b)$ by

$$\hat{I}(b) = \frac{1}{2\pi n(n-1)} \sum_{j \neq k} \mathrm{Re} \int K^{\mathrm{ft}}(th) \exp\big(\mathrm{i}t(Y_j - Y_k)\big) / \big|g^{\mathrm{ft}}(t)\big|^2 \, \mathrm{d}t,$$

where the notation indicates that the sum is to be taken over all $(j,k) \in \{1, \ldots, n\}^{(2)}$ without those pairs (j,k) satisfying $j = k$. We may derive that $E\hat{I}(b) = EI(b)$ for any $b > 0$. Note that this equality would not hold true if we take the sum over all $(j,k) \in \{1, \ldots, n\}^{(2)}$, including the diagonal terms as $\exp\big(\mathrm{i}t(Y_j - Y_j)\big) = 1$ almost surely. That allows us to construct an empirical fully accessible function $\mathrm{CV}(b)$ – called the cross validation function – which shall be minimized instead of the unknown ISE. We have

$$\mathrm{CV}(b) = \|\hat{f}_b\|_2^2 - 2\mathrm{Re}\,\hat{I}(b). \tag{2.63}$$

Therefore, we may define a fully data-driven bandwidth selector $b = \hat{b}$ by

$$\hat{b} = \arg\min_{b \in B} \mathrm{CV}(b), \tag{2.64}$$

where the set $B \subseteq (0, \infty)$ is still to be selected. The question about the existence and the uniqueness of the minimum above is deferred as it depends on the specific shape of the set B.

We realize that

$$\mathrm{ISE}(b) = \mathrm{CV}(b) + 2\mathrm{Re}\,\Delta I(b) + \|f\|_2^2. \tag{2.65}$$

where

$$\begin{aligned}
\Delta I(b) &= \hat{I}(b) - I(b) \\
&= \frac{1}{2\pi n(n-1)} \sum_{j \neq k} \int \frac{K^{\mathrm{ft}}(tb)}{|g^{\mathrm{ft}}(t)|^2} \exp(\mathrm{i}tY_j)\big(\exp(-\mathrm{i}tY_k) - E\exp(-\mathrm{i}tY_k)\big)\, \mathrm{d}t \\
&= \frac{1}{2\pi n(n-1)} \sum_{j \neq k} \int \frac{K^{\mathrm{ft}}(tb)}{|g^{\mathrm{ft}}(t)|^2} \psi_j(t)\psi_k(-t)\, \mathrm{d}t \\
&\qquad + \frac{1}{2\pi n} \sum_{j=1}^{n} \int \frac{K^{\mathrm{ft}}(tb)}{g^{\mathrm{ft}}(-t)} f^{\mathrm{ft}}(t)\psi_j(-t)\, \mathrm{d}t
\end{aligned}$$

$$= \frac{1}{2\pi n(n-1)} \sum_{j \neq k} \int \frac{K^{\mathrm{ft}}(tb)}{|g^{\mathrm{ft}}(t)|^2} \psi_j(t)\psi_k(-t)\,\mathrm{d}t$$
$$+ \frac{1}{2\pi n} \sum_{j=1}^{n} \int \frac{K^{\mathrm{ft}}(tb) - K^{\mathrm{ft}}(tb_B)}{g^{\mathrm{ft}}(-t)} f^{\mathrm{ft}}(t)\psi_j(-t)\,\mathrm{d}t$$
$$+ \frac{1}{2\pi n} \sum_{j=1}^{n} \int \frac{K^{\mathrm{ft}}(tb_B)}{g^{\mathrm{ft}}(-t)} f^{\mathrm{ft}}(t)\psi_j(-t)\mathrm{d}t, \tag{2.66}$$

where b_B denotes that bandwidth which minimizes the MISE restricted to $b \in B$, with B as in (2.64) and $\psi_j(t) = \exp(itY_j) - E\exp(itY_j)$. The first term in (2.66) is denoted by $\Delta I_1(b)$; the second one by $\Delta I_2(b)$, accordingly. The third term does not depend on b. Therefore, following from (2.65), we may write

$$\mathrm{ISE}(b) = \mathrm{CV}(b) + 2\mathrm{Re}\,\Delta I_1(b) + 2\mathrm{Re}\,\Delta I_1(b) + W_n, \tag{2.67}$$

with some random variable W_n, which may depend on n but not on b.

In the following lemma, we give an upper bound on any even absolute moment of $\Delta I_1(b)$.

Lemma 2.15. *Assume that $h = f * g \in L_2(\mathbb{R})$ and the bandwidth satisfies $b \in (0,1)$. For any integer $s > 0$ and any symmetric kernel $K \in L_2(\mathbb{R})$, where K^{ft} is supported on $[-1,1]$ and $|K^{\mathrm{ft}}(t)| \leq 1$ for all $t \in \mathbb{R}$, we have*

$$E|\Delta I_1(b)|^{2s} \leq C_s \cdot R_{n,1}^{2s}(b) \cdot \max\{n^{-2s}, b^{s/2}\},$$

with

$$R_{n,1}(b) = n^{-1}b^{-1} \max_{|s|\leq 1/b} |g^{\mathrm{ft}}(s)|^{-2},$$

and some constant C_s, which does not depend on n or b.

Proof. Consider that

$$E|\Delta I_1(b)|^{2s} = \Big(\frac{1}{2\pi n(n-1)}\Big)^{2s} \sum_{j_1 \neq k_1} \cdots \sum_{j_{2s} \neq k_{2s}} \int \cdots \int \Big(\prod_{l=1}^{2s} \frac{K^{\mathrm{ft}}(t_l b)}{|g^{\mathrm{ft}}(t_l)|^2}\Big)$$
$$\cdot E \prod_{l=1}^{2s} \psi_{j_l}\big(-(-1)^l t_l\big)\psi_{k_l}\big((-1)^l t_l\big)\,\mathrm{d}t_1 \cdots \mathrm{d}t_{2s}, \tag{2.68}$$

by Fubini's theorem. Also, we have applied that $\overline{\psi_j(t)} = \psi_j(-t)$. Therefore, we need to analyze the term

$$\Xi_{j'}(t) = E \prod_{l=1}^{2s} \psi_{j_l}\big(-(-1)^l t_l\big)\psi_{k_l}\big((-1)^l t_l\big),$$

with the vectors $t = (t_1, \ldots, t_{2l})$ and $j' = (j'_1, \ldots, j'_{4s}) = (j_1, k_1, \ldots, j_{2s}, k_{2s})$. As some components of j' may be identical, we must not assume that the $\psi_{j_\cdot}(-(-1)^\cdot t_\cdot)\psi_{k_\cdot}((-1)^\cdot t_\cdot)$ are all independent. Therefore, we collect the numbers of those identical components in equivalence classes driven by the relation

$$p \sim p' \iff j'_p = j'_{p'}.$$

This gives us a partition, that is, a system of disjoint covering sets of $\{1, \ldots, 4s\}$, consisting of the sets $A_1, \ldots, A_{m_{j'}}$. Writing $M_p = \min A_p$, we have the representation

$$A_p = \{p' \in \{1, \ldots, 4s\} : p' \sim M_p\}.$$

Also, with respect to the number of all equivalence classes, we derive that $m_{j'} = \#\{j'_1, \ldots, j'_{4s}\}$. Then, using the independence of those $\psi_{j'_l}$ with l from different sets A_p, we have

$$\begin{aligned} |\Xi_{j'}(t)| &= \Big| E \prod_{p=1}^{m_{j'}} \prod_{l \in A_p} \psi_{j'_l}(\varepsilon_{j',l} t_{\lceil l/2 \rceil}) \Big| = \prod_{p=1}^{m_{j'}} \Big| E \prod_{l \in A_p} \psi_{j'_l}(\varepsilon_{j',l} t_{\lceil l/2 \rceil}) \Big| \\ &= \prod_{p=1}^{m_{j'}} \Big| E \prod_{l \in A_p} \psi_{j'_{\lceil M_p/2 \rceil}}(\varepsilon_{j',l} t_{\lceil l/2 \rceil}) \Big| = \prod_{p=1}^{m_{j'}} \Big| E \prod_{l \in A_p} \psi_1(\varepsilon_{j',l} t_{\lceil l/2 \rceil}) \Big|, \end{aligned}$$

with some $\varepsilon_{j',l} \in \{-1, 1\}$, which are specified no further. Also, we have used that ψ_1 has the same distribution as any other ψ_l, $l = 2, \ldots, n$. We may write

$$\Psi_{j',p}(t_{\lceil M_p/2 \rceil}, \ldots, t_{2s}) = \Big| E \prod_{l \in A_p} \psi_1(\varepsilon_{j',l} t_{\lceil l/2 \rceil}) \Big|.$$

Whenever $m_{j'} > 2s$, there exists at least one A_p, which contains one element only because, otherwise, we would have

$$4s = \sum_{p=1}^{m'_j} \# A_p \geq 2m'_j > 4s.$$

In the case of $\# A_p = 1$, we fix that

$$\Psi_{j',p}(t_{\lceil M_p/2 \rceil}, \ldots, t_{2s}) = \big| E\psi_1(\varepsilon_{j',M_p} t_{\lceil M_p/2 \rceil}) \big| = 0,$$

as $E\psi_1(t) = 0$ for all $t \in \mathbb{R}$. Hence, the condition $m_{j'} > 2s$ implies that $\Xi_{j'}(t) = 0$.

On the other hand, if $m_{j'} \leq 2s$, the functions $\Psi_{j',p}$ have to be considered more precisely. We have

$$\begin{aligned} \Psi_{j',p}(t_{\lceil M_p/2 \rceil}, \ldots, t_{2s}) &= \Big| E \prod_{l \in A_p} \big[\exp(\mathrm{i} Y_1 \varepsilon_{j',l} t_{\lceil l/2 \rceil}) - E \exp(\mathrm{i} Y_1 \varepsilon_{j',l} t_{\lceil l/2 \rceil}) \big] \Big| \\ &= \Big| \sum_{J \subseteq A_p} (-1)^{\#(A_p \setminus J)} E \exp\Big(\mathrm{i} Y_1 \sum_{l \in J} \varepsilon_{j',l} t_{\lceil l/2 \rceil} \Big) \prod_{l \in A_p \setminus J} h^{\mathrm{ft}}(\varepsilon_{j',l} t_{\lceil l/2 \rceil}) \Big| \\ &\leq \sum_{J \subseteq A_p} \Big| h^{\mathrm{ft}}\Big(\sum_{l \in J} \varepsilon_{j',l} t_{\lceil l/2 \rceil} \Big) \Big| \prod_{l \in A_p \setminus J} \big| h^{\mathrm{ft}}(\varepsilon_{j',l} t_{\lceil l/2 \rceil}) \big|, \end{aligned}$$

where we recall the definition $h = f * g$. We have to check whether the same component of t may occur twice in the above term. If that was the case, then we would have $\lceil l/2 \rceil = \lceil l'/2 \rceil$ for some $l < l' \in A_p$. This implies that

$$k_{l'/2} = j'_{l'} = j'_l = j_{l'/2}.$$

However, this equality is prohibited by the restricted summation in (2.68). Then, for any fixed $t_{\lceil M_p/2 \rceil + 1}, \ldots, t_{2s}$, we have

$$\begin{aligned}
&\Big(\int \big| \Psi_{j',p}(u, t_{\lceil M_p/2 \rceil+1}, \ldots, t_{2s}) \big|^2 \, \mathrm{d}u \Big)^{1/2} \\
&\leq \sum_{J \subseteq A_p} \Big(\int \Big| h^{\mathrm{ft}} \Big(\sum_{l \in J} \varepsilon_{j',l} t_{\lceil l/2 \rceil} \Big) \Big|^2 \prod_{l \in A_p \setminus J} \big| h^{\mathrm{ft}} \big(\varepsilon_{j',l} t_{\lceil l/2 \rceil} \big) \big|^2 \, \mathrm{d}t_{\lceil M_p/2 \rceil} \Big)^{1/2} \\
&\leq \sum_{J \subseteq A_p} \Big(\int \big| h^{\mathrm{ft}}(u) \big|^2 \, \mathrm{d}u \Big)^{1/2} \leq \sqrt{2\pi} 2^{4s} \, \|h\|_2, \qquad (2.69)
\end{aligned}$$

where we have used Minkowski's inequality (triangle inequality with respect to $\| \cdot \|_2$), $\|h^{\mathrm{ft}}\|_\infty \leq 1$, and the fact that there exist exactly $2^{\#A_p} \leq 2^{4s}$ subsets of A_p. Finally, Parseval's identity (Theorem A.4) has been employed.

On the other hand, $\lceil M_p/2 \rceil$ may be identical for two different equivalence classes A_p; however, only for less than three classes, so that

$$m^*_{j'} = \#\big\{ \lceil M_p/2 \rceil \, : \, p = 1, \ldots, m_{j'} \big\} \geq m_{j'}/2 \, .$$

Therefore, we conclude the existence of subset $\mathcal{K} \subseteq \{1, \ldots, m'_j\}$ with $\#\mathcal{K} = m^*_{j'}$ so that all $\lceil M_p/2 \rceil$ differ from each other for the different $p \in \mathcal{K}$. For all $p \notin \mathcal{K}$, we apply 2^{4s} as an upper bound on $|\Psi_{j',p}|$ (since $\|h^{\mathrm{ft}}\|_\infty \leq 1$). Considering $m'_j \leq 4s$, we derive that

$$|\Xi_{j'}(t)| \leq \chi_{[0,2s]}(m_{j'}) \cdot 2^{16s^2} \prod_{p \in \mathcal{K}} \Psi_{j',p}\big(t_{\lceil M_p/2 \rceil}, \ldots, t_{2s}\big).$$

Inserting this inequality in (2.68) gives us

$$\begin{aligned}
E|\Delta I_1(b)|^{2s} &\leq O\big(n^{-4s}\big) \sum_{j_1 \neq k_1} \cdots \sum_{j_{2s} \neq k_{2s}} \chi_{[0,2s]}\big(m_{\{j_1,k_1,\ldots,j_{2s},k_{2s}\}}\big) \\
&\quad \int \cdots \int \Big(\prod_{l=1}^{2s} \frac{|K^{\mathrm{ft}}(t_l b)|}{|g^{\mathrm{ft}}(t_l)|^2} \Big) \cdot \prod_{p \in \mathcal{K}} \Psi_{j',p}\big(t_{\lceil M_p/2 \rceil}, \ldots, t_{2s}\big) \, \mathrm{d}t_1 \cdots \mathrm{d}t_{2s} \\
&\leq O\big(n^{-4s}\big) \sum_{j_1 \neq k_1} \cdots \sum_{j_{2s} \neq k_{2s}} \chi_{[0,2s]}\big(m_{\{j_1,k_1,\ldots,j_{2s},k_{2s}\}}\big) \cdot \Big(\max_{|s| \leq 1/b} |g^{\mathrm{ft}}(s)|^{-2} \Big)^{2s} \\
&\quad \cdot \int \cdots \int \Big[\prod_{l=1}^{2s} |K^{\mathrm{ft}}(t_l b)| \Big] \prod_{p \in \mathcal{K}} \Psi_{j',p}\big(t_{\lceil M_p/2 \rceil}, \ldots, t_{2s}\big) \, \mathrm{d}t_1 \cdots \mathrm{d}t_{2s}.
\end{aligned}$$

Let $k_1 < \cdots < k_{m^*_{j'}}$ denote the elements in $\{\lceil M_p/2 \rceil : k \in \mathcal{K}\}$ with increasing order. By Fubini's theorem, we may consider the integrals over $t_{k_1}, \ldots, t_{k_{m^*_{j'}}}$ as the inner integrals and the residual integrals as the outer ones in the above equation. Then, the Cauchy–Schwarz inequality may consecutively be applied on $m^*_{j'}$ different one-dimensional integrals. Leading to

$$\begin{aligned}
&\int \big|K^{\mathrm{ft}}(bt_{\lceil M_p/2\rceil})\big| \Psi_{\lceil M_p/2\rceil}(t_{\lceil M_p/2\rceil}, \ldots, t_{2s})\, \mathrm{d}t_{\lceil M_p/2\rceil} \\
&\le \Big(\int \big|K^{\mathrm{ft}}(bt_{\lceil M_p/2\rceil})\big|^2 \mathrm{d}t_{\lceil M_p/2\rceil}\Big)^{1/2} \\
&\qquad \cdot \Big(\int \big|\Psi_{\lceil M_p/2\rceil}(t_{\lceil M_p/2\rceil}, \ldots, t_{2s})\big|^2 \mathrm{d}t_{\lceil M_p/2\rceil}\Big)^{1/2} \\
&= O\big(b^{-1/2}\big),
\end{aligned}$$

according to (2.69). Then, we conclude that

$$\begin{aligned}
E\big|\Delta I(b)\big|^{2s} &\le O\big(n^{-4s}\big) \sum_{j_1 \neq k_1} \cdots \sum_{j_{2s} \neq k_{2s}} \chi_{[0,2s]}\big(m_{\{j_1,k_1,\ldots,j_{2s},k_{2s}\}}\big) \\
&\qquad \cdot \Big(\max_{|s|\le 1/b} |g^{\mathrm{ft}}(s)|^{-2}\Big)^{2s} O\big(b^{-m^*_{j'}/2}\big) \cdot b^{-2s+m^*_{j'}} \\
&\le O\big(n^{-4s}\big) \Big(\max_{|s|\le 1/b} |g^{\mathrm{ft}}(s)|^{-2}\Big)^{2s} \sum_{l=1}^{2s} \sum_{j' \in J(l)} O\big(b^{-2s+m^*_{j'}/2}\big),
\end{aligned}$$

where the set $J(l)$ collects all vectors $j' \in \{1, \ldots, n\}^{(4s)}$ with $m_{j'} = l$. Because of some combinational considerations, we derive that $\#J(l) = c_l \cdot n^l$, where c_l does not depend on b or n. Note that there are only l different elements within $\{j'_1, \ldots, j'_{4s}\}$, which may take their values from 1 to n. Then, as $m^*_{j'} \ge l/2$ for any $j' \in J(l)$, we obtain that

$$\begin{aligned}
E\big|\Delta I(b)\big|^{2s} &\le O\big(n^{-4s}\big)\Big(\max_{|s|\le 1/b} |g^{\mathrm{ft}}(s)|^{-2}\Big)^{2s} \sum_{l=1}^{2s} O\big(n^l b^{-2s+l/4}\big) \\
&\le O\big(n^{-4s}\big)\Big(\max_{|s|\le 1/b} |g^{\mathrm{ft}}(s)|^{-2}\Big)^{2s} \max\big\{b^{-2s}\,,\, n^{2s} b^{-3s/2}\big\}\,,
\end{aligned}$$

what completes the proof of the lemma. ■

Therefore, when b becomes small when n increases, Lemma 2.15 indicates that $|\Delta I_1(b)|^2$ tends to zero with a faster rate compared to $R_{n,1}(b)$, which is a lower bound on the MISE under some conditions. This will be made precise later. But before that, we derive a similar result for the term $\Delta I_2(b)$. From now on, we restrict our consideration to the sinc-kernel K as defined in (2.41), for simplicity.

Lemma 2.16. *Under the conditions of Lemma 2.15, put K equal to the sinc-kernel. Then, we have*

$$E\big|\Delta I_2(b)\big|^{2s} \le D_s \max\{R_{n,1}(b), R_{n,1}(b_B)\}^s \max\{R_n(b), R_n(b_B)\}^s \cdot \max\big\{n^{-s}, \big(b^{s/2} + b_B^{s/2}\big)\big\},$$

where the constant $D_s > 0$ does not depend on n or b, and

$$R_n(b) = R_{n,1}(b) + \int \big|K^{\mathrm{ft}}(tb) - 1\big|^2 |f^{\mathrm{ft}}(t)|^2 \, dt$$

Proof. We consider that

$$\begin{aligned} E\big|\Delta I_2(b)\big|^{2s} &= \Big(\frac{1}{2\pi n}\Big)^{2s} \sum_{j_1=1}^{n} \cdots \sum_{j_{2s}=1}^{n} \int \cdots \int \\ &\quad \Big(\prod_{k=1}^{2s} \frac{K^{\mathrm{ft}}(t_k b) - K^{\mathrm{ft}}(t_k b_B)}{g^{\mathrm{ft}}(-t_k)} f^{\mathrm{ft}}(t_k)\Big) \cdot E \prod_{k=1}^{2s} \psi_k\big((-1)^k t_k\big)\, dt_1 \cdots dt_{2s} \\ &\le O\big(n^{-2s}\big) \sum_{j_1=1}^{n} \cdots \sum_{j_{2s}=1}^{n} \\ &\quad \Big(\int \cdots \int \prod_{k=1}^{2s} |K^{\mathrm{ft}}(t_k b) - K^{\mathrm{ft}}(t_k b_B)|^2 |f^{\mathrm{ft}}(t_k)|^2 \, dt_1 \cdots dt_{2s}\Big)^{1/2} \\ &\quad \cdot \Big(\int \cdots \int_{\max|t_k| \le \max\{1/b, 1/b_B\}} \big|\Xi_j'(t)\big|^2 / \prod_{k=1}^{2s} |g^{\mathrm{ft}}(t_k)|^2 \, dt_1 \cdots dt_{2s}\Big)^{1/2} \end{aligned} \tag{2.70}$$

by the Cauchy–Schwarz inequality in $L_2(\mathbb{R}^{2s})$, where we write $j = (j_1, \ldots, j_{2s})$, $t = (t_1, \ldots, t_{2s})$, and

$$\Xi_j'(t) = E \prod_{k=1}^{2s} \psi_k\big((-1)^k t_k\big).$$

Also, we have used that K^{ft} is supported on $[-1, 1]$. Then, we derive that

$$\begin{aligned} &\int \cdots \int \prod_{k=1}^{2s} |K^{\mathrm{ft}}(t_k b) - K^{\mathrm{ft}}(t_k b_B)|^2 |f^{\mathrm{ft}}(t_k)|^2 \, dt_1 \cdots dt_{2s} \\ &\le 2 \int \cdots \int \prod_{k=1}^{2s} |K^{\mathrm{ft}}(t_k b) - 1|^2 |f^{\mathrm{ft}}(t_k)|^2 \, dt_1 \cdots dt_{2s} \\ &\qquad + 2 \int \cdots \int \prod_{k=1}^{2s} |K^{\mathrm{ft}}(t_k b_B) - 1|^2 |f^{\mathrm{ft}}(t_k)|^2 \, dt_1 \cdots dt_{2s} \\ &\le \big[R_n(b)\big]^{2s} + \big[R_n(b_B)\big]^{2s}. \end{aligned}$$

From (2.70), we derive that

$$E\big|\Delta I_2(b)\big|^{2s} \le O\big(n^{-2s}\max\{R_n(b),R_n(b_B)\}^s\big)\cdot\sum_{j_1=1}^{n}\cdots\sum_{j_{2s}=1}^{n}$$
$$\Big(\int\cdots\int_{\max|t_k|\le\max\{1/b,1/b_B\}} \big|\Xi_j'(t)\big|^2/\prod_{k=1}^{2s}|g^{\mathrm{ft}}(t_k)|^2\,\mathrm{d}t_1\cdots\mathrm{d}t_{2s}\Big)^{1/2}. \tag{2.71}$$

We analyze $\Xi_j'(t)$ in the spirit of the proof of Lemma 2.15. Therefore, the following consideration is reduced to a brief draft of the proof. Thus, we introduce the equivalence relation $k\sim k' \iff j_k=j_{k'}$. When $\mathcal{K}_j$ denotes the set of all equivalence classes, we may derive that

$$\big|\Xi_j'(t)\big| = \prod_{K\in\mathcal{K}_j}\Big|E\prod_{l\in K}\psi_1\big((-1)^l t_l\big)\Big|.$$

As in the proof of Lemma 2.15, we can show that $\big|\Xi_j'(t)\big|$ is the product of $m_j=\#\mathcal{K}_j$ functions depending on the components t_j, where j are taken from only one equivalence class for each function. As those functions are square-integrable over one component, we may use the Cauchy–Schwarz inequality m_j times in the integrals in (2.71). Again, we realize that $\Xi_j'(t)=0$ whenever $m_j>s$ as, then, at least one equivalence class consists of one element only. All that gives us

$$E\big|\Delta I_2(b)\big|^{2s} \le O\big(n^{-2s}\max\{R_n(b),R_n(b_B)\}^s\big)\cdot\sum_{j_1=1}^{n}\cdots\sum_{j_{2s}=1}^{n}\chi_{[0,s]}(m_j)$$
$$\cdot O\big(b^{-s+m_j/2}+b_B^{-s+m_j/2}\big)\cdot\Big(\min_{|s|\le\max\{1/b,1/b_B\}}|g^{\mathrm{ft}}(s)|^2\Big)^{-s}.$$

Again, recalling the techniques used in the proof of Lemma 2.15, we learn from combinatorics that

$$\begin{aligned}E\big|\Delta I_2(b)\big|^{2s} &\le O\Big(n^{-2s}\Big)\max\{R_n(b),R_n(b_B)\}^s\Big(\min_{|t|\le\max\{1/b,1/b_B\}}|g^{\mathrm{ft}}(t)|^2\Big)^{-s}\\ &\qquad\sum_{l=1}^{s}n^l\big(b^{-s+l/2}+b_B^{-s+l/2}\big)\\ &\le O\big(n^{-s}\big)\max\{b^{-s},b_B^{-s}\}\,\max\{R_n(b),R_n(b_B)\}^s\\ &\qquad\cdot\Big(\min_{|t|\le\max\{1/b,1/b_B\}}|g^{\mathrm{ft}}(t)|^2\Big)^{-s}\sum_{l=1}^{s}n^{l-s}\big(b^{l/2}+b_B^{l/2}\big),\end{aligned}$$

so that the lemma has been shown. ∎

Now, we restrict our consideration to ordinary smooth error densities, defined in (2.31). Under standard Sobolev conditions and supersmooth error densities, adaptivity is achieved by a nonempirical bandwidth selector, see Theorem 2.9. Hence, from an asymptotic point of view, there is no real need to apply cross validation under supersmooth contamination unless more restrictive conditions on f can be justified. Also, as we choose the sinc-kernel for K, we may fix that $R_n(b)$ is bounded above by the MISE times a constant by Proposition 2.2. Let us denote the MISE of the deconvolution kernel estimator by MISE(b). As b_B has been chosen so that the MISE is minimized we easily derive that

$$R_n(b_B) \leq O(1) \cdot \mathrm{MISE}(b_B) \leq O(1) \cdot \mathrm{MISE}(b)$$

for all $b \in B$.

As the main merit of Lemma 2.15 and 2.16, we obtain that $\left|\Delta I_1(b)\right|$ and $\left|\Delta I_2(b)\right|$ are asymptotically negligible in (2.67) as, for $n \to \infty$, they tend to zero with faster rates than the MISE. That can be seen as follows: We arrange that B is a finite set. At a computational point of view, we have to restrict the minimization to some finite set B anyway. Let us consider some fixed but arbitrary $\varepsilon > 0$. Then,

$$\begin{aligned}
& P\big[\sup_{b \in B} \big(|\Delta I_1(b)| + |\Delta I_2(b)|\big)/\mathrm{MISE}(b) > \varepsilon\big] \\
& \leq \sum_{b \in B} P\big[\big(|\Delta I_1(b)| + |\Delta I_2(b)|\big)/\mathrm{MISE}(b) > \varepsilon\big] \\
& \leq \sum_{b \in B} P\big[|\Delta I_1(b)|^{2s}/\mathrm{MISE}^{2s}(b) > (\varepsilon/2)^{2s}\big] \\
& \qquad\qquad + \sum_{b \in B} P\big[|\Delta I_2(b)|^{2s}/\mathrm{MISE}^{2s}(b) > (\varepsilon/2)^{2s}\big] \\
& \leq (\varepsilon/2)^{-2s} \sum_{b \in B} \mathrm{MISE}^{-2s}(b)\, \big(E|\Delta I_1(b)|^{2s} + E|\Delta I_2(b)|^{2s}\big) \\
& \leq O(1) \cdot (\#B) \cdot \big(n^{-s} + b_B^{s/2}\big) + O(1) \cdot \sum_{b \in B_\mu} b^{s/2} \\
& \qquad\qquad + O(1) \cdot \sum_{b \in B \backslash B_\mu} \mathrm{MISE}^{-2s}(b)\, \big(E|\Delta I_1(b)|^{2s} + E|\Delta I_2(b)|^{2s}\big) \\
& \leq O(1) \cdot (\#B) \cdot \big(n^{-s} + b_B^{s/2} + n^{-\mu s/2}\big) \\
& \qquad\qquad + O(1) \cdot \max_{k=1,2} \sum_{b \in B \backslash B_\mu} \big(R_n^{-s}(b) R_{n,1}^s(b)\big)^k \\
& \qquad\qquad + O(1) \cdot \sum_{b \in B \backslash B_\mu} R_n^{-s}(b) R_n^s(b_B),
\end{aligned} \tag{2.72}$$

where we have applied Markov's inequality and the Lemmas 2.15 and 2.16. We have introduced the set

$$B_\mu = \{b \in B : b \leq n^{-\mu}\}$$

for some $\mu > 0$. By Theorem 2.9, we know that under ordinary smooth contamination the bandwidth b shall decrease to zero at some specific algebraic rate to establish optimal convergence rates. If we assume that there exist both-side-bounds on f^{ft} in the following version,

$$\mathrm{const} \cdot \lambda^{-2\beta} \geq \int_{|t| \geq \lambda} \left|f^{\mathrm{ft}}(t)\right|^2 \mathrm{d}t \geq \mathrm{const} \cdot \lambda^{-2\beta^*}, \qquad \text{for all } \lambda > 1 \tag{2.73}$$

for some $\beta^* > \beta > 0$, then the bias term (second term) in Proposition 2.2 will not tend to zero with an algebraic rate unless b does so. Note that the upper bound contained in (2.73) corresponds to the Sobolev condition (2.30). So far, we are not guaranteed that b_B also tends to zero with an algebraic rate as we have not specified the set B yet. Therefore, we will just assume at this stage that

$$b_B = O(n^{-\gamma}) \tag{2.74}$$

for some $\gamma > 0$. Note that the integer $s > 0$ is still to be selected arbitrarily; however, its choice must not depend on n or b. Assuming that $\#B \leq O(n^\xi)$ for some fixed $\xi > 0$, we may choose s sufficiently large (for any $\mu > 0$ and $\beta^* > 0$) so that the first term in (2.72) converges to zero at some rate faster than $O(n^{-\nu})$ for some $\nu > 1$.

With respect to the second term in (2.72), we derive that

$$R_{n,1}(b) \leq \mathrm{const} \cdot n^{-1} b^{-1-2\alpha}$$

when inserting (2.31) as the assumed shape of the error density g. Further, we obtain

$$R_n(b) \geq \mathrm{const} \cdot \int_{|t| \geq 1/b} \left|f^{\mathrm{ft}}(t)\right|^2 \mathrm{d}t \geq \mathrm{const} \cdot b^{2\beta^*},$$

so that

$$R_{n,1}(b)/R_n(b) \leq \mathrm{const} \cdot n^{-1} b^{-1-2\alpha-2\beta^*},$$

where we have $b > n^{-\mu}$ as $b \in B \backslash B_\mu$. Hence, under the selection $0 < \mu < 1/(1 + 2\alpha + 2\beta^*)$ and, again, s sufficiently large, the term

$$\max_{k=1,2} \sum_{b \in B \backslash B_\mu} \left(R_n^{-s}(b) R_{n,1}^s(b)\right)^k$$

converges to zero at some rate $O(n^{-\nu})$, $\nu > 1$, where the condition $\#B \leq O(n^\xi)$ still holds true, of course. With respect to the third term in (2.72), we realize that $R_n(b_B)$ converges to zero with an algebraic rate so that first selecting $\mu > 0$ small enough and, then, choosing s sufficiently large give us the upper bound $O(n^{-\nu})$, $\nu > 1$, on that term, too. Combining those results, we derive that (2.72) is bounded above by $O\big(n^{-\nu}\big)$, $\nu > 1$. That implies

$$\sum_{n=1}^{\infty} P\big[\sup_{b\in B}\big(|\Delta I_1(b)| + |\Delta I_2(b)|\big)/\mathrm{MISE}(b) > \varepsilon\big] < \infty,$$

and, by the Borel–Cantelli lemma, we may derive that

$$\sup_{b\in B}\big(|\Delta I_1(b)| + |\Delta I_2(b)|\big)/\mathrm{MISE}(b) \stackrel{n\to\infty}{\longrightarrow} 0\,, \qquad \text{a.s.} \tag{2.75}$$

Revisiting (2.67), we may conclude that

$$\begin{aligned}
&\Big|\frac{\mathrm{ISE}(\hat b) - \mathrm{ISE}(b_B)}{\mathrm{MISE}(\hat b)} - \underbrace{\frac{\mathrm{CV}(\hat b) - \mathrm{CV}(b_B)}{\mathrm{MISE}(\hat b)}}_{\leq 0 \text{ a.s.}}\Big| \\
&\quad \leq 2\max_{k=1,2}\Big|\frac{\Delta I_k(\hat b)}{\mathrm{MISE}(\hat b)}\Big| + 2\max_{k=1,2}\Big|\frac{\Delta I_k(b_B)}{\mathrm{MISE}(b_B)}\Big| \cdot \underbrace{\Big|\frac{\mathrm{MISE}(b_B)}{\mathrm{MISE}(\hat b)}\Big|}_{\leq 1 \text{ a.s.}}
\end{aligned}$$

by the definition of $\hat b$ and b_B. Then, it follows from (2.75) that

$$\limsup_{n\to\infty} \frac{\mathrm{ISE}(\hat b) - \mathrm{ISE}(b_B)}{\mathrm{MISE}(\hat b)} \leq 0\,, \quad \text{a.s.} \tag{2.76}$$

Also, consider that

$$\begin{aligned}
1 \geq \frac{\mathrm{MISE}(b_B)}{\mathrm{MISE}(\hat b)} = \frac{\mathrm{MISE}(b_B) - \mathrm{ISE}(b_B)}{\mathrm{MISE}(\hat b)} &+ \frac{\mathrm{ISE}(b_B) - \mathrm{ISE}(\hat b)}{\mathrm{MISE}(\hat b)} \\
&+ \frac{\mathrm{ISE}(\hat b) - \mathrm{MISE}(\hat b)}{\mathrm{MISE}(\hat b)} + 1,
\end{aligned}$$

so that, given (2.76), we have

$$\frac{\mathrm{MISE}(b_B)}{\mathrm{MISE}(\hat b)} \stackrel{n\to\infty}{\longrightarrow} 1\,, \qquad \text{a.s.} \tag{2.77}$$

if we can prove that

$$\sup_{b\in B}\Big|\frac{\mathrm{MISE}(b) - \mathrm{ISE}(b)}{\mathrm{MISE}(b)}\Big| \stackrel{n\to\infty}{\longrightarrow} 0\,, \qquad \text{a.s.} \tag{2.78}$$

Therefore, we consider that

$$\mathrm{ISE}(b)-\mathrm{MISE}(b) = \frac{1}{2\pi}\int\frac{|K^{\mathrm{ft}}(tb)|^2}{|g^{\mathrm{ft}}(t)|^2}\Big(\Big|\frac{1}{n}\sum_{j=1}^n\psi_j(t)\Big|^2 - E\Big|\frac{1}{n}\sum_{j=1}^n\psi_j(t)\Big|^2\Big)\,\mathrm{d}t$$

$$= \frac{1}{2\pi}n^{-2}\int\frac{|K^{\mathrm{ft}}(tb)|^2}{|g^{\mathrm{ft}}(t)|^2}\sum_{j=1}^n\big(|\psi_j(t)|^2 - E|\psi_j(t)|^2\big)\,\mathrm{d}t$$

$$-\frac{1}{2\pi}n^{-2}\int\frac{|K^{\mathrm{ft}}(tb)|^2}{|g^{\mathrm{ft}}(t)|^2}\sum_{j_1\neq j_2}\big(\psi_{j_1}(t)\psi_{j_2}(-t) - \underbrace{E\psi_{j_1}(t)\psi_{j_2}(-t)}_{=0}\big)\,\mathrm{d}t$$

$$= -\frac{1}{\pi}n^{-2}\,\mathrm{Re}\int\frac{|K^{\mathrm{ft}}(tb)|^2}{|g^{\mathrm{ft}}(t)|^2}\sum_{j=1}^n h^{\mathrm{ft}}(t)\psi_j(-t)\,\mathrm{d}t$$

$$+\frac{1}{2\pi}n^{-2}\int\frac{|K^{\mathrm{ft}}(tb)|^2}{|g^{\mathrm{ft}}(t)|^2}\sum_{j_1\neq j_2}\psi_{j_1}(t)\psi_{j_2}(-t)\,\mathrm{d}t.$$

We recall that we have restricted our consideration to the sinc-kernel K. Then, the second term above corresponds to $\Delta I_1(b)$ as studied in Lemma 2.15 up to some factor $1+o(1)$. With respect to the first term, we consider its $(2s)$th moment.

$$E\Big|\frac{1}{2\pi}n^{-2}\int\frac{|K^{\mathrm{ft}}(tb)|^2}{|g^{\mathrm{ft}}(t)|^2}\sum_{j=1}^n h^{\mathrm{ft}}(t)\psi_j(-t)\,\mathrm{d}t\Big|^{2s}$$

$$\le O(n^{-4s})\int\cdots\int\Big(\prod_{j=1}^{2s}\frac{|K^{\mathrm{ft}}(t_jb)|^2}{|g^{\mathrm{ft}}(t_j)|^2}\Big)\sum_{j_1=1}^n\cdots\sum_{j_{2s}=1}^n$$

$$E\prod_{k=1}^{2s}E\psi_{j_k}(-t_k)h^{\mathrm{ft}}(t_k)\,\mathrm{d}t_1\cdots\mathrm{d}t_{2s}$$

$$\le O\big(R_{n,1}^{2s}(b)\big)\cdot n^{-s},$$

where we use that all the addends with $\#\{j_1,\ldots,j_{2s}\}\le s$ vanish; also we may use 2 as a rough upper bound on the $|\psi_{j.}(-t.)|$ almost surely. This result, combined with Lemma 2.15, proves (2.78) and (2.77).

Nevertheless, this result is less meaningful as the expectation of $\mathrm{ISE}(\hat b)$ is not identical with $E\,\mathrm{MISE}(\hat b)$. The expectations cannot be calculated separately from each other due to the dependence between $\hat b$ and the data $Y_1,\ldots,Y_n$. Despite, our consideration is useful since we may consider that

$$\frac{\mathrm{ISE}(\hat b)}{\mathrm{MISE}(b_B)} = \frac{\mathrm{ISE}(\hat b)-\mathrm{ISE}(b_B)}{\mathrm{MISE}(b_B)} + \frac{\mathrm{ISE}(b_B)-\mathrm{MISE}(b_B)}{\mathrm{MISE}(b_B)} + 1$$

$$= \frac{\mathrm{ISE}(\hat b)-\mathrm{ISE}(b_B)}{\mathrm{MISE}(\hat b)}\cdot\frac{\mathrm{MISE}(\hat b)}{\mathrm{MISE}(b_B)} + \frac{\mathrm{ISE}(b_B)-\mathrm{MISE}(b_B)}{\mathrm{MISE}(b_B)} + 1.$$

By (2.78), the second addend in the above equation tends to zero almost surely. Combining (2.76) and (2.77), we derive that the limit superior of the first addend is bounded above by zero, almost surely. From there, we conclude that

$$\limsup_{n\to\infty} \frac{\mathrm{ISE}(\hat{b})}{\mathrm{MISE}(b_B)} \leq 1\,, \quad \text{a.s.},$$

implying that $\mathrm{ISE}(\hat{b})$ converges to zero at least as fast as the minimized $\mathrm{MISE}(b_B)$, almost surely.

As the last step, we would like to replace $\mathrm{MISE}(b_B)$, which minimizes the MISE over the set B, by its minimum over the whole of $[0, \infty)$. All we need to show is that the deterministic term

$$\mathrm{MISE}(b_B) \,/\, \inf_{b>0} \mathrm{MISE}(b)$$

converges to 1 as $n \to \infty$. We specify the set B by the grid

$$B = \left\{ (1 + j/n)^{-1} \,:\, j = 0, \ldots, n^{1+1/(2\alpha)} \right\}, \tag{2.79}$$

with α as in (2.31). Then, the condition $\#B = O(n^{\xi})$ is satisfied. From Proposition 2.2, we conclude that

$$\begin{aligned}
&\mathrm{MISE}(b) - \mathrm{MISE}(b') \\
&\quad \leq O(1/n) + \frac{1}{2\pi} \int \left(\left| K^{\mathrm{ft}}(tb) - K^{\mathrm{ft}}(tb') \right|^2 \right) \left(n^{-1} \left| g^{\mathrm{ft}}(t) \right|^{-2} + |f^{\mathrm{ft}}(t)|^2 \right) dt \\
&\quad \leq O(1/n) + n^{-1} O(\max\{b^{-2\alpha}, (b')^{-2\alpha}\}) \cdot \left| b^{-1} - (b')^{-1} \right|,
\end{aligned}$$

from what follows Lipschitz continuity of the MISE, viewed as a function depending on $1/b$ on the domain $1/b \in [1, n^{1/(2\alpha)}]$. For large n, we may neglect the set $\mathbb{R} \backslash [1, n^{1/(2\alpha)}]$ with respect to the minimization problem as either the variance term contained in the MISE or the bias term, because (2.73) will not converge to zero, then. On the other hand, inside the interval $[1, n^{1/(2\alpha)}]$ the Lipschitz continuity implies that

$$\mathrm{MISE}(b_B) \leq \inf_{b>0} \mathrm{MISE}(b) + O(1/n),$$

as the distance between some $1/b \in [1, n^{1/(2\alpha)}]$ and its nearest neighbor contained in $\{b^{-1} \,:\, b \in B\}$ is bounded above by $1/n$. Combining that with (2.73), we also verify condition (2.74). As the optimal convergence rates of the deconvolution kernel estimator are slower than n^{-1}, due to (2.73), we have finally proved the following theorem.

Theorem 2.17. *Consider the density deconvolution model (2.1). Assume that $f \in L_2(\mathbb{R})$ and satisfies (2.73), and g is ordinary smooth, see (2.31). Choose the sinc-kernel K as defined in (2.41) for the density deconvolution kernel estimator $\hat{f}$ as in (2.7). Let $\hat{b}$ denote the CV-bandwidth defined in (2.64), and select the set B as in (2.79). Then, we have*

$$\limsup_{n\to\infty} \frac{ISE(\hat{b})}{\inf_{b>0} MISE(b)} \leq 1, \quad a.s.$$

Thus, Theorem 2.17 establishes individual strong adaptivity of the CV-procedure under some conditions.

2.6 Unknown Error Density

As already mentioned in the real-life examples in Sect. 2.2.1, the standard assumption of an exactly known error density g is unrealistic in many practical applications. The common procedures such as the kernel estimator (2.7), the wavelet-estimators considered in Proposition 2.4 or the ridge-parameter estimator (2.21) use g in their construction and, hence, are not applicable unless g is somehow accessible. In the most general case where any density is admitted to be the true f, this target density is not identifiable, that is, the following implication is violated:

$$f * g = \tilde{f} * \tilde{g} \Longrightarrow \mathrm{d}(f, \tilde{f}) = 0, \text{ for all } f, \tilde{f} \in \mathcal{F}, g, \tilde{g} \in \mathcal{G} \tag{2.80}$$

for some desired semi-metric d, where $\mathcal{F}$ and $\mathcal{G}$ denote the classes consisting of all admitted target densities f and error densities g, respectively. Therefore, (2.80) represents a generalization of (2.22) to the case of unknown g. The inconsistency proof, when (2.22) is violated, can analogously be extended to (2.80). The previous setting is still included by specifying $\mathcal{G} = \{g\}$, where g is the known true error density.

Whenever d is a metric so that $d(f,g) = 0$ implies $f = g$, and $\#\mathcal{G} \geq 2$, then condition (2.80) is violated in many cases. In the sequel, we consider some examples where f is not identifiable and, hence, there is no chance to construct a consistent estimator.

(a) $\#(\mathcal{F} \cap \mathcal{G}) \geq 2$.
Just take two different densities $f, g \in \mathcal{F} \cap \mathcal{G}$. On the one hand, f may be the target density and g the error density, but, on the other hand, g may be the target density and f the error density, as well. As convolution is symmetric, those two situations lead to the same observation density, but $d(f,g) \neq 0$.

(b) There exist some $f \in \mathcal{F}$ and $g \in \mathcal{G}$ so that the shifted densities $f(\cdot - a)$ and $g(\cdot + a)$ are also contained in $\mathcal{F}$ and $\mathcal{G}$, respectively, for some $a \neq 0$. Rather elementary calculation leads to $f(\cdot - a) * g(\cdot + a) = f * g$. On the other hand, f and $f(\cdot - a)$ do not coincide as f is a density function.

(c) Assume that g is not perfectly known, so $\#\mathcal{G} \geq 2$. We have

$$\mathcal{F} * \mathcal{G} = \big\{ f * g : f \in \mathcal{F}, \, g \in \mathcal{G} \big\} \subseteq \mathcal{F},$$

and there exists at least one $f \in \mathcal{F}$ such that f^{ft} vanishes nowhere. Then, (2.80) is violated.
To prove that assertion, we take two different $g, \tilde{g} \in \mathcal{G}$ and $f \in \mathcal{F}$ so that $f^{\mathrm{ft}}(t) \neq 0$ for all t. By assumption, $g * f$ and $\tilde{g} * f$ are admitted to be the true target density and $g, \tilde{g} \in \mathcal{G}$. Therefore, the equality

$$(g * f) * \tilde{g} = g * (\tilde{g} * f)$$

implies that $g * f \neq f * \tilde{g}$ if (2.80) holds. From there, we may conclude coincidence of the Fourier transforms $g^{\mathrm{ft}} f^{\mathrm{ft}} = \tilde{g}^{\mathrm{ft}} f^{\mathrm{ft}}$ and, hence, $g^{\mathrm{ft}} = \tilde{g}^{\mathrm{ft}}$; therefore, we have contradiction to $g \neq \tilde{g}$.

The Sobolev classes $\mathcal{F}_{\beta,C;L_2}$ as given in (2.30) satisfy the conditions imposed on $\mathcal{F}$ in the Example (c). That can be seen by the property $|f^{\mathrm{ft}} g^{\mathrm{ft}}| \leq |f^{\mathrm{ft}}|$ for any density g so that $f * g$ satisfies (2.30) whenever f does so. Also, normal densities with sufficiently large variances and nonvanishing Fourier transforms are included in $\mathcal{F}$. Hence, the common Sobolev conditions lead to nonidentifiability whenever $\mathcal{G}$ contains more than one density.

This inconsistency may be quantified by considering the asymptotic MISE for the deconvolution kernel estimator (2.7). This approach goes back to the paper of [87]. Assume that g is the true error density and some $\tilde{g}$ is incorrectly used as the error density in the construction of (2.7). Then, we obtain that

$$\begin{aligned}
&\mathrm{MISE}(\hat{f}, f) \\
&= E \int \Big| \frac{1}{2\pi} \exp(-\mathrm{i}tx) K^{\mathrm{ft}}(tb) \frac{1}{n} \sum_{j=1}^{n} \exp(\mathrm{i}tY_j)/\tilde{g}^{\mathrm{ft}}(t)\, \mathrm{d}t \;-\; f(x)\Big|^2 \mathrm{d}x \\
&= \frac{1}{2\pi} \int E \Big| K^{\mathrm{ft}}(tb) \frac{1}{n} \sum_{j=1}^{n} \exp(\mathrm{i}tY_j)/\tilde{g}^{\mathrm{ft}}(t) \;-\; f^{\mathrm{ft}}(t) \Big|^2 \\
&= \frac{1}{2\pi n} \int |K^{\mathrm{ft}}(tb)|^2 |\tilde{g}^{\mathrm{ft}}(t)|^{-2} (1 - |h^{\mathrm{ft}}(t)|^2)\, \mathrm{d}t \\
&\qquad + \frac{1}{2\pi} \int \Big| K^{\mathrm{ft}}(tb) E \exp(\mathrm{i}tY_1)/\tilde{g}^{\mathrm{ft}}(t) \;-\; f^{\mathrm{ft}}(t) \Big|^2 \mathrm{d}t \\
&= \frac{1}{2\pi n} \int |K^{\mathrm{ft}}(tb)|^2 |\tilde{g}^{\mathrm{ft}}(t)|^{-2} (1 - |h^{\mathrm{ft}}(t)|^2)\, \mathrm{d}t \\
&\qquad + \frac{1}{2\pi} \int |f^{\mathrm{ft}}(t)|^2 \, |K^{\mathrm{ft}}(tb) g^{\mathrm{ft}}(t)/\tilde{g}^{\mathrm{ft}}(t) \;-\; 1|^2 \, \mathrm{d}t,
\end{aligned}$$

where we have used Parseval's identity, Fubini's theorem, and the bias-variance decomposition as in the proof of Proposition 2.2. We realize that the bias term has changed significantly compared to Proposition 2.2. Whenever we want to keep uniform consistency in the case of correct specification, that is, $g = \tilde{g}$ – even if some deterioration of the convergence rates is accepted – the bandwidth must be selected so that

$$\begin{aligned}
&\frac{1}{2\pi n} \sup_{f \in \mathcal{F}} \int |K^{\mathrm{ft}}(tb)|^2 \big(|\tilde{g}^{\mathrm{ft}}(t)|^{-2} - |f^{\mathrm{ft}}(t)|^2\big)\, \mathrm{d}t \stackrel{n \to \infty}{\longrightarrow} 0, \\
&\frac{1}{2\pi} \sup_{f \in \mathcal{F}} \int |f^{\mathrm{ft}}(t)|^2 \, |K^{\mathrm{ft}}(tb) \;-\; 1|^2 \, \mathrm{d}t \stackrel{n \to \infty}{\longrightarrow} 0.
\end{aligned}$$

Meister [87] shows that those conditions suffice to derive that

$$\liminf_{n\to\infty} \sup_{f\in\mathcal{F}} \mathrm{MISE}(\hat{f}, f) = d_{\mathcal{F}}(g, \tilde{g})$$

for ordinary smooth and supersmooth error densities as in (2.31) and (2.32), respectively, where

$$d_{\mathcal{F}}(g,\tilde{g}) = \frac{1}{2\pi} \sup_{f\in\mathcal{F}} \int \left|g^{\mathrm{ft}}(t)/\tilde{g}^{\mathrm{ft}}(t) - 1\right|^2 \left|f^{\mathrm{ft}}(t)\right|^2 \mathrm{d}t$$

for any density class $\mathcal{F}$. The functional $d_{\mathcal{F}}(g,\tilde{g})$ may be viewed as a kind of distance between g and $\tilde{g}$, although it is no metric on $\mathcal{G}$. In particular, it is not symmetric. This means that exchanging the arguments g and $\tilde{g}$ may change the value of $d_{\mathcal{F}}(g,\tilde{g})$. Hence, $d_{\mathcal{F}}(g,\tilde{g}) < d_{\mathcal{F}}(\tilde{g},g)$ implies that inserting $\tilde{g}$ as the error density in our estimator is preferable to g in view of the asymptotic behavior of the MISE under possible misspecification of the error density if one has to decide between two potential error densities g and $\tilde{g}$ without any further a-priori knowledge. The fact that $|g^{\mathrm{ft}}(t)| \leq |\tilde{g}^{\mathrm{ft}}(t)|$ for all $t \in \mathbb{R}$ implies $d_{\mathcal{F}}(g,\tilde{g}) < d_{\mathcal{F}}(\tilde{g},g)$ indicates that less smooth $\tilde{g}$ are less dangerous with respect to misspecification. That can be illustrated as follows:

Assume that the true error density g is Laplace $g(x) = \exp(-|x|)$ with $g(t) = 1/(1+t^2)$ and we mistakenly insert the standard normal density $\tilde{g}$. Then, $d_{\mathcal{F}}(g,\tilde{g}) = \infty$ implies that the supremum of the MISE tends to infinity as $n \to \infty$. That is a disaster. Even the standard kernel density estimator which ignores the contamination of the data provides better asymptotic results than the deconvolution estimator with a misspecified error density.

On the other hand, assume the reverse situation where the true error density g is standard normal and the inserted density $\tilde{g}$ is Laplace. Then, we have $d_{\mathcal{F}}(g,\tilde{g}) < \infty$, at least, so that the supremum of the MISE is asymptotically bounded. This fact emphasizes that the Laplace density is preferable. In general, ordinary smooth error densities (see (2.31)) are preferable to supersmooth ones (2.32).

A similar result can be derived in the case of a continuous class of normal errors with unknown variances. The variance shall be chosen rather too small than too large as any variance, which is larger than the true one may also lead to $d_{\mathcal{F}}(g,\tilde{g}) = \infty$.

2.6.1 Deterministic Constraints

As derived in the previous subsection, the common Sobolev conditions do not allow for any kind of misspecification of the error density g without losing consistency. That has been shown by Example (c), where the important condition is the inclusion $\mathcal{F} * \mathcal{G} \subseteq \mathcal{F}$. Clearly, the Sobolev condition (2.30) imposes only an upper bound on the decay of f^{ft} and, hence, on the smoothness of f. As convolving some function with any density makes the function smoother, the one-side bound given by (2.30) cannot violate the validity of $\mathcal{F} * \mathcal{G} \subseteq \mathcal{F}$, which is essential to keep some chance for identifiability. This inspires us to impose the following two-side bound on f^{ft},

$$D_2(1+|t|^{\beta+1/2})^{-1} \geq \left|f^{\mathrm{ft}}(t)\right| \geq D_1(1+|t|^{\beta+1/2})^{-1}, \text{ for all } t \in \mathbb{R} \quad (2.81)$$

for some $D_2 > D_1 > 0$ and $\beta > 1/2$. Note that those conditions are more restrictive compared to the inequality (2.30), which can be viewed as an integrated version of the upper bound contained in (2.81); however, they have also been used in the definition of ordinary smooth densities (2.31), referring to the error density g. In the present context, we restrict our consideration to normal contamination, that is, we assume that g is a normal density with known mean μ but unknown variance σ^2. Further, we propose that the variance is located to some known interval $(0, \sigma_1^2]$. Hence, the class of all admitted error densities is equal to

$$\mathcal{G} = \left\{N(\mu, \sigma^2) : \sigma \in (0, \sigma_1^2]\right\}.$$

In the sequel, we will show that the lower bound imposed in (2.81) makes the density f identifiable and consistently estimable, indeed. To derive an estimator for σ^2, we consider that

$$\left|h^{\mathrm{ft}}(t)\right| = \left|f^{\mathrm{ft}}(t)\right|\left|g^{\mathrm{ft}}(t)\right| \geq D_1(1+|t|^{\beta+1/2})^{-1} \exp\left(-\sigma_1^2 t^2/2\right), \text{ for all } t \in \mathbb{R}.$$

Therefore, we may define a truncated empirical version of $|h^{\mathrm{ft}}(t)|$, which takes into account the above lower bound, by

$$\hat{\varphi}_n(t) = \begin{cases} \left|\frac{1}{n}\sum_{j=1}^n \exp(itY_j)\right|, & \text{if } \left|\frac{1}{n}\sum_{j=1}^n \exp(itY_j)\right| \\ & \geq D_1(1+|t|^{\beta+1/2})^{-1}\exp\left(-\sigma_1^2 t^2/2\right), \\ D_1(1+|t|^{\beta+1/2})^{-1}\exp\left(-\sigma_1^2 t^2/2\right), & \text{otherwise.} \end{cases}$$

Solving the equation

$$h^{\mathrm{ft}}(t) = f^{\mathrm{ft}}(t)\exp\left(-\sigma^2 t^2/2\right)$$

by the error variance σ^2, we have

$$\sigma^2 = -\frac{2}{t^2}\cdot\log\left(|h^{\mathrm{ft}}(t)/f^{\mathrm{ft}}(t)|\right),$$

and, hence,

$$\sigma^2 \in \left[\frac{2}{t^2}\cdot\left|\log\left(|h^{\mathrm{ft}}(t)|D_1^{-1}(1+|t|^{\beta+1/2})\right)\right|, \frac{2}{t^2}\cdot\left|\log\left(|h^{\mathrm{ft}}(t)|D_2^{-1}(1+|t|^{\beta+1/2})\right)\right|\right] \quad (2.82)$$

for all t. Since

$$\Big|\log\big(|h^{\mathrm{ft}}(t)|D_1^{-1}(1+|t|^{\beta+1/2})\big)\Big| \Big/ \Big|\log\big(|h^{\mathrm{ft}}(t)|D_2^{-1}(1+|t|^{\beta+1/2})\big)\Big|$$
$$= \left|\frac{-\sigma^2 t^2/2 + \log\big|f^{\mathrm{ft}}(t)D_1^{-1}(1+|t|^{\beta+1/2})\big|}{-\sigma^2 t^2/2 + \log\big|f^{\mathrm{ft}}(t)D_2^{-1}(1+|t|^{\beta+1/2})\big|}\right| \stackrel{|t|\to\infty}{\longrightarrow} 1,$$

the length of the interval in (2.82) converges to 0 as t tends to infinity. Now we realize that σ^2 is reconstructable from (2.82). Therefore, we replace $|h^{\mathrm{ft}}(t)|$ by $\hat{\varphi}_n(t)$ and choose the lower bound on $|f^{\mathrm{ft}}(t)|$ to estimate the error variance by

$$\hat{\sigma}_n^2 = \max\Big\{-\frac{2}{t_n^2}\log\big(\hat{\varphi}_n(t_n)D_1^{-1}(1+|t_n|^{\beta+1/2})\big)\,,\,0\Big\}, \tag{2.83}$$

where the sequence $(t_n)_n \uparrow \infty$ is still to be selected. Because of the construction of $\hat{\varphi}_n$, we may conclude that $\hat{\sigma}_n^2 \leq \sigma_1^2$ holds almost surely. Hence, we may fix that $|\hat{\sigma}_n^2 - \sigma^2| \leq \sigma_1^2$ for the true $\sigma^2 \in (0, \sigma_1^2]$. This gives us the final density estimator

$$\hat{f}_n(x) = \frac{1}{2\pi}\int \exp(-itx)K^{\mathrm{ft}}(tb)\frac{1}{n}\sum_{j=1}^{n}\exp(itY_j)\,\exp\Big(-it\mu + \frac{1}{2}\hat{\sigma}_n^2 t^2\Big)\,dt. \tag{2.84}$$

The convergence rates of the MISE considered uniformly over the class $\mathcal{F}'_{\beta,D_1,D_2}$ collecting all densities satisfying (2.81) will be studied in the sequel. Applying Parseval's identity and Fubini's theorem as in Proposition 2.2, we obtain, for the estimator (2.84), that

$$\begin{aligned} E\|\hat{f} - f\|_2^2 &\\ &\leq \frac{1}{\pi}\int \big|K^{\mathrm{ft}}(tb)\big|^2 E\,\exp\big(\hat{\sigma}_n^2 t^2\big)\Big|\frac{1}{n}\sum_{j=1}^{n}\big(\exp(itY_j) - E\exp(itY_j)\big)\,dt\Big|^2 dt \\ &\quad + \frac{1}{\pi}\int \big|K^{\mathrm{ft}}(tb)\big|^2 E\big|\exp[(\hat{\sigma}_n^2 - \sigma^2)t^2/2] - 1\big|^2 |f^{\mathrm{ft}}(t)|^2\,dt \\ &\quad + \frac{1}{2\pi}\int \big|K^{\mathrm{ft}}(tb) - 1\big|^2 |f^{\mathrm{ft}}(t)|^2\,dt. \end{aligned}$$

Since $\hat{\sigma}_n^2 \leq \sigma_1^2$ almost surely, the first and the third term in the above equation are bounded above by the MISE for standard deconvolution with the normal error density $N(0,\sigma_1^2)$. According to Theorem 2.9(b), we choose the bandwidth $b = c\big(\log n\big)^{-1/2}$ with some sufficiently large constant c and the sinc-kernel K as in (2.41). Then, the sum of the first and the third term decays as $O\big((\log n)^{-\beta}\big)$. The second term vanishes in deconvolution problems with known error variance and, therefore, needs more careful consideration in the present framework. For some sequence $(d_n)_n \downarrow 0$, this second term is bounded above by

$$\frac{1}{\pi}E\int \big|K^{\rm ft}(tb)\big|^2\big|\exp\big[\big|\hat\sigma_n^2-\sigma^2\big|t^2/2\big]-1\big|^2|f^{\rm ft}(t)|^2\,{\rm d}t\cdot\chi_{[0,d_n]}(|\hat\sigma_n^2-\sigma^2|)$$
$$+\frac{1}{\pi}E\int \big|K^{\rm ft}(tb)\big|^2\big|\exp\big[\big|\hat\sigma_n^2-\sigma^2\big|t^2/2\big]-1\big|^2|f^{\rm ft}(t)|^2\,{\rm d}t\cdot\chi_{(d_n,\infty)}(|\hat\sigma_n^2-\sigma^2|)$$
$$\leq \frac{1}{\pi}\int |K^{\rm ft}(tb)|^2|f^{\rm ft}(t)|^2\big|\exp\big(d_nt^2/2\big)-1\big|^2\,{\rm d}t$$
$$+O\big(n^{\sigma_1^2/c^2}\big)\cdot P\big[|\hat\sigma_n^2-\sigma^2|>d_n\big],$$

where we have used that $\|f^{\rm ft}\|_2$ is uniformly bounded over $f\in\mathcal{F}'_{\beta,D_1,D_2}$. Therefore, the constants contained in $O(\cdots)$ do not depend on the specific $f\in\mathcal{F}'_{\beta,D_1,D_2}$. Note that the sequence $(d_n)_n$ need not be chosen by the person who applies the estimator but is only to be selected within the current proof. Let us derive an upper bound on the following probability

$$P\big[\hat\sigma_n^2>\sigma^2+d_n\big]$$
$$\leq P\big[\hat\varphi_n(t_n)\leq D_1\big(1+t_n^{\beta+1/2}\big)^{-1}\exp\big(-t_n^2\sigma^2/2\big)\exp\big(-t_n^2d_n/2\big)\big]$$
$$\leq P\big[\hat\varphi_n(t_n)\leq\big|h^{\rm ft}(t_n)\big|\cdot\exp\big(-t_n^2d_n/2\big)\big]$$
$$\leq\big(1-\exp(-t_n^2d_n/2)\big)^{-2}\big|h^{\rm ft}(t_n)\big|^{-2}\cdot E\big|\hat\varphi_n(t_n)-\big|h^{\rm ft}(t_n)\big|\big|^2$$
$$\leq O\big(1/n\big)\cdot\big(1+t_n^{\beta+1/2}\big)^2\exp\big(\sigma_1^2t_n^2\big),$$

with uniform constants under the condition $t_n^2d_n\geq{\rm const}>0$, where Markov's inequality has been applied. Analogously, we have

$$P\big[\hat\sigma_n^2<\sigma^2-d_n\big]$$
$$\leq P\big[\hat\varphi_n(t_n)-|h^{\rm ft}(t_n)|\geq\big((D_1/D_2)\exp(t_n^2d_n/2)-1\big)\cdot|h^{\rm ft}(t_n)|\big]$$
$$\leq\frac{4}{n}\cdot(1+t_n^{\beta+1/2})^2\exp\big(\sigma_1^2t_n^2\big),$$

when fixing that

$$d_n=2t_n^{-2}\log\big(2D_2/D_1\big)\,.$$

Then, summarizing, the MISE is uniformly bounded above by

$$O\big((\log n)^{-\beta}\big)+\frac{1}{\pi}\int|K^{\rm ft}(tb)|^2|f^{\rm ft}(t)|^2\big|\exp\big(d_nt^2/2\big)-1\big|^2\,{\rm d}t,$$

whenever $t_n=c_t\cdot\big(\log n\big)^{1/2}$ with $c_t>0$ and $c^{-2}+c_t^2<\sigma_1^{-2}$. By the substitution $s=tb$ and the fact that the function $\big(1-\exp(\cdot)\big)/\cdot^2$ is bounded on each compact interval, we derive that the supremum of the MISE has the upper bound,

$$O\big((\log n)^{-\beta}\big)+O\big(d_n^2b^{-5}\big)\cdot\int|K^{\rm ft}(s)|^2|f^{\rm ft}(s/b)|^2s^4\,{\rm d}s.$$

Now we have to distinguish between three cases. In the first case, we assume $\beta<2$ and apply the upper bound ${\rm const}\cdot|s|^{-2\beta+3}b^{2\beta+1}\chi_{[-1,1]}(s)$ to the function to be integrated. Hence, the MISE is uniformly bounded from above by

$$O\big((\log n)^{-\beta}\big) + O\big(d_n^2 b^{2\beta-4}\big) = O\big((\log n)^{-\beta}\big).$$

In the second case $\beta > 2$, we substitute the integral again by $t = s/b$ so that we face the bound

$$O\big((\log n)^{-\beta}\big) + O\big(d_n^2\big) \cdot \int |K^{\mathrm{ft}}(tb)|^2 |f^{\mathrm{ft}}(t)|^2 t^4 \, dt = O\big(d_n^2\big) = O\big((\log n)^{-2}\big).$$

Finally, in the third case $\beta = 2$, an iterated logarithmic factor gets in. All in all, we have shown the following proposition.

Proposition 2.18. *Consider the density deconvolution problem where $f \in \mathcal{F}'_{\beta,D_1,D_2}$ and $\mathcal{G} = \{N(\mu,\sigma^2) : \sigma^2 \leq \sigma_1^2\}$. Use estimator (2.84) where $t_n = c_t\big(\log n\big)^{1/2}$ and $b = c\big(\log n\big)^{-1/2}$ with $c, c_t > 0$ and $c^{-2} + c_t^2 < \sigma_1^{-2}$. Then we have*

$$\sup_{g\in\mathcal{G}} \sup_{f\in\mathcal{F}'_{\beta,D_1,D_2}} E\|\hat{f} - f\|_2^2 = \begin{cases} O\big((\log n)^{-\beta}\big), & \text{if } \beta \in (1/2, 2), \\ O\big((\log n)^{-2}\big), & \text{if } \beta > 2, \\ O\big((\log n)^{-2} \log\log n\big), & \text{otherwise.} \end{cases}$$

Therefore, the ignorance of σ causes some deterioration of the convergence rates only if $\beta > 2$. One can also show that those convergence rates are optimal (or nearly optimal in the case $\beta = 1/2$) in the underlying model, see [89] for the theory and numerical simulations for this estimation problem. A similar approach to such deconvolution problems is given in [8], where some further density classes are considered. Still, those approaches suffer from the slow rates and the fact that the lower bound on f^{ft} as in (2.81) is also hard to justify in practice – just as the exact knowledge of the error variance in the standard approach.

Another model where the assumption of a perfectly known error density g can be relaxed is suggested in [92]. In the corresponding experiment, we assume that the target densities f are supported on some fixed compact interval and satisfy the usual Sobolev assumption (2.30). Then, with respect to the class of the error densities $\mathcal{G}$, it suffices to assume that the restriction of their Fourier transform to some compact interval, $g^{\mathrm{ft}}(t)$, $|t| \leq c$ is known. Outside $[-c, c]$, no information about g^{ft} is required. This is due to the fact that compactly supported densities possess analytic Fourier transforms; more precisely, their Fourier transforms are representable by the pointwise limit of their Taylor series on the whole real line. Hence, one may estimate the functions $f^{\mathrm{ft}}(t)$ on their restriction to $t \in [-c, c]$ by (2.3) as usual and, then, apply some polynomial continuation of f^{ft} in the Fourier domain on some interval $[-1/b, 1/b]$, where b denotes some bandwidth parameter. Meister [92] suggests to use orthogonal projection involving Legendre polynomials. In this paper, consistency along with optimal rates is shown. Although the convergence rates are logarithmic, we notice the surprising result that the rates are faster than those whose optimality is established in Theorem 2.14(b). Hence, the condition of compact support of the target densities improves the optimal convergence rates; this is a very unusual situation in density estimation in general. Note that there is no improvement of the rates for known ordinary smooth error densities.

2.6.2 Additional Data

In this subsection, we focus on the density deconvolution problem where the error density g is unknown but estimable from direct i.i.d. data $\varepsilon'_1, \ldots, \varepsilon'_m$, which are collected in a separate independent experiment. This model is studied in [40, 101] and the book of Efromovich published in 1999 [41]. Of course, its applicability is restricted to cases where the system of measurement can be calibrated somehow. In particular, the model should be considered when, in some cases, the same individual or quantity can be observed both in an error-free way (call this measurement $X'_{j,1}$) and by a measurement procedure, which is affected by nonnegligible noise (denote this observation by $X'_{j,2}$). Then put $\varepsilon'_j = X'_{j,2} - X'_{j,1}$, where ε'_j is indeed a direct observation from the error distribution. Usually, noisy measurement procedures are less expensive so that there likely exists a large amount of measurements where no exact observation is possible. The previously estimated error density may be employed in the deconvolution step for that latter dataset.

Recalling the practical Example 1, as described in Sect. 2.1, one can arrange some training experiment where the same run of one athlete is measured by an electronic system (error-free observations) as well as by a human (noisy hand-measured observations). This acquired information about the error distribution can be used to analyze solely hand-measured data.

Considering the structure of the deconvolution kernel estimator (2.7), we recognize that $g^{\mathrm{ft}}(t)$ must be replaced by some empirical version based on the additional data $\varepsilon'_1, \ldots, \varepsilon'_m$. Obviously, it is reasonable to propose the empirical characteristic function

$$\hat{g}^{\mathrm{ft}}(t) = \frac{1}{m} \sum_{j=1}^{m} \exp(\mathrm{i}t\varepsilon'_j).$$

However, inserting $\hat{g}^{\mathrm{ft}}(t)$ into the deconvolution kernel estimator without any regularization is rather dangerous as we divide by this quantity. Because of random deviation, $\hat{g}^{\mathrm{ft}}(t)$ could vanish although its true counterpart $g^{\mathrm{ft}}(t)$ is not equal to zero. Therefore, a ridge-parameter technique is recommended as in (2.21) even if g^{ft} does not have any zeros and satisfies (2.31) or (2.32). Such a method is also used in [101]. Applying a combined ridge-kernel approach, we propose the following density estimator for f,

$$\hat{f}(x) = \frac{1}{2\pi} \int \exp(-\mathrm{i}tx) K^{\mathrm{ft}}(tb) \frac{1}{n} \sum_{j=1}^{n} \exp(\mathrm{i}tY_j) \frac{\hat{g}^{\mathrm{ft}}(-t)}{\max\{|\hat{g}^{\mathrm{ft}}(t)|, \rho_n\}^2} \,\mathrm{d}t. \quad (2.85)$$

where the bandwidth b and the ridge-parameter $\rho_n > 0$ are still to be selected. We employ the conventional sinc-kernel as in (2.41) for K. Then the expectation of the squared $L_2(\mathbb{R})$-distance between estimator (2.85) and the deconvolution kernel density estimator (2.7) with correctly specified g is bounded above by

$$\frac{1}{2\pi}E\int \big|K^{\mathrm{ft}}(tb)\big|^2 \Big|\frac{1}{n}\sum_{j=1}^n \exp(itY_j)\Big|^2 \Big|\frac{\hat{g}^{\mathrm{ft}}(-t)}{\max\{|\hat{g}^{\mathrm{ft}}(t)|,\rho_n\}^2} - \big[g^{\mathrm{ft}}(t)\big]^{-1}\Big|^2 dt$$

$$= \frac{1}{2\pi}E\int \big|K^{\mathrm{ft}}(tb)\big|^2 \Big|\frac{1}{n}\sum_{j=1}^n \exp(itY_j)\Big|^2 / |g^{\mathrm{ft}}(t)|^2$$

$$E\Big|\frac{\hat{g}^{\mathrm{ft}}(-t)g^{\mathrm{ft}}(t)}{\max\{|\hat{g}^{\mathrm{ft}}(t)|,\rho_n\}^2} - 1\Big|^2 dt$$

$$\leq \frac{1}{2\pi n}\int \big|K^{\mathrm{ft}}(tb)\big|^2 / |g^{\mathrm{ft}}(t)|^2 \cdot E\Big|\frac{\hat{g}^{\mathrm{ft}}(-t)g^{\mathrm{ft}}(t)}{\max\{|\hat{g}^{\mathrm{ft}}(t)|,\rho_n\}^2} - 1\Big|^2 dt$$

$$+ \frac{1}{2\pi}\int \big|K^{\mathrm{ft}}(tb)\big|^2 \big|f^{\mathrm{ft}}(t)\big|^2 \cdot E\Big|\frac{\hat{g}^{\mathrm{ft}}(-t)g^{\mathrm{ft}}(t)}{\max\{|\hat{g}^{\mathrm{ft}}(t)|,\rho_n\}^2} - 1\Big|^2 dt,$$

where we have used Parseval's identity and the independence of the datasets $Y_1,\ldots,Y_n$ and $\varepsilon'_1,\ldots,\varepsilon'_m$. Furthermore, we derive that

$$E\Big|\frac{\hat{g}^{\mathrm{ft}}(-t)g^{\mathrm{ft}}(t)}{\max\{|\hat{g}^{\mathrm{ft}}(t)|,\rho_n\}^2} - 1\Big|^2$$

$$\leq E\Big|\frac{g^{\mathrm{ft}}(t)\hat{g}^{\mathrm{ft}}(-t)}{\rho_n^2} - 1\Big|^2 \cdot \chi_{[0,\rho_n)}(|\hat{g}^{\mathrm{ft}}(t)|)$$

$$+ E\Big|\frac{g^{\mathrm{ft}}(t) - \hat{g}^{\mathrm{ft}}(t)}{\hat{g}^{\mathrm{ft}}(t)}\Big|^2 \cdot \chi_{[\rho_n,\infty)}(|\hat{g}^{\mathrm{ft}}(t)|)$$

$$\leq E\Big|\frac{g^{\mathrm{ft}}(t)\hat{g}^{\mathrm{ft}}(-t)}{\rho_n^2} - 1\Big|^2 \cdot \chi_{[0,\rho_n)}(|\hat{g}^{\mathrm{ft}}(t)|)$$

$$+ E\big|\Delta(t)\big|^2 \Big|\frac{g^{\mathrm{ft}}(t)}{\hat{g}^{\mathrm{ft}}(t)}\Big|^2 \cdot \chi_{[\rho_n,\infty)}(|\hat{g}^{\mathrm{ft}}(t)|)$$

$$\leq \Big(\Big|\frac{g^{\mathrm{ft}}(t)}{\rho_n}\Big| + 1\Big)^2 \cdot \big(P\big[|\hat{g}^{\mathrm{ft}}(t)| < \rho_n\big] + P\big[|\Delta(t)| > 1/2\big]\big) + 4E\big|\Delta(t)\big|^2, \tag{2.86}$$

where $\Delta(t) = \big(g^{\mathrm{ft}}(t) - \hat{g}^{\mathrm{ft}}(t)\big)/g^{\mathrm{ft}}(t)$. Note that $|\Delta(t)| \leq 1/2$ implies

$$\big|g^{\mathrm{ft}}(t)/\hat{g}^{\mathrm{ft}}(t)\big| \leq 2\,.$$

Rather easily, we derive that

$$E\big|\Delta(t)\big|^2 \leq 1/\big(m|g^{\mathrm{ft}}(t)|^2\big).$$

Now we restrict the framework of this study to ordinary smooth g satisfying (2.31). Then, the parameter ρ_n shall be selected so that

$$\rho_n \asymp n^{-\gamma},$$

with some $\gamma > 1$. As any bandwidth selection under which $nb^{1+2\alpha}$ does not tend to infinity will consequently lead to inconsistency, we may assume that

$$\inf_{|t|\leq 1/b} |g^{\mathrm{ft}}(t)|^2 \geq 2\rho_n$$

for all n sufficiently large, as the left side of the above equation decays to zero with an algebraic rate. Hence, for $|t| \leq 1/b$,

$$\begin{aligned}
&P\big[|\hat{g}^{\mathrm{ft}}(t)| < \rho_n\big] + P\big[|\Delta(t)| > 1/2\big] \\
&\leq 2P\Big[\big|\hat{g}^{\mathrm{ft}}(t) - g^{\mathrm{ft}}(t)\big| > \inf_{|s|\leq 1/b} |g^{\mathrm{ft}}(s)|/2\Big] \\
&\leq 2P\Big[\Big|\frac{1}{m}\sum_{j=1}^{m} \big(\exp(it\varepsilon_j') - E\exp(it\varepsilon_j')\big)\Big|^2 > \mathrm{const}\cdot b^{2\alpha}\Big] \\
&\leq \mathrm{const}\cdot \exp\big(-\,\mathrm{const}\cdot mb^{2\alpha}\big),
\end{aligned}$$

where Hoeffding's inequality (see [66]) has been applied. It is used in the following version (not the most general one): Assume that $Z_1, \ldots, Z_n$ are i.i.d. real-valued random variables whose distribution is concentrated on some compact interval $[a, b]$. Then, we have

$$P\Big[\Big|\frac{1}{n}\sum_{j=1}^{n}(Z_j - EZ_j)\Big| > \varepsilon\Big] \leq 2\exp\Big(-\,2(b-a)^{-2}\cdot n\varepsilon^2\Big).$$

In general, this inequality is used to derive exponential upper bounds on the centered average of compactly distributed i.i.d. random variables. In our context, by this exponential bound, we may derive any algebraic convergence rate $n^{-\delta}$, $\delta > 0$, as an upper bound on the first term in (2.86) whenever $mb^{2\alpha}/\log n \to \infty$ as $n \to \infty$. Summarizing, we have shown that the mean squared $L_2(\mathbb{R})$-distance between the kernel deconvolution estimator (2.7) and estimator (2.85) has the following upper bound,

$$\begin{aligned}
o(n^{-1}) + \frac{1}{2\pi nm}\int |K^{\mathrm{ft}}(tb)|^2/|g^{\mathrm{ft}}(t)|^4\,dt& \\
+ \frac{1}{2\pi m}\int |K^{\mathrm{ft}}(tb)|^2|f^{\mathrm{ft}}(t)|^2/|g^{\mathrm{ft}}(t)|^2\,dt&.
\end{aligned}$$

When introducing the common Sobolev condition (2.30) with respect to f, the third term in the above equation is bounded above by $O\big(\max\{1, b^{2(\beta-\alpha)}\}/m\big)$. Obviously, if the mean squared distance between both estimators is not larger than the MISE of the deconvolution kernel estimator, rate optimality, which is shown for the kernel estimator in Theorem 2.9, can be extended to the case of unknown error density. Therefore, we have proved the following proposition.

Proposition 2.19. *Assume the common density deconvolution model (2.1) where the error density g is ordinary smooth (see (2.31)) but unknown, and assume the availability of the additional i.i.d. data $\varepsilon'_1, \ldots, \varepsilon'_m$ having the error density g. Then, estimator (2.85) with $\rho_n \asymp n^{-\gamma}$, $\gamma > 1$ and the bandwidth b as in Theorem 2.9(a) satisfies*

$$\sup_{f \in \mathcal{F}_{\beta,C;L_2}} E\|\hat{f} - f\|_2^2 = O\big(n^{-2\beta/(2\beta+2\alpha+1)}\big),$$

as long as

$$m \,/\, \big(n^{2\beta/(2\beta+2\alpha+1)} + n^{2\alpha/(2\beta+2\alpha+1)} \log n\big) \overset{n\to\infty}{\longrightarrow} \infty.$$

The necessity of the availability of enough direct data to keep the convergence rates from the known error case is rather intuitive. Critically, one may address that, although the ridge parameter ρ_n is chosen without assuming any property of the error density g to be known, the estimator is not adaptive because the choice of the bandwidth parameter requires knowledge of β as well as of the smoothness degree α of g, leading to a further problem in the unknown error case. One could think of cross validation techniques as introduced in Sect. 2.5.1 to select the bandwidth. An adaptive wavelet estimator for the underlying problem has recently been introduced in [17].

2.6.3 Replicated Measurements

In this subsection, we address the model where the same incorrupted but unobserved random variable X_j is independently measured for several times, but each measurement is affected by error. Therefore, we change the standard additive measurement error model (2.1) into that statistical experiment where we observe the data $Y_{j,k}$, $j = 1, \ldots, n$; $k = 1, \ldots, m_j$, $m_j \geq 2$ defined by

$$Y_{j,k} = X_j + \varepsilon_{j,k}. \tag{2.87}$$

Our goal is still to estimate the density f of the X_j where all the involved random variables $X_\cdot, \varepsilon_{\cdot,\cdot}$ are independent. Each $\varepsilon_{j,k}$ has the error density g.

This model, which includes repeated measurements, has broad applications in practice as already indicated in the Examples 2(b),2(c),4(a),4(b) in Sect. 2.1.

Let us explain why the error density g is identifiable in the current model unlike in the standard experiment (2.1). For simplicity, we assume that $m_j \equiv 2$ for all j so that the minimal valid assumption is studied. Then, we consider the fully accessible differences

$$\Delta Y_j = Y_{j,1} - Y_{j,2} = \varepsilon_{j,1} - \varepsilon_{j,2}, \quad j \in \{1, \ldots, n\} \quad \text{a.s.}$$

As an essential fact, the random variable X_j has been eliminated in this simple calculation so that ΔY_j is measurable in the σ-algebra generated by $\varepsilon_{j,1}$ and $\varepsilon_{j,2}$. The characteristic function of ΔY_j is then equal to

$$\psi_{\Delta Y_j}(t) = E\exp(it\Delta Y_j) = E\exp(it\varepsilon_{j,1}) \cdot E\exp(-it\varepsilon_{j,2}) = g^{\mathrm{ft}}(t)g^{\mathrm{ft}}(-t)$$
$$= \left|g^{\mathrm{ft}}(t)\right|^2.$$

If we assume that g is symmetric with respect to the y-axis, that is, $g(x) = g(-x)$ for all $x \in \mathbb{R}$, then g^{ft} is real-valued and we conclude that

$$g^{\mathrm{ft}}(-t) = g^{\mathrm{ft}}(t), \qquad \text{for all } t \in \mathbb{R}$$

by Lemma A.1(g) from the appendix. If, in addition, we assume the standard condition of nonvanishing g^{ft} as in (2.18), we may conclude that

$$g^{\mathrm{ft}}(t) > 0, \qquad \text{for all } t \in \mathbb{R}$$

by elementary analysis, as g^{ft} is a continuous, nonvanishing, real-valued function on the whole real line with $g^{\mathrm{ft}}(0) = 1 > 0$ (see Lemma A.1(d)). Therefore, it appears to be reasonable to use

$$\hat{g}^{\mathrm{ft}}(t) = \left|\frac{1}{n}\sum_{j=1}^{n}\exp(it\Delta Y_j)\right|^{1/2}$$

as an estimator for $g^{\mathrm{ft}}(t)$, where the empirical characteristic function of the ΔY_j has been inserted (see (2.2)). Considering that g^{ft} is real-valued, we may restrict to its real part, that is, replace the terms $\exp(it\Delta Y_j)$ by $\cos(t\Delta Y_j)$. Also, we shall use all given measurements of X_j so that we generalize to the case $m_j > 2$. Therefore, we define

$$\hat{g}^{\mathrm{ft}}(t) = \left|\frac{1}{N}\sum_{j=1}^{n}\sum_{(k_1,k_2)\in K_j}\cos\left(t\Delta Y_j\right)\right|^{1/2}, \tag{2.88}$$

where K_j denotes the set of all pairs $(k_1, k_2) \in \mathbb{Z}^2$ with $1 \leq k_1 < k_2 \leq m_j$. Hence, by an elementary combinational result, we fix that K_j contains $m_j(m_j-1)/2$ elements. Then, put $N = \sum_{j=1}^{n} m_j(m_j-1)/2$, the total number of all addends in the definition of estimator (2.88). Still, it is possible that some of the m_j are equal to one, implying that only one measurement of that X_j is given. Those measurements are simply ignored in the construction of (2.88).

While the use of replications in density deconvolution has first been mentioned in [70] in an econometric context, many recent papers have adopted this setting to make the error density identifiable, for example, [29, 113, 114]; where the estimator (2.88) is mainly taken from the latter work. In [83], the assumption of symmetry of g is relaxed and a rather sophisticated reconstruction algorithm of f^{ft} is introduced. However, the convergence rates can obviously not be kept from the known error case but suffer from some deterioration. Another reconstruction algorithm of f is studied in [62]. Neumann [103] shows that f is not identifiable if, for some density g, both g and the shifted function $g(\cdot - a)$, $a > 0$ are admitted to be the true error density. With respect to our estimator, we recognize that ΔY_j has the same distribution for both competing error densities so that g really cannot be identified based on the difference of two measurements. On the other hand, [103] establishes consistency for the rather general case where the $\varepsilon_{j,k}$ are centered, that is, their expectation and median are equal to zero and satisfy some regularity condition. However, convergence rates are not studied in that work.

Delaigle et al. [29] suggest as the density estimator for f,

$$\hat{f}(x) = \frac{1}{2\pi} \int \exp(-itx) K^{\mathrm{ft}}(tb) \frac{1}{n} \sum_{j=1}^{n} \exp(itY_{j,1}) / \big(\hat{g}^{\mathrm{ft}}(t) + \rho\big)\, \mathrm{d}t, \qquad (2.89)$$

with the bandwidth parameter b, a kernel function K, and an additional ridge regularization parameter $\rho > 0$. As in the previous subsection, the ridge-parameter is used to avoid random effects, which may cause that the denominator $\hat{g}^{\mathrm{ft}}$ becomes too close to zero.

Under appropriate selection of b and ρ, the optimal convergence rates can be preserved from Theorem 2.9(a) if the smoothness degree β of the target density is larger than the characteristic α belonging to g and $m_j \geq 2$ for each $j \in \{1, \ldots, n\}$ in the case of ordinary smooth contamination (see (2.31)). The proof requires more complicated stochastic proximity arguments between the estimator (2.89) and the standard deconvolution kernel estimator (2.7) compared to the proof in the setting of the previous subsection. Here, we have to realize that the dataset for estimating g^{ft} by (2.88) and those observations applied in the final deconvolution step in estimator (2.89) are not independent, which makes the proof technically more involved. Also, we note that different types of smoothness classes are used for f. In particular, we have a uniform upper bound on f^{ft} with polynomial decay rather than the Sobolev condition (2.30). In the countercase, where β is smaller than α, one has to pay a price for the estimation of $\hat{g}^{\mathrm{ft}}$ by accepting slower convergence rates. Then, some loss in the convergence rates is unavoidable. Under supersmooth contamination, the logarithmic convergence rates derived in Theorem 2.9(b) may be kept. The numerical simulations contained in [29] also indicate that the procedure performs well in practice.

Another great advantage of the replicated data model concerns the treatment of error densities g whose Fourier transforms show some oscillations and have some zeros. That effect has recently been investigated by [95]. In the model (2.1) without any repeated measurements, isolated zeros of g^{ft} will unavoidably cause slower convergence rates in many cases unless the order of the zeros is small compared to the smoothness β under more restrictive smoothness assumptions or moment restrictions on f are assumed (see [58, 94] or – as a summary – the discussion in the current book at the end of Sect. 2.4.3). However, in the replicated measurement model, the situation changes dramatically. For instance, in deconvolution from uniform error densities and their self-convolutions, one is able to attain the same convergence rates as for ordinary smooth g with the corresponding Fourier tails – without the requirement of any stronger conditions on the target density; in fact, the usual Sobolev assumption (2.30) can be used again. The main reason for that phenomenon is the characteristic function of the bivariate observation $(Y_{j,1}, Y_{j,2})$, which is equal to $f^{\mathrm{ft}}(t_1+t_2)g^{\mathrm{ft}}(t_1)g^{\mathrm{ft}}(t_2)$, or equivalently $f^{\mathrm{ft}}(t)g^{\mathrm{ft}}(t_1)g^{\mathrm{ft}}(t-t_1)$ so that we have to divide by $g^{\mathrm{ft}}(t_1)g^{\mathrm{ft}}(t-t_1)$ in the Fourier domain where we are still free to select the parameter $t_1 \in \mathbb{R}$. For error densities with isolated zeros only, t_1 may be selected for each t so that the denominator is sufficiently far away from zero. Of course, when only one measurement Y_j is given, its characteristic function is equal to $f^{\mathrm{ft}}(t)g^{\mathrm{ft}}(t)$ so that no parameter is selectable to estimate $f^{\mathrm{ft}}(t)$. That is an intuitive explanation for the difference between the two models.

In repeated measurement models, it is frequently suggested to reduce the observation scheme by averaging. In our context, it means that we base the estimator on the observations $(\overline{Y}_1, \ldots, \overline{Y}_n)$ with

$$\overline{Y}_j = \frac{1}{m_j}\sum_{k=1}^{m_j} Y_{j,k}\,, \qquad k \in \{1,\ldots,m_j\},$$

instead of the original data $Y_{\cdot,\cdot}$. We have

$$\overline{Y}_j = X_j + \frac{1}{m_j}\sum_{k=1}^{m_j} \varepsilon_{j,k}.$$

Although averaging reduces the variance of the error components, it also makes the error density smoother when g is ordinary smooth as in (2.31). For instance, if $m_j \equiv 2$, then the characteristic function of the random variables $\frac{1}{2}\sum_{k=1}^{2}\varepsilon_{j,k}$ decays as $|t|^{-2\alpha}$. Therefore, the smoothness degree is multiplied by the number of measurements, over which the average has been taken, as long as the m_j are bounded with respect to the sample size n. It follows from there that the convergence rates are made worse by averaging, according to Theorem 2.14(a). Therefore, it is preferable to use just one row, for example, $Y_{\cdot,1}$, of the data in the deconvolution step.

On the other hand, if the error density g is normal, the situation changes as normal densities are closed with respect to convolution, that is, the convolution of two normal densities is normal again. Then, the smoothness degree of g does not increase when averaging, but the variance is reduced. Delaigle and Meister [31] show that the dataset $(\overline{Y}_1, \ldots, \overline{Y}_n)$ is statistically sufficient for the target density f. This justifies the use of the averaging method. However, the methods used in the sufficiency proof required some special properties of normal densities so that there exists no straightforward extension to more general densities g. As shown earlier, in the case of ordinary smooth contamination, sufficiency cannot hold true. The paper of [31] also addresses the heteroscedastic case where the numbers m_j are not put equal to two; but they are indeed different among j.

2.7 Special Problems

In this section, we want to address some rather specific topics in the field of density deconvolution.

2.7.1 Heteroscedastic Contamination

In some real-life applications in the field of density deconvolution, the data are collected under different conditions or taken from different surveys or measurement systems. Then, the measurement error might differ among the data, which may no longer be assumed to be i.i.d. in that case. This problem has recently been highlighted in [31]. In that work, the authors apply the model (2.1) under the significant modification that the random variables ε_j are no longer assumed to be identically distributed, but each ε_j may have its own known density $g_{j,n}$, which may depend on both the sample size n and the observation number j.

As a real data example, the authors mention a dataset taken from [97] and discussed in [19], where fibre ingestion is investigated based on 333 individuals, and replicated measurements are available for 76 of those individuals. Again, the different outcome for the replications shows that measurement error must not be neglected. On the other hand, we assume that all 333 data suffer from the identical error density. The authors assume that the error density is normal with mean zero, and take the average of the replicated observations so that the error variance is reduced for the 76 repeatedly measured data. However, the averaged data do not have the same distribution as the residual 257 data. Then, we have a normal measurement error model where the errors have different variances; hence, a model, which includes heteroscedastic contaminated observations. Furthermore, the authors mention that knowledge of the original error variance need not be assumed in advance; instead, it can be estimated by considering the difference between the two replicates, estimating the variance of these differences, and, then, dividing by two.

How can the standard deconvolution kernel density estimator (2.7) be modified so that we can handle the heteroscedastic deconvolution problem? As a general kernel estimator, we specify

$$\hat{f}(x) = \frac{1}{2\pi} \int \exp(-\mathrm{i}tx) K^{\mathrm{ft}}(tb) \hat{\Psi}_{X_1}(t)\, \mathrm{d}t,$$

with a kernel function $K \in L_2(\mathbb{R})$, where K^{ft} is compactly supported, and a bandwidth parameter $b > 0$. The quantity $\hat{\Psi}_{X_1}(t)$, which is still to be chosen, denotes an estimator of the Fourier transform $f^{\mathrm{ft}}(t)$. In the homoscedastic case, that is, $g_{j,n} = g$ for all n, $j = 1, \ldots, n$, we simply employ

$$\hat{\Psi}_{X_1}(t) = \frac{1}{n} \sum_{j=1}^{n} \exp(\mathrm{i}tY_j)/g^{\mathrm{ft}}(t),$$

In the heteroscedastic setting, where the $g_{j,n}$ are allowed to differ from each other, the most obvious generalization of $\hat{\Psi}_{X_1}(t)$ may be

$$\hat{\Psi}_{X_1;1}(t) = \frac{1}{n}\sum_{j=1}^{n}\exp(itY_j)/g_{j,n}^{\mathrm{ft}}(t).$$

We may establish unbiasedness of this estimator, that is, $E\hat{\Psi}_{X_1;1}(t) = f^{\mathrm{ft}}(t)$. However, imagine the situation where half of the data are contaminated by ordinary smooth error (see (2.31)) and the residual data are corrupted by supersmooth error as in (2.32). Then, the variance of $\hat{\Psi}_{X_1;1}(t)$ is calculable by

$$\operatorname{var}\hat{\Psi}_{X_1;1}(t) = n^{-2}\sum_{j=1}^{n}\left(\left|g_{j,n}^{\mathrm{ft}}(t)\right|^{-2} - \left|f^{\mathrm{ft}}(t)\right|^{2}\right),$$

so that the variance is bounded below by $\mathrm{const}\cdot n^{-1}\exp\left(2d_1|t|^{\gamma}\right)$, and, for the sake of consistency, the bandwidth b must only tend to zero with a logarithmic rate. Whenever the target density is assumed to have polynomial Fourier tails, this implies logarithmic convergence rates with respect to the MISE, see, for example, the proof of Theorem 2.9(b). On the other hand, if we apply estimator (2.7) to those (homoscedastic) data, which suffer from ordinary smooth contamination, and leave the other data group unused, then we attain algebraic rates, according to Theorem 2.9(a), as the sample size is now equal to $n/2$. Hence, we learn that under some conditions, using $\hat{\Psi}_{X_1;1}(t)$ as the empirical version of $f^{\mathrm{ft}}(t)$ does not lead to optimal rates of convergence.

As an alternative choice of $\hat{\Psi}_{X_1}(t)$, [31] suggest to consider

$$\hat{\Psi}_{X_1;2}(t) = \sum_{j=1}^{n}\exp(itY_j)\Big/\Big(\sum_{j=1}^{n}g_{j,n}^{\mathrm{ft}}(t)\Big).$$

Unbiasedness can be verified for this estimator of $f^{\mathrm{ft}}(t)$, too. Nevertheless, its variance is equal to

$$\begin{aligned}\operatorname{var}\hat{\Psi}_{X_1;2}(t) &= \Big(\sum_{j=1}^{n}g_{j,n}^{\mathrm{ft}}(t)\Big)^{-2}\sum_{j=1}^{n}\operatorname{var}\exp(itY_j)\\ &= \Big(\sum_{j=1}^{n}g_{j,n}^{\mathrm{ft}}(t)\Big)^{-2}\sum_{j=1}^{n}\left(1-\left|g_{j,n}^{\mathrm{ft}}(t)\right|^{2}\left|f^{\mathrm{ft}}(t)\right|^{2}\right).\end{aligned}$$

If the $g_{j,n}$ are not symmetric, then their Fourier transforms are complex-valued. Then, however, the denominator occurring in the variance bound may be equal to zero although all the $\left|g_{j,n}^{\mathrm{ft}}\right|$ satisfy (2.31). For instance, consider $g_{j,n}(x) = g(x-1)$ for $j = 1,\ldots,n/2$ and $g_{j,n}(x) = g(x+1)$ for the other data. Then, the denominator equals $n\cos(t)\cdot g^{\mathrm{ft}}(t)$.

As a third method, [31] propose

$$\hat{\Psi}_{X_1;3}(t) = \sum_{j=1}^{n} g_{j,n}^{\mathrm{ft}}(-t)\exp(\mathrm{i}tY_j) \Big/ \Big(\sum_{j=1}^{n} \big|g_{j,n}^{\mathrm{ft}}(t)\big|^2\Big).$$

We derive that

$$\begin{aligned} E\hat{\Psi}_{X_1;3}(t) &= \sum_{j=1}^{n} g_{j,n}^{\mathrm{ft}}(-t) E\exp(\mathrm{i}tY_j) \Big/ \Big(\sum_{j=1}^{n} \big|g_{j,n}^{\mathrm{ft}}(t)\big|^2\Big) \\ &= \sum_{j=1}^{n} g_{j,n}^{\mathrm{ft}}(-t) g_{j,n}^{\mathrm{ft}}(t) f^{\mathrm{ft}}(t) \Big/ \Big(\sum_{j=1}^{n} \big|g_{j,n}^{\mathrm{ft}}(t)\big|^2\Big) \\ &= f^{\mathrm{ft}}(t), \end{aligned}$$

so that unbiasedness of estimator $\hat{\Psi}_{X_1;3}(t)$ holds true. With respect to the variance, we obtain that

$$\begin{aligned} \operatorname{var}\hat{\Psi}_{X_1;3}(t) &= \Big(\sum_{j=1}^{n} \big|g_{j,n}^{\mathrm{ft}}(t)\big|^2\Big)^{-2} \sum_{j=1}^{n} \big|g_{j,n}^{\mathrm{ft}}(t)\big|^2 \operatorname{var}\exp(\mathrm{i}tY_j) \\ &\le \Big(\sum_{j=1}^{n} \big|g_{j,n}^{\mathrm{ft}}(t)\big|^2\Big)^{-2} \sum_{j=1}^{n} \big|g_{j,n}^{\mathrm{ft}}(t)\big|^2 \\ &= \Big(\sum_{j=1}^{n} \big|g_{j,n}^{\mathrm{ft}}(t)\big|^2\Big)^{-1}. \end{aligned}$$

As mentioned in [31], the estimator derived from $\hat{\Psi}_{X_1;3}(t)$, which is defined by

$$\hat{f}(x) = \frac{1}{2\pi}\int \exp(-\mathrm{i}tx)K^{\mathrm{ft}}(tb)\sum_{j=1}^{n} g_{j,n}^{\mathrm{ft}}(-t)\exp(\mathrm{i}tY_j) \Big/ \Big(\sum_{j=1}^{n} \big|g_{j,n}^{\mathrm{ft}}(t)\big|^2\Big), \tag{2.90}$$

is favorable because of the disadvantages of the alternative estimators derived from $\hat{\Psi}_{X_1;k}(t)$, $k = 1,2$. Also, in [31], optimality of the convergence rates with respect to the MISE is established for estimator (2.90), and data-driven choice of the bandwidth is suggested and computed by plug-in approaches. However, to apply estimator (2.90) the density of each error must be known, while we can abandon the information which observation is contaminated by which error density when using the estimator based on $\hat{\Psi}_{X_1;2}$, but that might be seen as an advantage of minor importance.

The paper of [31] also discusses the case of unknown $g_{j,n}$ where replicated measurements should be available to avoid nonidentifiability. In the case where the $g_{j,n}$ are known and identical up to some unknown scaling parameter, this parameter is estimable without any loss of the speed of convergence. However, under fully nonparametric conditions, it is possible to construct a consistent estimator, where the convergence rates may become slower in the case of ordinary smooth contamination.

Finally, we mention that heteroscedastic contamination has also been considered in an errors-in-variables regression problem, see [30].

2.7.2 Distribution and Derivative Estimation

Now let us return to the homoscedastic measurement model (2.1). So far, we have focussed on the estimation of the density f of the X_j. In this subsection, we are interested in the distribution function F of the X_j. This problem has been studied in, for example, [43, 56, 63, 95, 131]. At some point of view, estimating the distribution function is more convenient than density estimation, at least in the error-free setting. There, we are not required to assume the existence of a Lebesgue density, and the estimators enjoy better convergence rates. In many cases, even the parametric rates n^{-1} are attainable. The canonical estimator for the distribution function F based on direct data $X_1, \ldots, X_n$ is the empirical distribution function, defined by

$$\hat{F}(x) = \frac{1}{n} \sum_{j=1}^{n} \chi_{(-\infty,x]}(X_j), \quad x \in \mathbb{R},$$

which does not contain any smoothing parameter, which has to be selected, in its construction. Therefore, it is very easy to compute. In fact, it is a convenient estimator to make the likelihood of some intervals accessible, for instance, we can estimate the probability $P\big[X_1 \in (a,b]\big]$ by $\hat{F}(b) - \hat{F}(a)$.

A more challenging problem, however, is the estimation of the likelihood of a general Lebesgue measurable set A. Still, we can modify the definition of the empirical distribution function so that the likelihood of such a general A is empirically accessible. The resulting estimator is defined by

$$\hat{P}(A) = \frac{1}{n} \sum_{j=1}^{n} \chi_A(X_j),$$

which is usually referred to as the empirical probability measure. The typical distance to evaluate the accuracy of the estimator $\hat{P}$ is the total variation distance (TV) between two probability measures P and Q, defined by

$$\mathrm{TV}(P,Q) = \sup_{A \in \mathfrak{B}(\mathbb{R})} \big|P(A) - Q(A)\big|,$$

where $\mathfrak{B}(\mathbb{R})$ denotes the collection of all Lebesgue-measurable subsets of $\mathbb{R}$. The TV-distance will be investigated further in Sect. 3.3.2. Let us now consider the TV-distance between the empirical probability measure $\hat{P}$ and its true counterpart $P(A) = \int_A f(x)\,\mathrm{d}x$.

$$\begin{aligned} \mathrm{TV}(\hat{P},P) &\geq \big|\hat{P}(\{X_1,\ldots,X_n\}) - P(\{X_1,\ldots,X_n\})\big| \\ &= |1-0| = 1 \quad \text{a.s.}, \end{aligned}$$

whenever P is a continuous probability measure, that is, a Lebesgue density exists. This implies that the probability of any finite set under P is equal to zero. However, this means that the empirical probability measure fails to estimate the true probability measure consistently with respect to the TV-distance. One may think of changing the measure estimator. Nevertheless, [36] show that no empirical probability measure can converge to the true probability measure almost surely when any probability measure is admitted. On the other hand, if only continuous distributions are admitted to be P, the empirical probability measure is still inconsistent as seen above, but we can apply that probability estimator which is generated by the standard kernel density estimator (2.4) when using a density function for the kernel. In this case, the estimator (2.4) is a density itself. We conclude from Lemma 3.4 that strong consistency (i.e., almost sure consistency) of some density estimator with respect to the $L_1(\mathbb{R})$-metric is equivalent to strong consistency of the corresponding empirical probability measure with respect to the TV-distance. Finally, Devroye showed in his so-called equivalence theorem in 1983 that the standard kernel estimator with some conditions on the kernel and the bandwidth, which are rather common and can easily be fulfilled, achieves strong consistency in the $L_1(\mathbb{R})$-metric (see the books of [35] and [33]). From this point of view, the kernel estimator beats the simple empirical probability measure, where the assumption of the existence of the density f is reasonable in many problems.

Now we focus on the measurement error model (2.1) again. There, we realize big trouble already at the stage when trying to adopt the empirical probability measure from the error-free case. The main problem concerns the fact that the incorrupted random variables X_j are not available. Hence, we shall rather focus on estimation methods derived from the deconvolution kernel density estimator (2.7).

Let us assume that the distribution function F possesses a Lebesgue density f. Then, we can estimate the probability $P\big[X_1 \in (c,d]\big]$ for some $c < d$ by

$$\hat{F}_{c,d} \;=\; \int_c^d \hat{f}(x)\,\mathrm{d}x \;=\; \int \chi_{(c,d]}(x)\hat{f}(x)\,\mathrm{d}x \;=\; \frac{1}{2\pi}\int \chi^{\mathrm{ft}}_{(c,d]}(t)\hat{f}^{\mathrm{ft}}(t)\,\mathrm{d}t,$$

where $\hat{f}$ denotes the deconvolution kernel density estimator (2.7). In the last step, the Plancherel isometry has been used (see Theorem A.4 in the appendix). Therein, the membership of $\hat{f}$ in $L_2(\mathbb{R})$ can be ensured whenever $K \in L_2(\mathbb{R})$ and K^{ft} is compactly supported. By Fourier inversion (also see Theorem A.4), we derive that

$$\hat{f}^{\mathrm{ft}}(t) \;=\; K^{\mathrm{ft}}(tb)\frac{1}{n}\sum_{j=1}^{n}\exp(\mathrm{i}tY_j)/g^{\mathrm{ft}}(t).$$

By elementary integration we obtain that

$$\chi^{\mathrm{ft}}_{(c,d]}(t) \;=\; 2\exp\big(\mathrm{i}t(c+d)/2\big)\sin\big(t(d-c)/2\big)/t.$$

Hence,

$$\hat{F}_{c,d} = \frac{1}{\pi n} \sum_{j=1}^{n} \int \exp\big(it(c+d)/2\big) \sin\big(t(d-c)/2\big) K^{\mathrm{ft}}(tb) \exp(itY_j) \\ /\big[tg^{\mathrm{ft}}(t)\big]\, dt. \tag{2.91}$$

Let us calculate the expectation of $\hat{F}_{c,d}$,

$$\begin{aligned} E\hat{F}_{c,d} &= \frac{1}{\pi} \int \exp\big(it(c+d)/2\big) \sin\big(t(d-c)/2\big) K^{\mathrm{ft}}(tb) E \exp(itY_1)/\big[tg^{\mathrm{ft}}(t)\big]\, \mathrm{d}t \\ &= \frac{1}{2\pi} \int \chi_{[c,d]}^{\mathrm{ft}}(t) K^{\mathrm{ft}}(tb) f^{\mathrm{ft}}(t)\, \mathrm{d}t \\ &= \int_c^d \big[K_b * f\big](x)\, \mathrm{d}x, \end{aligned}$$

where Fubini's theorem and the Plancherel isometry (see Theorem A.4) have been applied as before but in the reverse direction. Again, we define $K_b = K(\cdot/b)/b$. Then, the (squared) bias term is bounded above by

$$\begin{aligned} \Big|E\hat{F}_{c,d} - \int_c^d f(x)\, \mathrm{d}x\Big|^2 &= \Big|\frac{1}{b}\int_c^d \int K\big((x-y)/b\big) f(y)\, \mathrm{d}y \mathrm{d}x - \int_c^d f(x)\, \mathrm{d}x\Big|^2 \\ &= \Big|\int_c^d \int K(z)\big(f(x-zb) - f(x)\big)\, \mathrm{d}z \mathrm{d}x\Big|^2 \\ &= \Big|\int K(z)\Big(\int_{c-zb}^{d-zb} f(x)\, \mathrm{d}x - \int_c^d f(x)\, \mathrm{d}x\Big)\, \mathrm{d}z\Big|^2 \\ &= \Big|\int K(z)\big(F(d-zb) - F(d) - F(c-zb) + F(c)\big)\, \mathrm{d}z\Big|^2, \end{aligned} \tag{2.92}$$

with the distribution function F of the random variables X_j under a choice of the kernel function such that $\int K(z)\, \mathrm{d}z = 1$.

First, our assumptions on f are restricted to the existence of a uniform upper bound on f on the whole real line, say $\|f\|_\infty \le C$. Then, F satisfies the Lipschitz condition,

$$\big|F(x) - F(y)\big| \le \Big|\int_x^y f(t)\, \mathrm{d}t\Big| \le \int_x^y |f(t)|\, \mathrm{d}t \le \|f\|_\infty \cdot |x-y|.$$

By (2.92), the bias term is bounded above by

$$\Big|E\hat{F}_{c,d} - \int_c^d f(x)\, \mathrm{d}x\Big|^2 \le 4C^2 b^2 \cdot \Big(\int |K(z)||z|\, \mathrm{d}z\Big)^2,$$

and, hence, is of the order $O\big(b^2\big)$ whenever we select a kernel function K whose first absolute moment is finite.

Now, we assume more restrictive conditions on f; in particular, we impose global Hölder conditions by assuming that f is uniformly bounded by C and $\lfloor\beta\rfloor$-fold continuously differentiable and satisfies

$$\left|f^{(\lfloor\beta\rfloor)}(x) - f^{(\lfloor\beta\rfloor)}(y)\right| \leq C|x-y|^{\beta-\lfloor\beta\rfloor} \tag{2.93}$$

for all $x, y \in \mathbb{R}$. As f is the derivative of F, we may equivalently write $F^{(\beta+1)}$ instead of $f^{(\beta)}$. Let us choose a $(\beta+1)$-order kernel K as defined in (2.38). Then we have

$$\begin{aligned}
&\Big|\int K(z)\big(F(d-zb) - F(d) - F(c-zb) + F(c)\big)\,\mathrm{d}z\Big| \\
&\leq \Big|\sum_{j=0}^{\lfloor\beta\rfloor+1} \frac{1}{j!}F^{(j)}(d)\cdot(-b)^j \int K(z)z^j\,\mathrm{d}z\Big| \\
&\qquad + \frac{1}{(\lfloor\beta\rfloor+1)!}b^{\lfloor\beta\rfloor+1}\int |K(z)||z|^j\big|f^{(\lfloor\beta\rfloor)}(d) - f^{(\lfloor\beta\rfloor)}(\xi_{d,z,b})\big|\,\mathrm{d}z \\
&+ \Big|\sum_{j=0}^{\lfloor\beta\rfloor+1} \frac{1}{j!}F^{(j)}(c)\cdot(-b)^j \int K(z)z^j\,\mathrm{d}z\Big| \\
&\qquad + \frac{1}{(\lfloor\beta\rfloor+1)!}b^{\lfloor\beta\rfloor+1}\int |K(z)||z|^j\big|f^{(\lfloor\beta\rfloor)}(c) - f^{(\lfloor\beta\rfloor)}(\xi_{c,z,b})\big|\,\mathrm{d}z,
\end{aligned}$$

where $\xi_{\cdot,z,b}$ denotes some real number between $\cdot$ and $\cdot - zb$. Therefore, we have applied the Taylor approximation as in (2.36). Also, by the condition (2.93) and the $(\beta+1)$-order property of the kernel K, we may show that the bias term is of order $O(b^{2\beta+2})$ with uniform constants. The previously derived upper bound for those f, which are only assumed to be bounded above by some C, is included by putting $\beta = 0$.

With respect to the variance, we derive that

$$\begin{aligned}
\operatorname{var}\hat{F}_{c,d} &\leq \frac{1}{4\pi^2 n}E\Big|\int \exp\big(\mathrm{i}t(d+c)/2\big)2\sin\big(t(d-c)/2\big)K^{\mathrm{ft}}(tb)\exp(\mathrm{i}tY_j) \\
&\qquad\qquad / \big[tg^{\mathrm{ft}}(t)\big]\,\mathrm{d}t\Big|^2 \\
&= \frac{1}{4\pi^2 n}\int\Big|\int \exp\big(\mathrm{i}t(d+c)/2\big)\sin\big(t(d-c)/2\big)K^{\mathrm{ft}}(tb)\exp(\mathrm{i}ty) \\
&\qquad\qquad / \big[tg^{\mathrm{ft}}(t)\big]\,\mathrm{d}t\Big|^2\big[f*g\big](y)\,\mathrm{d}y \\
&= \frac{1}{2\pi n}\int\Big|\exp\big(\mathrm{i}t(d+c)/2\big)2\sin\big(t(d-c)/2\big)K^{\mathrm{ft}}(tb)/\big[tg^{\mathrm{ft}}(t)\big]\,\mathrm{d}t\Big|^2\mathrm{d}t \\
&\qquad\qquad \cdot \|f*g\|_\infty \\
&= \frac{2}{\pi n}\int\big|\sin\big(t(d-c)/2\big)K^{\mathrm{ft}}(tb)\big|^2|t|^{-2}\big|g^{\mathrm{ft}}(t)\big|^{-2}\,\mathrm{d}t\cdot\|f*g\|_\infty,
\end{aligned} \tag{2.94}$$

where we have used Parseval's identity on the hidden Fourier transform occurring as the inner integral as in the proof of Lemma 2.1(b). To classify the error density g, we consider ordinary smooth g as in (2.31) and supersmooth g (see (2.32)) again. For supersmooth g, we may apply the same upper bound on the variance as in density deconvolution, that is, $O\big(n^{-1}b^{-1}\exp\big(-2d_1b^{-\gamma}\big)\big)$ for any kernel satisfying $\|K^{\rm ft}\|_\infty \le \|K\|_1 \le$ const and $\operatorname{supp} K^{\rm ft} \subseteq [-1,1]$, which are included in, for example, (2.38). Note that the uniform boundedness of f by C implies $\|f*g\|_\infty \le C$.

On the other hand, for ordinary smooth g, we have to distinguish between some cases with respect to α. If $\alpha < 1/2$ then the integral in (2.94) is finite even when the term $K^{\rm ft}(tb)$ is removed so that the variance is uniformly bounded above by $O\big(1/n\big)$. Formally, we may put $b = 0$, implying that $K^{\rm ft}(\cdot b) \equiv 1$ so that we can remove that term from the definition of estimator $\hat{F}_{c,d}$ in (2.91) so that we can estimate $\int_c^d f(x)\,\mathrm{d}x$ without involving a kernel function or a bandwidth parameter. That is similar to estimation by the empirical distribution function when direct data $X_1,\ldots,X_n$ are available.

If $\alpha > 1/2$, the kernel and the bandwidth are needed again. Imposing the same conditions on the selected kernel function as in the supersmooth error case, the uniform upper bound $O\big(n^{-1}b^{2-2\alpha}\big)$ occurs.

Finally, in the case of $\alpha = 1/2$, the integrand behaves as $|t|^{-1}$ for large $|t|$ so that the variance is uniformly bounded above by $O\big(n^{-1}\big|\log b\big|\big)$.

Then, we have to choose the bandwidth so that the convergence of the variance and the bias term are well balanced – considering the usual variance and bias decomposition of the mean squared error. The theoretically rate-optimal selection is given in the following proposition.

Proposition 2.20. *Consider the problem of estimating the likelihood* $F_{c,d} = \int_a^b f(x)\,\mathrm{d}x$ *in the measurement error model (2.1). We apply estimator* $\hat{F}_{c,d}$ *as defined in (2.91). Collect all those* f*, which are uniformly bounded by* C*;* $\lfloor\beta\rfloor$*-fold continuously differentiable and satisfy (2.93), in the density class* $\mathcal{F}^{gl}_{C,\beta}$*. We choose a* $(\beta+1)$*-kernel* K *as defined in (2.38) except in case (a).*
(a) For ordinary smooth g *satisfying (2.31) with* $\alpha \in (0,1/2)$*, put* $K^{\rm ft}(tb)$ *equal to* 1 *in the definition of estimator (2.91). Then, we have*

$$\sup_{f\in\mathcal{F}^{gl}_{C,\beta}} E\big|\hat{F}_{c,d} - F_{c,d}\big|^2 = O\big(1/n\big)\,.$$

(b) For ordinary smooth g *with* $\alpha = 1/2$*, select the bandwidth* $b \asymp n^{-1}$*. Then, we have*

$$\sup_{f\in\mathcal{F}^{gl}_{C,\beta}} E\big|\hat{F}_{c,d} - F_{c,d}\big|^2 = O\big((\log n)/n\big)\,.$$

(c) For ordinary smooth g with $\alpha > 1/2$, select the bandwidth so that $b \asymp n^{-1/(2\beta+2\alpha+1)}$. Then, we have

$$\sup_{f \in \mathcal{F}^{gl}_{C,\beta}} E\big|\hat{F}_{c,d} - F_{c,d}\big|^2 = O\big(n^{-(2\beta+2)/(2\beta+2\alpha+1)}\big).$$

(d) For supersmooth g satisfying (2.32), select the bandwidth so that $b = \big(c_b \log n\big)^{-1/\gamma}$ with $c_b \in (0, 1/(2d_1))$. Then, we have

$$\sup_{f \in \mathcal{F}^{gl}_{C,\beta}} E\big|\hat{F}_{c,d} - F_{c,d}\big|^2 = O\big((\log n)^{-(2\beta+2)/\gamma}\big),$$

where $F_{c,d} = \int_c^d f(x)\,\mathrm{d}x$. Note that the results are also valid in the case of $\beta = 0$. Then, $\mathcal{F}^{gl}_{C,0}$ shall be interpreted as the set of all densities f, which are uniformly bounded above by C. Note that we treated this case separately with respect to its bias term. Obviously, in likelihood estimation, convergence rates are available even if no smoothness constraints are assumed. On the other hand, additional smoothness assumptions ($\beta > 0$) allow us to improve the speed of convergence.

The bandwidth selection is adaptive in the sense that it does not require information about the smoothness degree β except in part (c) of Proposition 2.20. In part (c) and (d), the conditions on the bandwidth are identical with those in deconvolution density estimation, see Theorems 2.8 and 2.9. Hence, it seems reasonable to consider the cross validation procedure for bandwidth selection as introduced in Sect. 2.5.1 for likelihood estimation, too.

Proposition 2.20 solves the problem of deconvolution distribution estimation to some extent. In particular, if f is compactly supported, the distribution function is fully accessible. However, if the left interval boundary a is equal to $-\infty$, we face a more difficult problem. In [43], the condition saying that f is compactly supported is relaxed to decay conditions. However, the problem of rate optimality is obviously open in this problem.

Formally, the distribution function can be viewed as the -1-derivative of the density f. That is also mentioned in [43]. We realize that the convergence rates become faster in distribution estimation compared to density estimation. From that point of view, we understand that the problem of estimating the derivatives of f is a harder statistical problem than estimating the density itself. This will be made precise in the following.

Whenever the Hölder condition (2.29) holds, the existence of $f^{(l)}$ for $l = 0, \ldots, \lfloor\beta\rfloor$ is guaranteed. As an obvious approach to estimate $f^{(l)}$, we may consider the lth derivative of the deconvolution kernel density estimator as defined in (2.7). Note that estimator (2.7) is l-fold continuously differentiable in x whenever K^{ft} is supported on $[-1,1]$. Then, we have

$$\hat{f}^{(l)}(x) = \frac{1}{2\pi}\int \exp(-\mathrm{i}tx)(-\mathrm{i}t)^l K^{\mathrm{ft}}(tb)\frac{1}{n}\sum_{j=1}^n \exp(\mathrm{i}tY_j)/g^{\mathrm{ft}}(t)\,\mathrm{d}t. \qquad (2.95)$$

Following the usual scheme of evaluating the quality of an estimator, we consider the expectation

$$\begin{aligned} E\hat{f}^{(l)}(x) &= \frac{1}{2\pi}\int \exp(-\mathrm{i}tx)(-\mathrm{i}t)^l K^{\mathrm{ft}}(tb)\frac{1}{n}\sum_{j=1}^n E\,\exp(\mathrm{i}tY_j)/g^{\mathrm{ft}}(t)\,\mathrm{d}t \\ &= \frac{1}{2\pi}\int \exp(-\mathrm{i}tx)(-\mathrm{i}t)^l K^{\mathrm{ft}}(tb)f^{\mathrm{ft}}(t)\,\mathrm{d}t \\ &= \frac{\mathrm{d}^l}{\mathrm{d}x^l}\Big\{\frac{1}{2\pi}\int \exp(-\mathrm{i}tx)K^{\mathrm{ft}}(tb)f^{\mathrm{ft}}(t)\,\mathrm{d}t\Big\} \\ &= \frac{\mathrm{d}^l}{\mathrm{d}x^l}\{K_b * f\} = \int f^{(l)}(x-y)K_b(y)\,\mathrm{d}y, \end{aligned}$$

where we have used the notation $K_b = K(\cdot/b)/b$ and dominated convergence to exchange the integral and the derivative. Note that convolution is commutative so that

$$\big[f * K_b\big](x) = \int f(x-y)K_b(y)\,\mathrm{d}y = \int K_b(x-y)f(y)\,\mathrm{d}y = \big[K_b * f\big](x)\,,$$

which can be seen by simple substitution of the integral. Further, we shall study the proximity of the expectation and the true $f^{(l)}(x)$, that is, we consider the bias term

$$\big|E\hat{f}^{(l)}(x) - f^{(l)}(x)\big|^2 = \Big|\int \big[f^{(l)}(x-zb) - f^{(l)}(x)\big]K(z)\,\mathrm{d}z\Big|^2,$$

where the kernel satisfies (2.38) so that the bias term for derivative estimation has the same shape as the bias B_n when estimating the density f in (2.35). Note that we can apply the Taylor expansion of $f^{(l)}$ analogously to (2.36); however, the degree of the approximating Taylor polynomial is restricted to $\lfloor\beta\rfloor - l$ in the current setting. In addition, we assume that $f^{(l)}$ is bounded above by the uniform constant C on the whole real line so that the term $B_{n,2}$ can also be bounded above in the same way as when estimating the density f. That gives us the upper bound $O\big(b^{2\beta-2l}\big)$ on the bias.

Now we consider the variance. We have

$$\begin{aligned} \operatorname{var}\hat{f}^{(l)}(x) &\le \frac{1}{(2\pi)^2 n}E\Big|\int \exp\big(-\mathrm{i}t(x-Y_1)\big)(\mathrm{i}t)^l K^{\mathrm{ft}}(tb)/g^{\mathrm{ft}}(t)\,\mathrm{d}t\Big|^2 \\ &= \frac{1}{(2\pi)^2 n}\int\Big|\int \exp\big(-\mathrm{i}t(x-y)\big)(\mathrm{i}t)^l K^{\mathrm{ft}}(tb)/g^{\mathrm{ft}}(t)\,\mathrm{d}t\Big|^2\big[f*g\big](y)\,\mathrm{d}y \\ &\le \frac{1}{(2\pi)^2 n}\|f*g\|_\infty\int\Big|\int \exp(-\mathrm{i}ty)(\mathrm{i}t)^l K^{\mathrm{ft}}(tb)/g^{\mathrm{ft}}(t)\,\mathrm{d}t\Big|^2\mathrm{d}y \\ &= \frac{1}{2\pi n}\|f*g\|_\infty\int |t|^{2l}\big|K^{\mathrm{ft}}(tb)\big|^2/\big|g^{\mathrm{ft}}(t)\big|^2\,\mathrm{d}t \end{aligned}$$

by Parseval's identity. We have used the same techniques as in the proof of Proposition 2.1. As K^{ft} is supported on $[-1,1]$, the variance term is bounded above by

$$\operatorname{var} \hat{f}^{(l)}(x) \leq O\big(n^{-1}b^{-2l-1}\big) \cdot \max\big\{\big|g^{\mathrm{ft}}(t)\big|^{-2} : t \in [-1/b, 1/b]\big\},$$

as f is assumed to be bounded by C on $\mathbb{R}$; hence, $\|f * g\|_\infty \leq C$. Thus, the variance bound in deconvolution derivative estimation deteriorates by the term $O\big(b^{-2l}\big)$ compared to density deconvolution. Using the usual bias-variance decomposition again, we derive that

$$\sup_{f \in \mathcal{F}^{(l)}_{\beta,C,\delta;x}} E\big|\hat{f}^{(l)}(x) - f^{(l)}(x)\big|^2$$
$$\leq O\big(b^{2\beta-2l}\big) + O\big(n^{-1}b^{-2l-1}\big) \cdot \max\big\{\big|g^{\mathrm{ft}}(t)\big|^{-2} : t \in [-1/b, 1/b]\big\},$$

where $\mathcal{F}^{(l)}_{\beta,C,\delta;x}$ denotes the class of all densities, which satisfy (2.29) along with uniform boundedness of f and $f^{(l)}$ by C on the whole of $\mathbb{R}$. We have to assume that $l \leq \lfloor \beta \rfloor$. Again, we have to distinguish between ordinary smooth and supersmooth error densities to specify the asymptotic order of $\max\big\{\big|g^{\mathrm{ft}}(t)\big|^{-2} : t \in [-1/b, 1/b]\big\}$ and, again, the bias term increases when b becomes larger; while the variance term decreases. Solving the equation between those terms, that is,

$$b^{2\beta-2l} = n^{-1}b^{-2l-1} \max\big\{\big|g^{\mathrm{ft}}(t)\big|^{-2} : t \in [-1/b, 1/b]\big\},$$

with respect to b gives us a choice of the bandwidth b, under which we may establish optimal convergence rates, at least. Inserting that selection into the upper bound on the MISE leads to the following proposition.

Proposition 2.21. *Consider the problem of estimating the* l*th derivative of* f *in the additive measurement error model (2.1). We apply estimator (2.95) with an* β*-order kernel as defined in (2.38) where* $\lfloor \beta \rfloor \geq l$. *Then,*
(a) when g *is ordinary smooth as defined in (2.31), select the bandwidth so that* $b \asymp n^{-2\beta/(2\beta+2\alpha+1)}$. *Then,*

$$\sup_{f \in \mathcal{F}^{(l)}_{\beta,C,\delta;x}} E\big|\hat{f}^{(l)}(x) - f^{(l)}(x)\big|^2 = O\big(n^{-(2\beta-2l)/(2\beta+2\alpha+1)}\big).$$

(b) when g *is supersmooth as defined in (2.32), select the bandwidth so that* $b = \big(c_b \log n\big)^{-1/\gamma}$ *with* $c_b \in (0, 1/(2d_1))$. *Then,*

$$\sup_{f \in \mathcal{F}^{(l)}_{\beta,C,\delta;x}} E\big|\hat{f}^{(l)}(x) - f^{(l)}(x)\big|^2 = O\big((\log n)^{-(2\beta-2l)/\gamma}\big).$$

Therefore, the rate-optimal bandwidth selection follows the same criteria as in density estimation in both cases; however, the uniform convergence rates of the MSE are slower than those derived for density deconvolution in Theorem 2.8, and we realize that this deterioration becomes worse when the degree of the desired derivative l increases. In particular, l must not be larger than $\lfloor \beta \rfloor$, which is rather clear because the assumption $f \in \mathcal{F}^{(l)}_{\beta,C,\delta;x}$ guarantees differentiability only to the degree $\lfloor \beta \rfloor$. We give the remark that Proposition 2.20, part (c) and (d), are included in the general framework of Proposition 2.21 by formally considering the distribution function F as the -1-derivative of f.

The problem of derivative estimation under the statistical experiment (2.1) along with the rates as derived in Proposition 2.21 has been addressed in [43]. In that paper, the convergence rates have also been shown to be optimal where any estimator based on the given empirical information is admitted as in Sect. 2.4.5 for deconvolution density estimation.

2.7.3 Other Related Topics

We include the current subsection to give a survey on some further problems, which are related to density deconvolution.

The problem of estimating the mode (i.e., the global maximum of a continuous density function f) when the data are drawn from the measurement error model (2.1) is studied in [109]. In that paper, the authors propose to fix the point where the deconvolution kernel density estimator (2.7) takes its maximum as the estimator of the mode. In general, mode estimation has been studied for very long time in nonparametric statistics. In the error-free case, it was already highlighted in [105], which represents one of the first papers where kernel techniques have been used. Rachdi and Sabre [109] utilize smoothness conditions, which correspond to boundedness of the second-order derivative, where this function must also be bounded away from zero at the mode. This local sharpness constraint is not required in density estimation but in mode estimation since we recognize that the localization problem of the mode becomes very hard when the density has a very flat shape around the mode. In the extreme case where the density is locally constant around the mode, we cannot determine the mode even under full knowledge of the density. We realize that solely smoothness constraints as used for density estimation do not rule out such bad behavior of the density. Rachdi and Sabre [109] derive upper bounds on the convergence rates in an individual sense, where algebraic rates occur for ordinary smooth error densities (2.31) and logarithmic rates are attained under supersmooth contamination (2.32), as in density deconvolution.

Other problems where one is not interested in the density itself but in some of its characteristics occur in the field of deconvolution support estimation. In general, we are interested in estimating the support of the density f, where noisy data are observed under the model (2.1). In the error-free case, there exists a quite obvious method of estimating the upper and the lower support boundaries (also known as the endpoints of the distribution) by just taking the maximum and the minimum of the observations, respectively, as the estimators of the endpoints. As no direct data are available under model (2.1), this simple procedure is not applicable. In general, the support of the density f can be defined as

$$\operatorname{supp} f = \bigcap \Big\{ G \subseteq \mathbb{R} : \int_G f(x)\,\mathrm{d}x = 1,\, G \text{ closed} \Big\}.$$

The paper of [90] addresses this general setting in a multivariate case. However, in most problems, the support is assumed to be simply connected and f is assumed to be continuous up to its support boundaries so that, in the univariate case, the support denotes the closure of the set $\{x \in \mathbb{R} : f(x) > 0\}$, which is an interval in fact. Therefore, the support may be uniquely characterized by its boundaries.

Hall and Simar [61] study the problem of frontier estimation under the modified measurement error model when a scaling factor of the error density g tends to zero as the sample size increases. This condition, which makes the estimation problem easier, is not assumed in the following papers.

Goldenshluger and Tsybakov [51] propose to manipulate those estimators by appropriate scaling factors and a shift so that the outcome estimator is consistent for the upper endpoint. Also, the authors derive optimal convergence rates for their procedure when the error density g is assumed to be either compactly supported or to have tails as a normal density.

Delaigle and Gijbels [26] study the problem of support estimation under the condition that the density function is not smooth at the endpoints, while smoothness is assumed for f inside the support. Their method leans on empirical detection of the discontinuities.

Meister [91] introduces a method based on moment estimation. The moments of the X_1 may be reconstructed from those of Y_1 by inverting a system of linear equations, given by

$$EY_1^k = E(X_1 + \varepsilon_1)^k = \sum_{j=1}^{n} \binom{n}{k} EX_1^k E\varepsilon_1^{n-k}.$$

Then, as $\big(EX_1^{2k}\big)^{1/(2k)}$, for integer k, converges to the right endpoint of the density f, for $k \to \infty$, that endpoint is estimable by those moment estimators for large k. This method is consistent under general conditions.

A related problem of change-point estimation from noisy data is considered in [102].

Another deconvolution problem apart from the model (2.1) concerns the aggregated or grouped data problem. It has first been mentioned by [84], motivated by a real data application from econometrics. It is applicable when the data, in whose density we are interested, are not reported as single observations but only the averages over some data are available. Therefore, we have a self-convolution problem, where the observed data have the structure

$$Y_j = \sum_{k=1}^{m} X_{j,k} + \varepsilon_j\,, \quad j = 1, \ldots, n.$$

where the i.i.d. components $X_{j,k}$ have the target density f, and the error component ε_j has a known density g. Apparently,

$$E \exp(itY_j) = \left[f^{\mathrm{ft}}(t)\right]^m g^{\mathrm{ft}}(t),$$

so that nonparametric estimation of f requires to take the mth root in the Fourier domain rather than simple division by g^{ft} as in common deconvolution problems. Linton and Whang [84] introduce a procedure for estimating the density f in this nonlinear statistical inverse problem; they also give an extension to regression estimation. Meister [93] considers that problem when the error component is zero and develops a minimax theory in this setting.

3

Nonparametric Regression with Errors-in-Variables

3.1 Errors-in-Variables Problems

As a broad field in statistics in general, the investigation of the link or the dependence between some quantity, which is affected by random noise, and some circumstances, under which the quantity is observed, is referred to as regression. We assume that those circumstances may be represented by a real number x, which is called the covariate or the independent variable. Thus, we restrict our consideration to the univariate case, that is, the opposite of the multivariate case where x is assumed to be a finite-dimensional vector with real-valued components. The observed quantity is denoted by some real-valued random variable Y, the so-called dependent variable. In some cases, x also suffers from random effects so that the covariate should be modeled as a random variable X. That latter setting is referred to as the random design case, while those problems where the covariate is deterministic are classified as fixed design models.

As a basic example for a regression problem, we consider the investigation of the size of children with respect to their age. Therefore, in the above notation, Y denotes the size of a child and X represents his age. Observations are collected from n individuals; they are reported by the data sample

$$(X_1, Y_1), \ldots, (X_n, Y_n).$$

Then, the question whether the fixed or the random design model is preferable in that application must be answered with respect to how the individuals have been chosen. If the survey is planned so that only children at some fixed ages $x_1, \ldots, x_n$ are eligible, the fixed design model should be favored. On the other hand, if the children are chosen arbitrarily and independently of each other from some large totality without any respect to their age, we choose the random design model where the children's ages occur as random variables $X_1, \ldots, X_n$. In the sequel, we only study the random design model, which seems more appropriate for the deconvolution topics, which will be

A. Meister, *Deconvolution Problems in Nonparametric Statistics*, Lecture Notes in Statistics 193, DOI 10.1007/978-3-540-87557-4_3,

introduced later. Then, we may assume that the observations (X_j, Y_j) are i.i.d. (independent and identically distributed). The X_j and Y_j may be viewed as independent copies of the random variables X and Y as in the previous notation. However, note that X_j and Y_j are not stochastically independent in general. Contrarily, their dependence is subject of our investigation.

A mathematical tool from probability theory used to study the link between X_j and Y_j is provided by the conditional expectation of Y_j, given $X_j = x$. Thus, the regression function m is defined by

$$m(x) = E\big(Y_j \mid X_j = x\big), \quad x \in I \subseteq \mathbb{R}. \tag{3.1}$$

Transferred to the described application, the regression function m at the point x denotes the expected size of an arbitrarily chosen child at the age of x.

The following lemma underlines that the regression function is indeed the best predictor of the dependent variable Y_j based on observing the covariate X_j in L_2-sense.

Lemma 3.1. *Define the regression function as in (3.1). Then, we have*

$$E\big|Y_j - m(X_j)\big|^2 \leq E\big|Y_j - f(X_j)\big|^2$$

for all measurable functions $f : \mathbb{R} \to \mathbb{R}$ with $Ef^2(X_j) < \infty$.

Proof. For any function f satisfying the conditions required in the lemma, we derive that

$$\begin{aligned}
& E\big|Y_j - f(X_j)\big|^2 = E\big|Y_j - m(X_j) + m(X_j) - f(X_j)\big|^2 \\
& = E\big|Y_j - m(X_j)\big|^2 + E\big|m(X_j) - f(X_j)\big|^2 \\
& \qquad\qquad\qquad + 2\mathrm{Re}E\big(Y_j - m(X_j)\big) \cdot \big(m(X_j) - f(X_j)\big) \\
& \geq E\big|Y_j - m(X_j)\big|^2 + 2\mathrm{Re}\, E\big(m(X_j) - f(X_j)\big) \cdot E\big(\big[Y_j - m(X_j)\big] \mid X_j\big) \\
& = E\big|Y_j - m(X_j)\big|^2 + 2\mathrm{Re}\, E\big(m(X_j) - f(X_j)\big) \cdot \underbrace{\big[E\big(Y_j \mid X_j\big) - m(X_j)\big]}_{=0},
\end{aligned}$$

where we have used the formula $E(XY) = E\big(X \cdot E(Y \mid Z)\big)$ from probability theory where X, Y, Z are some integrable random variables, which are not necessarily independent, but X is measurable in the σ-algebra generated by Z. ∎

The random variables

$$\varepsilon_j = Y_j - m(X_j), \quad j \in \{1, \ldots, n\},$$

are called the regression errors and reflect the difference between the size of the jth individual and its expected size based on the information of his age. Quite obviously, $E\varepsilon_j = 0$ holds true as

$$EY_1 = E\,E(Y_1 \mid X_1) = Em(X_1).$$

Furthermore, we assume that the $X_1, \ldots, X_n$ are continuously distributed and have the (Lebesgue) density function f_X, which is called the design density. The design density is typically unknown but estimable from the empirical information, as too is the regression function.

Our goal is to estimate the regression function m based on the data structure $(X_1, Y_1), \ldots, (X_n, Y_n)$. According to (3.1), $m(x)$ is the expectation of the conditional distribution of Y_1, given $X_1 = x$, which is denoted by $F_{(Y_1|X_1=x)}$. In the field of nonparametric regression, we do not assume that the regression function has some known shape, which determines m up to finitely many real-valued parameters – just as, in nonparametric density estimation, the density to be estimated does not satisfy such restrictive assumptions. Therefore, we are dealing with problems of function estimation rather than parameter estimation. Generally, in statistics, expectations are usually estimated by the average of the underlying data. Imagine the simpler situation where we observe the i.i.d. data $Y_1^*, \ldots, Y_n^*$ from the distribution $F_{(Y_1|X_1=x)}$. Then, by the strong law of large numbers we derive that

$$\frac{1}{n}\sum_{j=1}^{n} Y_j^* \stackrel{n\to\infty}{\longrightarrow} m(x), \quad \text{a.s.}$$

Nevertheless, the real dataset is more complicated. Since the X_j are continuously distributed, we must not assume that any X_j is equal to x for some arbitrary but fixed $x \in \mathbb{R}$. Despite, we are interested in m at that specific point x. If m is continuous at x, those data (X_j, Y_j) whose components X_j are near x contain some information about $m(x)$. In particular, the smaller the distance between X_j and x, the more informative the observation (X_j, Y_j) for $m(x)$. Therefore, we shall not treat any observation equal as is done by averaging, but Y_j should be weighted according to the proximity of X_j and x. That leads to the introduction of the general regression weight estimator

$$\hat{m}_0(x) = \sum_{j=1}^{n} Y_j \cdot W_{j,n}(x; X_1, \ldots, X_n),$$

where the weights shall satisfy the scaling condition,

$$\sum_{j=1}^{n} W_{j,n}(x; X_1, \ldots, X_n) = 1, \text{ for all } x \in \mathbb{R}, \quad \text{a.s.},$$

so that the average is still included as the equally weighted estimator with $W_{j,n}(x; X_1, \ldots, X_n) \equiv 1/n$. A popular choice of the weights will be discussed in the following section.

Now let us focus on the specific characteristics of an errors-in-variables problem. In the standard nonparametric measurement error model, we assume that the covariates X_j can only be observed with some additive independent noise. Therefore, we change the observation scheme into the i.i.d. dataset

$$(W_1, Y_1), \ldots, (W_n, Y_n),$$

where

$$\begin{aligned} W_j &= X_j + \delta_j, \\ Y_j &= m(X_j) + \varepsilon_j. \end{aligned} \tag{3.2}$$

The covariate errors δ_j are i.i.d. unobservable random variables having the error density g. Note that they are different from the regression errors ε_j. The δ_j are stochastically independent of the X_j and the Y_j. As in the previous chapter on density deconvolution, we assume that g is known in the standard setting, while the distribution of the ε_j need not be known. That distinction can be explained by the different quantities we are interested in: we aim at estimating $m(x)$, the conditional expectation of the Y_j rather than the whole of the distribution function of the Y_j. On the other hand, to make m accessible, we have to estimate the design density f_X first, that is, the density of the X_j, where a deconvolution procedure must be applied. Under nonparametric conditions, that model has first been studied in [47].

In the sequel, we discuss some practical applications of the errors-in-variables model (3.2). A lot of real-data problems can be found in the book of [16]. For instance, in the field of medical statistics, some dataset is given, which consists of measurements of the systolic blood pressure (SBP) taken by two different exams. The logarithm of one of those measurements represents the contaminated covariate W_j for the jth individual. That logarithmic transform is used to turn the type of the error from an independent multiplicative error into an independent additive one, in a similar way as in Example 4(a) in Sect. 2.1. The fact that two different results are obtained in both exams shows that the true SBP of the jth individual, denoted by X_j in the above notation, can be measured with some noise δ_j only. Both observations of the SBP may be interpreted as replicated measurements of X_j so that the density g of the δ_j is empirically accessible in some cases, see Sect. 2.6.3. For the same individual, the incidence of coronary heart disease (CHD) is reported by a digital random variable Y_j, indicating 1 in the case of a CHD and 0 otherwise. Researchers in medical science are interested in the link between the SBP and the likelihood of a CHD so that errors-in-variables regression analysis is required in the framework of (3.2).

Another real-data example is given in [13]. The authors apply their methods to data from the Nevada Test Site (NTS) Thyroid Disease Study, where the goal is to derive some link between the radiation exposure and thyroid disease outcomes, based on contaminated data on radiation exposures measured in the food of an individual and absence or presence of thyroid disease of the same individual. However, the authors choose the model where not only so-called classical errors δ_j occur, but also additional Berkson errors are taken into account. The study of those type of contamination will be deferred to Sect. 3.4.

Errors-in-variables regression problem also occur in the fields of econometrics. We revisit Example 4(a) in Sect. 2.1, where the income of an individual is measured by repeated observation of his expenditure, hence by corrupted observations. When, in addition, for each individual, the university education is reported, for example, by some random variables Y_j, indicating some integers representing some university degrees, the link between the income and the level of university education is empirically accessible as an errors-in-variables regression problem (3.2).

3.2 Kernel Methods

Now our goal is the derivation of some procedure to estimate the regression function in the errors-in-variables regression model (3.2). But first we focus on the standard regression case where the covariates X_j are not affected by contamination. In the previous section, we have already introduced the general weight estimator $\hat{m}_0$ but left the weight functions $W_{j,n}$ undefined. As in density estimation, kernel methods are widely used in nonparametric regression problems. The standard kernel regression estimator has become known as the Nadaraya–Watson estimator and mainly goes back to the papers of [98] and [125]. There, the weights are chosen as

$$W_{j,n}(x; X_1, \ldots, X_n) = K\big((x - X_j)/b\big) / \sum_{k=1}^{n} K\big((x - X_k)/b\big).$$

It easily follows that the weights add to one. We select some kernel function $K : \mathbb{R} \to \mathbb{R}$ and some bandwidth parameter $b > 0$. Therefore, the Nadaraya–Watson estimator is defined by

$$\hat{m}(x) = \Big(\sum_{j=1}^{n} Y_j \cdot K\big((x - X_j)/b\big)\Big) / \Big(\sum_{j=1}^{n} K\big((x - X_j)/b\big)\Big). \tag{3.3}$$

In a simple case, the kernel function can be put equal to the indicator function of the interval $[-1, 1]$, thus $K = \chi_{[-1,1]}$. Then, the kernel estimator is equal to the average of those observations Y_j where the distance between the corresponding X_j and x is smaller or equal to the bandwidth b. All other data are left unused. We recall that we postulated in the previous subsection that more weight should be given to those data whose components X_j are closer to x. In the light of this, the construction of the Nadaraya–Watson estimator seems reasonable. We generalize the method by allowing for more general kernel functions K.

As we may equivalently multiply the numerator and the denominator by $1/(nb)$, we realize the close relation to the kernel density estimator (2.4). Indeed, the kernel density estimator of $f_X(x)$ based on the i.i.d. data $X_1, \ldots, X_n$ occurs as the denominator of $\hat{m}(x)$. The numerator may be analyzed by considering its expectation

$$
\begin{aligned}
E\frac{1}{nb}\sum_{j=1}^{n} Y_j \cdot K\big((x-X_j)/b\big) &= \frac{1}{b}EY_1K\big((x-X_1)/b\big) \\
&= \frac{1}{b}E\,E(Y_1 \mid X_1)\cdot K\big((x-X_1)/b\big) \\
&= \int E(Y_1 \mid X_1 = t)\cdot K\big((x-t)/b\big) f_X(t)\,\mathrm{d}t/b \\
&= \int m(t)\cdot K\big((x-t)/b\big) f_X(t)\,\mathrm{d}t/b \\
&= \int m(x-sb) f_X(x-sb) K(s)\,\mathrm{d}s,
\end{aligned}
$$

where the definition of the regression function m as well as the substitution $s=(x-t)/b$ have been utilized. Writing

$$
p(x) = m(x)\cdot f_X(x),
$$

for the regression function times the design density, we learn from dominated convergence that the expectation of the numerator converges to $p(x)$ as $b\downarrow 0$ whenever K integrates to one and p is bounded and continuous at x. Hence, the numerator mimics $p(x)$. Therefore, this intuitive consideration also indicates that the Nadaraya–Watson estimator is a reasonable estimator of $m(x)=p(x)/f_X(x)$.

Now we focus on the question how to extend the kernel method to the more challenging errors-in-variables setting (3.2). Our given empirical information is restricted to the contaminated data (W_j,Y_j), $j=1,\ldots,n$. The denominator of the Nadaraya–Watson estimator may be replaced by the deconvolution kernel density estimator (2.7) using the data $W_1,\ldots,W_n$, which are additively corrupted by unobservable random variables with the density g. Then the denominator is an empirical version of the density $f_X(x)$ as in the error-free setting above. Also, the numerator of the Nadaraya–Watson estimator must be prepared so that it does not require knowledge of the unobservable $X_1,\ldots,X_n$ but only uses the noisy data $W_1,\ldots,W_n$ in its construction. In the spirit of (2.7), we suggest to use

$$
\hat{p}(x) = \frac{1}{2\pi n}\sum_{j=1}^{n} Y_j \cdot \int \exp(-itx)K^{\mathrm{ft}}(tb)\exp(itW_j)/g^{\mathrm{ft}}(t)\,\mathrm{d}t, \tag{3.4}
$$

as the estimator for $p(x)$. This leads to the final deconvolution kernel regression estimator

$$
\hat{m}(x) = \Big(\sum_{j=1}^{n} Y_j \cdot \int \exp(-itx)\exp(itW_j)K^{\mathrm{ft}}(tb)/g^{\mathrm{ft}}(t)\,\mathrm{d}t\Big) \Big/ \Big(\sum_{j=1}^{n}\int \exp(-itx)\exp(itW_j)K^{\mathrm{ft}}(tb)/g^{\mathrm{ft}}(t)\,\mathrm{d}t\Big), \tag{3.5}
$$

which has first been derived by [47]. To fix that $\hat{p}(x)$ suits well to mimic $p(x)$, we consider its expectation

$$
\begin{aligned}
E\hat{p}(x) &= \frac{1}{2\pi n}\sum_{j=1}^{n} EY_j \cdot \int \exp(-itx)\,\exp(itW_j)K^{\mathrm{ft}}(tb)/g^{\mathrm{ft}}(t)\,\mathrm{d}t \\
&= \frac{1}{2\pi}\int \exp(-itx)\big(EY_1\,\exp(itX_1)\big)E\,\exp(it\delta_1)K^{\mathrm{ft}}(tb)/g^{\mathrm{ft}}(t)\,\mathrm{d}t \\
&= \frac{1}{2\pi}\int \exp(-itx)\big(E\,E(Y_1 \mid X_1)\,\exp(itX_1)\big)g^{\mathrm{ft}}(t)K^{\mathrm{ft}}(tb)/g^{\mathrm{ft}}(t)\,\mathrm{d}t \\
&= \frac{1}{2\pi}\int \exp(-itx)\big(\int m(u)\,\exp(itu)f_X(u)\,\mathrm{d}u\big)K^{\mathrm{ft}}(tb)\,\mathrm{d}t \\
&= \frac{1}{2\pi}\int \exp(-itx)p^{\mathrm{ft}}(t)K^{\mathrm{ft}}(tb)\,\mathrm{d}t,
\end{aligned}
$$

using the independence between X_j, ε_j, on the one hand, and δ_j, on the other hand. We have assumed that $p \in L_1(\mathbb{R})$ so that its Fourier transform is well-defined. This is satisfied when m is bounded, for instance. Arranging K so that it lies in $L_1(\mathbb{R})$, Lemma A.1(b) and (e) from the appendix lead to the conclusion that

$$\big(p * K(\cdot/b)/b\big)^{\mathrm{ft}}(t) \;=\; p^{\mathrm{ft}}(t)K^{\mathrm{ft}}(tb).$$

Usually, we arrange $\int K(x)\,\mathrm{d}x = 1$ and that K^{ft} is bounded and supported on $[-1,1]$ so that $\big(p * K(\cdot/b)/b\big)^{\mathrm{ft}}$ is integrable over the whole real line since p^{ft} is bounded above by $\|p\|_1$. Also, $p * K(\cdot/b)/b \in L_1(\mathbb{R})$ follows from the previous assumptions $p, K \in L_1(\mathbb{R})$. Then, by Fourier inversion (Theorem A.2), we conclude that

$$E\hat{p}(x) \;=\; \big[p * K_b\big](x), \tag{3.6}$$

if, in addition, K is assumed to be bounded and continuous on the whole real line. We write $K_b(x) = K(x/b)/b$. Again, by dominated convergence, the assumed boundedness and continuity of p at x combined with the integrability of K suffices to establish that $E\hat{p}(x)$ converges to $p(x)$ as $b \downarrow 0$. A more detailed study of the asymptotic properties will be given in the following section.

3.3 Asymptotic Properties

In this section, we investigate the asymptotic behavior of the deconvolution kernel regression estimator (3.5), as the number of observations n tends to infinity – as we did for the density deconvolution kernel estimator in the previous chapter.

3.3.1 Consistency

As explained with respect to density deconvolution in Sect. 2.3, the criterion of consistency can be made precise in several mathematically reasonable ways. Roughly speaking, consistency means that the estimator converges to the quantity to be estimated as the sample size tends to infinity.

We will focus on the pointwise squared risk. Note that the MISE criterion, as studied in Sect. 2.4.3 in the field of density deconvolution, is not appropriate to evaluate the quality of regression estimators, as we are not guaranteed that, in general, the true regression function is integrable or square integrable, while, for density functions, integrability holds by definition and square integrability can be verified under slight additional assumptions, see Lemma 2.3. Also, when considering the pointwise squared estimation error, defined by

$$R(x) = \left|\hat{m}(x) - m(x)\right|^2,$$

the MSE-criterion, that is, the convergence of the expectation of $R(x)$ to zero, is rather difficult to show and might require some additional regularization due to the construction of estimator (3.5) as the fraction of two random variables. For instance, some ridge parameter for the denominator is needed to prevent this term from being too close to zero, following from some random deviation. That is closely related to the general result from probability theory, that $X_n \to x$ a.s. and $Y_n \to y$ a.s. for two sequences $(X_n)_n$ and $(Y_n)_n$ of real-valued random variables and $x, y \in \mathbb{R}$ with $y \neq 0$, implies that $X_n/Y_n \to x/y$ a.s.. On the other hand, this implication does not hold true in general when considering convergence of the corresponding means rather than strong convergence. Therefore, our investigation concentrates on weak and strong consistency, that is, convergence of $R(x)$ in probability and almost sure convergence, respectively. Recalling the definition of $\hat{p}$ in (3.4) we realize that the deconvolution kernel regression estimator (3.5) may be written as

$$\hat{m}(x) = \hat{p}(x) \,/\, \hat{f}_X(x),$$

where $\hat{f}_X$ denotes the deconvolution kernel density estimator (2.7) when inserting the data $W_1, \ldots, W_n$ rather than $Y_1, \ldots, Y_n$.

We give the following theorem.

Theorem 3.2. (Strong consistency) *Consider the errors-in-variables regression problem defined in (3.2) under the conditions that $g^{\mathrm{ft}}(t) \neq 0$ for all $t \in \mathbb{R}$; the functions $p = m \cdot f_X$ and f_X are bounded on the whole real line and continuous at some $x \in \mathbb{R}$; $p \in L_1(\mathbb{R})$; the continuous and bounded kernel function $K \in L_1(\mathbb{R})$ satisfies $\int K(z)\,\mathrm{d}z = 1$ and K^{ft} is supported on $[-1,1]$. Furthermore, select the bandwidth b so that $b = b_n \downarrow 0$ and*

$$b \cdot \min_{|t| \leq 1/b} |g^{\mathrm{ft}}(t)| \geq n^{-\xi} \tag{3.7}$$

for some $\xi \in (0, 1/2)$ and all n sufficiently large; assume that the $(2s)$th moment of Y_1 exists for some $s > 1/(1-2\xi)$, and that

$$|m(x)| \leq C_1 \qquad \text{and} \qquad f_X(x) \geq C_2 \tag{3.8}$$

holds for some constants $C_1, C_2 > 0$.

Then, estimator (3.5) satisfies

$$\hat{m}(x) \stackrel{n\to\infty}{\longrightarrow} m(x), \quad a.s.$$

Proof. The inequality $R(x) > \epsilon$, for some $\epsilon > 0$, implies that

$$\begin{aligned}
\sqrt{\epsilon} &< \big|\hat{p}(x)/\hat{f}_X(x) - p(x)/f_X(x)\big| \\
&\leq \big|\hat{p}(x)/\hat{f}_X(x) - p(x)/\hat{f}_X(x)\big| + \big|p(x)/\hat{f}_X(x) - p(x)/f_X(x)\big| \\
&\leq \big|\hat{p}(x) - p(x)\big| \cdot \big|1/\hat{f}_X(x) - 1/f_X(x)\big| + \big|\hat{p}(x) - p(x)\big|/f_X(x) \\
&\qquad + \big|p(x)\big| \cdot \big|1/\hat{f}_X(x) - 1/f_X(x)\big| \\
&= \big|\Delta p(x)\big|\,\big|f_X(x)/\hat{f}_X(x) - 1\big| + \big|\Delta p(x)\big| + \big|m(x)\big|\,\big|f_X(x)/\hat{f}_X(x) - 1\big| \\
&= \big|\Delta p(x)\Delta f_X(x)\big|/\big|1 + \Delta f_X(x)\big| + \big|\Delta p(x)\big| \\
&\qquad + \big|m(x)\Delta f_X(x)\big|/\big|1 + \Delta f_X(x)\big|,
\end{aligned}$$

where

$$\begin{aligned}
\Delta p(x) &= \big(\hat{p}(x) - p(x)\big) / f_X(x), \\
\Delta f_X(x) &= \big(\hat{f}_X(x) - f_X(x)\big) / f_X(x).
\end{aligned}$$

Hence, it follows from the condition $\big|R(x)\big| > \epsilon$ that $\big|\Delta p(x)\big| > \sqrt{\epsilon}/3$ or $\big|\Delta p(x)\Delta f_X(x)\big|/\big|1 + \Delta f_X(x)\big| > \sqrt{\epsilon}/3$ or $\big|m(x)\Delta f_X(x)\big|/\big|1 + \Delta f_X(x)\big| > \sqrt{\epsilon}/3$. The third assertion implies that $\big|\Delta f_X(x)\big| > 1/2$ or

$$\big|\Delta f_X(x)\big| > \sqrt{\epsilon}/(6|m(x)|).$$

The second assertion implies that $\big|\Delta f_X(x)\big| > 1/2$ or

$$\big|\Delta p(x)\big|/2 \geq \big|\Delta p(x)\Delta f_X(x)\big| > \sqrt{\epsilon}/6.$$

Summarizing, we have shown that

$$\big|\Delta p(x)\big| > \sqrt{\epsilon}/3 \vee \big|\Delta f_X(x)\big| > \min\big\{1/2, \sqrt{\epsilon}/(6|m(x)|)\big\}$$

follows from $\big|R(x)\big| > \epsilon$. Hence, we have

$$\begin{aligned}
P\big[\big|R(x)\big| > \epsilon\big] &\leq P\big[\big|\hat{p}(x) - p(x)\big| > \sqrt{\epsilon} f_X(x)/3\big] \\
&\quad + P\big[\big|\hat{f}_X(x) - f_X(x)\big| > f_X(x) \min\big\{1/2, \sqrt{\epsilon}/(6|m(x)|)\big\}\big],
\end{aligned} \tag{3.9}$$

for all $\epsilon > 0$. Under the assumption (3.8), we have

$$\begin{aligned}
P\big[\big|R(x)\big| > \epsilon\big] &\leq P\big[\big|\hat{p}(x) - p(x)\big| > \text{const} \cdot \sqrt{\epsilon}\big] \\
&\quad + P\big[\big|\hat{f}_X(x) - f_X(x)\big| > \text{const} \cdot \sqrt{\epsilon}\big],
\end{aligned} \tag{3.10}$$

for $\epsilon > 0$ small enough. By the usual splitting

$$\big|\hat{p}(x) - p(x)\big| \le \big|\hat{p}(x) - E\hat{p}(x)\big| + \big|E\hat{p}(x) - p(x)\big|$$

into the bias term and stochastic part, we obtain that

$$P\big[\big|\hat{p}(x) - p(x)\big| > \text{const} \cdot \sqrt{\epsilon}\big] \le P\big[\big|\hat{p}(x) - E\hat{p}(x)\big| > \text{const} \cdot \sqrt{\epsilon}\big] + \chi_{(\text{const}\cdot\sqrt{\epsilon},\infty)}\big(\big|p(x) - E\hat{p}(x)\big|\big),$$

the analogous inequality holds true when replacing $\hat{p}(x)$, $p(x)$ by $\hat{f}_X(x)$, $f_X(x)$, respectively. Applying Markov's inequality to the stochastic term, we conclude that

$$P\big[\big|\hat{p}(x) - p(x)\big| > \text{const} \cdot \sqrt{\epsilon}\big] \le \text{const} \cdot \epsilon^{-s}\, E\big|\hat{p}(x) - E\hat{p}(x)\big|^{2s} + \chi_{(\text{const}\cdot\sqrt{\epsilon},\infty)}\big(\big|p(x) - E\hat{p}(x)\big|\big), \tag{3.11}$$

for any integer $s > 0$. That motivates us to derive an upper bound on the expectation

$$\begin{aligned} E\big|\hat{p}(x) - E\hat{p}(x)\big|^{2s} &= (2\pi n)^{-2s}\, E\Big|\sum_{j=1}^{n} \int \exp(-itx)K^{\mathrm{ft}}(tb)\Psi_j(t)/g^{\mathrm{ft}}(t)\,\mathrm{d}t\Big|^{2s} \\ &\le (2\pi n)^{-2s} \sum_{j_1=1}^{n} \cdots \sum_{j_{2s}=1}^{n} \int \cdots \int \prod_{l=1}^{s} K^{\mathrm{ft}}(t_{2l-1}b)K^{\mathrm{ft}}(-t_{2l}b) \\ &\qquad \big[g^{\mathrm{ft}}(t_{2l-1}b)g^{\mathrm{ft}}(-t_{2l}b)\big]^{-1} E\prod_{l=1}^{s} \Psi_{j_{2l-1}}(t_{2l-1})\Psi_{j_{2l}}(-t_{2l})\,\mathrm{d}t_1 \cdots \mathrm{d}t_{2s}, \end{aligned}$$

when putting $\Psi_j(t) = Y_j \exp(itW_j) - EY_j \exp(itW_j)$ and choosing a symmetric kernel. Whenever the set $\{j_1, \ldots, j_{2s}\}$ contains more than s elements, at least one of the Ψ_{j_l} is stochastically independent of all other $\Psi_{j_l'}$, $l' \neq l$, so that

$$E\prod_{l=1}^{s} \Psi_{j_{2l-1}}(t_{2l-1})\Psi_{j_{2l}}(-t_{2l}) = 0\,.$$

On the other hand, if $\#\{j_1, \ldots, j_{2s}\} \le s$, then we use the rough bound

$$\Big|E\prod_{l=1}^{s} \Psi_{j_{2l-1}}(t_{2l-1})\Psi_{j_{2l}}(-t_{2l})\Big| \le 4^s \cdot E|Y_1|^{2s},$$

where we use the inequality

$$\begin{aligned} &E\big(|Y_1|^{\lambda_1} \cdot \ldots \cdot |Y_m|^{\lambda_m}\big) = E|Y_1|^{\lambda_1} \cdot \ldots \cdot E|Y_m|^{\lambda_m} \\ &\le \big(E|Y_1|^{\lambda_1+\cdots+\lambda_m}\big)^{\lambda_1/(\lambda_1+\cdots+\lambda_m)} \cdot \ldots \cdot \big(E|Y_1|^{\lambda_1+\cdots+\lambda_m}\big)^{\lambda_m/(\lambda_1+\cdots+\lambda_m)} \\ &= E|Y_1|^{\lambda_1+\cdots+\lambda_m}, \end{aligned}$$

for all positive integers m and $\lambda_1, \ldots, \lambda_m$ by Jensen's inequality. Then, we may employ the same combinational techniques as in the proof of Theorem 2.6 to establish the upper bound

$$E\big|\hat{p}(x) - E\hat{p}(x)\big|^{2s} \leq O\big(n^{-s}\big) \cdot \Big(b \cdot \min_{|t|\leq 1/b} |g^{\mathrm{ft}}(t)|\Big)^{-2s} E|Y_1|^{2s},$$

where the constants contained in $O(\cdots)$ depend on neither g nor f_X. As in the previous chapter, we assume that $K \in L_2(\mathbb{R})$ and K^{ft} is supported on $[-1,1]$. However, in addition, we have to assume that the $(2s)$th moment of Y_1 exists. When g is assumed to be uniformly bounded on the whole real line, this condition can be satisfied by the corresponding assumption of the existence of the $(2s)$th moment of the regression error ε_1.

Applying the standard condition saying that g^{ft} does not vanish (see (2.18)), we may select the bandwidth b as a sequence $b = b_n \downarrow 0$ so that (3.7) is satisfied. Otherwise, the sequence $(d_n)_n$, defined by

$$d_n = \inf\big\{b > 0 \,:\, b \cdot \min_{|t|\leq 1/b} |g^{\mathrm{ft}}(t)| \geq n^{-\xi}\big\}$$

would be bounded away from zero. That would imply the existence of a constant $d > 0$ so that

$$d \cdot \min_{|t|\leq 1/d} |g^{\mathrm{ft}}(t)| < n^{-\xi},$$

for all n and, hence, $\min_{|t|\leq 1/d} |g^{\mathrm{ft}}(t)| = 0$, contradicting assumption (2.18) along with the continuity of g^{ft}.

Therefore, for any selection rule for the bandwidth that satisfies (3.7), we have

$$E\big|\hat{p}(x) - E\hat{p}(x)\big|^{2s} = O\big(n^{(2\xi-1)s}\big).$$

Inserting this in (3.11), we obtain that

$$\begin{aligned} P\big[\big|\hat{p}(x) - p(x)\big| > \mathrm{const} \cdot \sqrt{\epsilon}\big] \\ \leq O\big(n^{(2\xi-1)s}\big) + \chi_{(\mathrm{const}\cdot\sqrt{\epsilon},\infty)}\big(\big|p(x) - \big(p * K_b\big)(x)\big|\big), \end{aligned}$$

where (3.6) has been applied. Consequently, we have to assume that p is bounded and integrable on the whole real line and continuous at x. Since

$$\begin{aligned} \big|p(x) - \big(p * K_b\big)(x)\big| &= \Big|p(x) - \int p(y) K\big((x-y)/b\big)/b \, \mathrm{d}y\Big| \\ &= \Big|\int K(z)\big[p(x) - p(x - zb)\big] \, \mathrm{d}z\Big|, \end{aligned}$$

when assuming that $K \in L_1(\mathbb{R})$ and $\int K(z) dz = 1$. Finally, by the boundedness and the local continuity of p, we can show that

$$\big|p(x) - \big(p * K_b\big)(x)\big| \stackrel{b\to 0}{\longrightarrow} 0,$$

by dominated convergence. Hence, for n sufficiently large, the term $\chi_{(\mathrm{const}\cdot\sqrt{\epsilon},\infty)}\big(|p(x) - (p * K_b)(x)|\big)$ vanishes so that

$$P\big[|\hat{p}(x) - p(x)| > \mathrm{const}\cdot\sqrt{\epsilon}\big] \leq O\big(n^{(2\xi-1)s}\big)$$

for n large enough. That gives us

$$\sum_{n=1}^{\infty} P\big[|\hat{p}(x) - p(x)| > \mathrm{const}\cdot\sqrt{\epsilon}\big] < \infty$$

under some selection $s > 1/(1-2\xi)$. Then, it follows from the Borel–Cantelli lemma from probability theory that

$$\hat{p}(x) \stackrel{n\to\infty}{\longrightarrow} p(x), \quad \text{a.s.} \tag{3.12}$$

Revisiting the formula (3.10), we realize that it remains to be shown that

$$\hat{f}_X(x) \stackrel{n\to\infty}{\longrightarrow} f_X(x), \quad \text{a.s.}$$

However, that convergence immediately follows from (3.12) when putting $Y_j \equiv 1$ almost surely. Under that tricky choice of Y_j, we have $g \equiv 1$ and $\varepsilon_j \equiv 0$ a.s., of course. Still, the assumptions about p have to be taken over with respect to f_X. Thus, we assume that f_X is bounded and continuous at x. ■

As in the case of density deconvolution (see Theorem 2.6), there is some potential to extend the theorem to those densities g whose Fourier transforms have some isolated zeros. Again, this requires to change the concept of kernel estimators to, for example, ridge-parameter approaches. Hall and Meister [58] introduce the deconvolution ridge-parameter regression estimator by

$$\begin{aligned}\hat{m}(x) &= \int g^{\mathrm{ft}}(-t)\sum_{j=1}^{n} Y_j \exp(itW_j)/\max\{|g^{\mathrm{ft}}(t)|^2, \rho_n(t)\}\exp(-itx)\,\mathrm{d}t \\ &\quad /\Big(\int g^{\mathrm{ft}}(-t)\sum_{j=1}^{n}\exp(\mathrm{i}tW_j)/\max\{|g^{\mathrm{ft}}(t)|^2, \rho_n(t)\}\exp(-\mathrm{i}tx)\,\mathrm{d}t\Big)\end{aligned}$$

– transferred to our notation – with some ridge function $\rho_n(t)$ as in estimator (2.21) in the density deconvolution case. Hall and Meister [58] also introduced an additional regularization parameter $r \geq 0$. In errors-in-variables regression, such estimators have rarely been studied so far.

3.3.2 Optimal Convergence Rates

As in density deconvolution, we are not only interested in consistency itself but we also focus on the question how fast $\hat{m}(x)$ converges to the true value of the regression function at some $x \in \mathbb{R}$. In regression estimation, we mainly focus on convergence rates in the weak sense rather than using criteria such as the MSE or the MISE for the reasons explained at the beginning of the previous subsection. The concept of weak consistency as applied in the previous subsection is modified by changing the constant $\epsilon > 0$ into a sequence $(\epsilon_n)_n \downarrow 0$. Then, $(\epsilon_n)_n$ represents the (weak) convergence rate of an estimator. More precisely, a regression estimator $\hat{m}(x)$ is said to attain the weak convergence rate $(\epsilon_n)_n$ if

$$\lim_{C\to\infty} \Big(\limsup_{n\to\infty} \sup_{(m,f_X)\in\mathcal{P}} P\big[\big|\hat{m}(x) - m(x)\big|^2 \geq C \cdot \epsilon_n\big]\Big) = 0.$$

Again, the convergence rates is actually achieved by a sequence of estimators with respect to the sample size n rather than by a single estimator. However, for simplicity, it has become common to say that an estimator attains that rate. With respect to the true regression function m as well as the true design density f_X, we choose a uniform version of the rates, that is, we define some set $\mathcal{P} = \mathcal{M} \times \mathcal{F}$, where $\times$ denotes the common set product, and we write $\mathcal{M}$ and $\mathcal{F}$ for some appropriate classes consisting of the admitted regression functions and design densities, respectively.

The correct order of the limits (first, put n to infinity and, then, $C \to \infty$) is essential in the above definition. To give more intuitive consideration about the concept of weak convergence rates, we show that the definition is indeed implied by the condition that the corresponding convergence rates are attained under the MSE-criterion, which means that

$$\sup_{(m,f_X)\in\mathcal{P}} E\big|\hat{m}(x) - m(x)\big|^2 = O\big(\epsilon_n\big).$$

That can be seen as follows: whenever a regression estimator $\hat{m}(x)$ achieves the convergence rate $(\epsilon_n)_n$ with respect to the MSE-criterion, then Markov's inequality gives us

$$\begin{aligned}\sup_{(m,f_X)\in\mathcal{P}} P\big[\big|\hat{m}(x) - m(x)\big|^2 \geq C \cdot \epsilon_n\big] &\leq C^{-1}\epsilon_n^{-1} \sup_{(m,f_X)\in\mathcal{P}} E\big|\hat{m}(x) - m(x)\big|^2 \\ &\leq \text{const} \cdot C^{-1},\end{aligned}$$

for all integer $n > 0$, where the constant factor depends on neither n nor C. Therefore, we have

$$\limsup_{n\to\infty} \sup_{(m,f_X)\in\mathcal{P}} P\big[\big|\hat{m}(x) - m(x)\big|^2 \geq C \cdot \epsilon_n\big] \leq \text{const} \cdot C^{-1}.$$

Then, putting $C \to \infty$, we verify that estimator $\hat{m}(x)$ also achieves the weak convergence rate $(\epsilon_n)_n$.

Studying the weak convergence rates seems more convenient for investigating the asymptotics of estimator (3.5) as no further regularization is required to keep the denominator away from zero. Also, we may use the inequality

(3.9) from the previous subsection when changing ϵ to $C \cdot \epsilon_n$. Then, condition (3.8) may be included in the definition of the combined regression and density class $\mathcal{P}$ as a uniform version. This means that the constants C_1 and C_2 may not depend on the individual pair (m, f_X) but they may be viewed as a significant characteristic parameter of the class $\mathcal{P}$. Applying Markov's inequality, we obtain that

$$\sup_{(m,f_X)\in\mathcal{P}} P\big[\big|\hat{m}(x) - m(x)\big|^2 \geq C \cdot \epsilon_n\big]$$
$$\leq \text{const} \cdot \big(C\epsilon_n\big)^{-1} \sup_{(m,f_X)\in\mathcal{P}} \big(E\big|\hat{p}(x) - p(x)\big|^2 + E\big|\hat{f}_X(x) - f_X(x)\big|^2\big). \quad (3.13)$$

While the moments of $\big|\hat{m}(x) - m(x)\big|$ have already been considered in the previous subsection, a more detailed study is required now to derive the convergence rates. By the bias-variance decomposition for the second moment, we derive that

$$E\big|\hat{p}(x) - p(x)\big|^2 = \operatorname{var}\hat{p}(x) + \big|E\hat{p}(x) - p(x)\big|^2,$$

and, further,

$$\operatorname{var}\hat{p}(x) \leq \frac{1}{(2\pi)^2 n} E\big|Y_1\big|^2\Big|\int \exp\big(-\mathrm{i}t(x - W_1)\big)K^{\mathrm{ft}}(tb)/g^{\mathrm{ft}}(t)\,\mathrm{d}t\Big|^2$$
$$\leq \frac{1}{(2\pi)^2 n}\int E\big(\big|Y_1\big|^2 \mid X_1 = \xi\big)$$
$$\cdot E\Big|\int \exp\big(-\mathrm{i}t(x - \xi)\big)\exp(\mathrm{i}t\delta_1)K^{\mathrm{ft}}(tb)/g^{\mathrm{ft}}(t)\,\mathrm{d}t\Big|^2 f_X(\xi)\,\mathrm{d}\xi$$
$$\leq \frac{1}{(2\pi)^2 n}\big\|f_X(\cdot)\,E\big(\big|Y_1\big|^2 \mid X_1 = \cdot\big)\big\|_\infty$$
$$\cdot\int E\Big|\int \exp(-\mathrm{i}t\xi)\exp(\mathrm{i}t\delta_1)K^{\mathrm{ft}}(tb)/g^{\mathrm{ft}}(t)\,\mathrm{d}t\Big|^2\mathrm{d}\xi$$
$$= \frac{1}{2\pi n}\big\|f_X(\cdot)\,E\big(\big|Y_1\big|^2 \mid X_1 = \cdot\big)\big\|_\infty \cdot \int E\big|\exp(\mathrm{i}t\delta_1)K^{\mathrm{ft}}(tb)/g^{\mathrm{ft}}(t)\big|^2\,\mathrm{d}t$$
$$= \frac{1}{2\pi n}\big\|f_X(\cdot)\,E\big(\big|Y_1\big|^2 \mid X_1 = \cdot\big)\big\|_\infty \cdot \int \big|K^{\mathrm{ft}}(tb)/g^{\mathrm{ft}}(t)\big|^2\,\mathrm{d}t,$$

where we have used Parseval's identity (Theorem A.4). Obviously, we shall assume that the product of f_X and the conditional expectation of Y_1^2 given $X_1 = x$ is uniformly bounded for all $x \in \mathbb{R}$. This expectation may be decomposed as follows,

$$f_X(x)E\big(Y_1^2 \mid X_1 = x\big) = \tau^2(x) + m^2(x)f_X(x),$$

where $\tau^2(x) = f_X(x)\operatorname{var}\big(Y_1 \mid X_1 = x\big) = f_X(x)E\big(|Y_1 - m(X_1)|^2 \mid X_1 = x\big)$. Hence, it suffices to assume that the functions τ^2 and $m^2 f_X$ are uniformly bounded on the whole real line. Then, we may fix that

$$\operatorname{var} \hat{p}(x) \leq \text{const} \cdot (nb)^{-1} \max_{|t| \leq 1/b} \left|g^{\text{ft}}(t)\right|^{-2},$$

with some constant which is uniform with respect to the specific regression function m and the design density f_X. Again, we have used some kernel function K whose Fourier transform is bounded and supported on $[-1, 1]$.

Recalling the two standard types of error densities, ordinary smooth g as defined in (2.31) and supersmooth g as in (2.32), we may derive that

$$\operatorname{var} \hat{p}(x) = O\big(n^{-1} b^{-1-2\alpha}\big)$$

for ordinary smooth g, and

$$\operatorname{var} \hat{p}(x) = O\big(n^{-1} b^{-1} \exp(2 d_1 b^{-\gamma})\big),$$

in the case of supersmooth contamination, uniformly in m and f_X.

Now we focus on the bias term $\left|E\hat{p}(x) - p(x)\right|^2$. Under the conditions imposed in Sect. 3.2, we may utilize (3.6) so that our goal is to derive an upper bound on

$$\left|(p * K_b)(x) - p(x)\right|^2 = \left|\int K(z)\big[p(x) - p(x - zb)\big]\, \mathrm{d}z\right|^2$$

when the kernel K integrates to one. Unlike in the previous subsection, which has been dedicated to general consistency, we are not satisfied with the sole convergence of the bias term to zero, but its convergence rates, considered uniformly over the class $(m, f_X) \in \mathcal{P}$, shall be studied.

In Sect. 2.4.2, we obtain the uniform convergence rates for the bias term in the field of density deconvolution. In fact, the bias term B_n occurring in that subsection coincides with the above bias term for errors-in-variables regression when replacing the function f by p. That motivates us to impose Hölder conditions on p, which correspond to (2.29) when writing p instead of f. Also, we assume that f_X is bounded above by some uniform constants. Then, p adopts all assumptions contained in $f \in \mathcal{F}_{\beta,C,\delta;x}$ except the condition that f is a density function. Of course, those conditions will be included in the definition of the combined regression-density-class $\mathcal{P}$. However, this latter condition is not required to establish upper bounds on the bias. Still, we assume integrability of p over the whole of $\mathbb{R}$ to guarantee well-definiteness of its Fourier transform. With respect to the kernel function K, we assume condition (2.38), that is, we employ some β-order kernel K. Then, the arguments follow the line as in density deconvolution so that we finally obtain

$$\sup_{(m, f_X) \in \mathcal{P}} \left|E\hat{p}(x) - p(x)\right|^2 = O\big(b^{2\beta}\big).$$

Considering (3.13) again, we shall show the same bound on $E\left|\hat{f}_X(x) - f_X(x)\right|^2$ as on $E\left|\hat{p}(x) - p(x)\right|^2$ up to some constant factors which do not depend on the specific f_X. This, however, is a standard problem of density deconvolution, which is readily solved in Sect. 2.4.2 when assuming the same Hölder conditions on f_X as on m.

Hence, we conclude that

$$\sup_{(m,f_X)\in\mathcal{P}} \left(E\big|\hat{p}(x)-p(x)\big|^2 + E\big|\hat{f}_X(x)-f_X(x)\big|^2\right) = O\big(n^{-1}b^{-1-2\alpha}+b^{2\beta}\big)$$

holds true for ordinary smooth g; analogously, for supersmooth g, we have

$$\sup_{(m,f_X)\in\mathcal{P}} \left(E\big|\hat{p}(x)-p(x)\big|^2 + E\big|\hat{f}_X(x)-f_X(x)\big|^2\right)$$
$$= O\big(n^{-1}b^{-1}\exp(2d_1 b^{-\gamma})+b^{2\beta}\big).$$

The optimal bandwidth choice may be taken from Theorem 2.8 for both types of error densities. Then, put

$$\epsilon_n = \begin{cases} n^{-2\beta/(2\beta+2\alpha+1)}, & \text{for ordinary smooth } g, \\ \big(\log n\big)^{-2\beta/\gamma}, & \text{for supersmooth } g, \end{cases}$$

that is, the rates occurring in Theorem 2.8. Finally, by (3.13), we derive that

$$\sup_{(m,f_X)\in\mathcal{P}} P\big[\big|\hat{m}(x)-m(x)\big|^2 \geq C\cdot\epsilon_n\big] \leq \text{const}\cdot C^{-1},$$

for any $C>0$, so that the weak rate $(\epsilon_n)_n$ is evident for both ordinary smooth and supersmooth error densities.

Let us summarize all conditions imposed on m and f_X in the following. Then, the class $\mathcal{P}$ is defined as the collection of all those pairs (m,f_X) that satisfy

$\|m^2 f_X\|_\infty, \|f_X\|_\infty \leq C_1$, $m\cdot f_X, (m\cdot f_X)^{\mathrm{ft}}, f_X^{\mathrm{ft}} \in L_1(\mathbb{R})$,

$m\cdot f_X$ and f_X satisfy (2.29) when putting f equal to $m\cdot f_X$ and f_X, respectively,

$(\beta\geq 1)$,

$f_X(x)\geq C_2>0$,

for some uniform positive constants C_1, C_2. Note that the uniform boundedness of mf_X follows immediately since

$$\big|m(x)f_X(x)\big| \leq \big(1+m^2(x)\big)|f_X(x)| \leq 2C_1,$$

and continuity of m and f_X at x is included in the Hölder conditions. With respect to the regression errors ε_j we assume that

$$\|\tau^2\|_\infty = \|\mathrm{var}\,(Y_j^2 \mid X_j=\cdot)\|_\infty = \|E(\varepsilon_j^2\mid X_j=\cdot)\|_\infty \leq C_3, \tag{3.14}$$

with some constant $C_3>0$. Hence, this condition can be generalized to a moderately heteroscedastic setting where the regression errors ε_j are not necessarily identically distributed but their conditional second moments possess a joint upper bound C_3.

All in all, we have proved the following theorem.

Theorem 3.3. *Consider the errors-in-variables regression model (3.2) and assume that condition (3.14) holds. Use estimator (3.5) with a bounded and continuous kernel function satisfying (2.38).*
(a) For ordinary smooth g (see (2.31)), select the bandwidth $b \asymp n^{-1/(2\beta+2\alpha+1)}$. *Then, we have*

$$\lim_{C\to\infty}\Big(\limsup_{n\to\infty}\ \sup_{(m,f_X)\in\mathcal{P}} P\big[\big|\hat{m}(x)-m(x)\big|^2 \geq C\cdot n^{-2\beta/(2\beta+2\alpha+1)}\big]\Big) = 0\,.$$

(b) For supersmooth g (see (2.32)), select the bandwidth $b = c_b^{-1/\gamma}\cdot(\log n)^{-1/\gamma}$ *with* $c_b \in (0, 1/(2d_1))$ *so that*

$$\lim_{C\to\infty}\Big(\limsup_{n\to\infty}\ \sup_{(m,f_X)\in\mathcal{P}} P\big[\big|\hat{m}(x)-m(x)\big|^2 \geq C\cdot\big(\log n\big)^{-2\beta/\gamma}\big]\Big) = 0\,.$$

That result has first been proven in [47].

Therefore, we notice coincidence of the convergence rates with those derived in density deconvoluion in Theorems 2.8 and 2.9 under the corresponding smoothness constraints. Also, the theoretical selection of the bandwidth parameter and the assumptions about the kernel function may be adopted.

However, as in density deconvolution, we realize that the selection rule for the bandwidth in (a) is not adaptive, that is, it requires knowledge of the joint smoothness degree β of $m\cdot f_X$ and f_X. Hence, there is some necessity to use data-driven bandwidth selectors as introduced in Sect. 2.5 for the density deconvolution problem. In the error-free nonparametric regression problem, cross-validation (CV) procedures are frequently applied with the following concept: as the estimate for the regression function m should be as close to m as possible, it seems reasonable to select the bandwidth $b > 0$ so that the empirically fully accessible function

$$\sum_{j=1}^{n}\big(Y_j - \hat{m}_{-j}(X_j)\big)^2 p(X_j),$$

with some positive weight function p is minimized. Therein, $\hat{m}_{-j}$ denotes the Nadaraya–Watson estimator (3.3) based on the observations $(X_1, Y_1), \ldots,$ $(X_{j-1}, Y_{j-1}), (X_{j+1}, Y_{j+1}), \ldots, (X_n, Y_n)$. Thus, the observation (X_j, Y_j) is omitted to avoid stochastic dependence between the estimator and the inserted value X_j. Therefore, the CV-procedure, in the regression context, is a kind of least square method.

Nevertheless, there is no straightforward extension of this method to the errors-in-variables regression experiment (3.2), as the covariates X_j are only observed with noise and none of the X_j can consistently be reconstructed from the data. To our best knowledge, the first approach to data-driven bandwidth selection for estimator (3.5) is given by [28], where the authors use SIMEX methods to determine the optimal b empirically, while the criterion of pointwise estimation of $m(x)$ must be changed into an integrated version of the risk.

To give a brief overview about SIMEX methods in general, we mention that SIMEX (SIMulation and EXtrapolation) has been introduced as an algorithm for estimation problems under measurement error by [22]. At some point of views, it is a competing method with respect to deconvolution. The main idea is to generate new observations with increased noise levels. Then, fitting (extrapolating) those data at different noise levels resulting in some appropriate curve with respect to the noise level σ in some real interval, one obtains an empirical version of the quantity to be estimated by the value of the extrapolated curve at $\sigma = 0$. SIMEX methods are favored by some applied scientists, although SIMEX is consistent and rate-optimal in rare cases only. Its applicability seems to be restricted to the availability of original data observed at small noise levels σ.

In [28], the authors suggest to add some additional independent noise to the covariates. Transferred to the notation used in model (3.2), this means one generates the data $W_j^* = W_j + \delta_j^*$ and $W_j^{**} = W_j + \delta_j^{**}$, $j = 1, \ldots, n$, where the W_j are the originally observed covariates and the $\delta_j^*, \delta_j^{**}$ are independent replicates of δ_j having the (known) error density g. Then, one may use the deconvolution regression estimator where we insert first the W_j^* and then the W_j^{**} instead of the W_j in the construction of (3.5) so that we obtain the estimators $\hat{m}^*$ and $\hat{m}^{**}$, respectively. The authors propose to derive CV-bandwidths $\hat{b}^*$ and $\hat{b}^{**}$ by minimizing the functionals

$$CV^*(b) = \frac{1}{n} \sum_{j=1}^{n} \big(Y_j - \hat{m}^*_{-j}(W_j)\big)^2 p(W_j),$$

$$CV^{**}(b) = \frac{1}{n} \sum_{j=1}^{n} \big(Y_j - \hat{m}^{**}_{-j}(W_j^*)\big)^2 p(W_j^*),$$

respectively, over $b > 0$. Therein, $\hat{m}^*_{-j}$ denotes that modification of estimator $\hat{m}^*$ where the jth observation is omitted and $\hat{m}^{**}_{-j}$, analogously. Furthermore, p denotes some weight function, which should be integrable on the real line to guarantee well-definiteness of the functionals CV^* and CV^{**}. As we may generate as many independent replicates as we like – in practice, the number of replicates which is realistic is only a question of time – we are able to approximate the conditional expectations of $CV^*(b)$ and $CV^{**}(b)$, given the original data $(W_1, Y_1), \ldots, (W_n, Y_n)$ with arbitrary precision by averaging over several versions of $CV^*(b)$ and $CV^{**}(b)$ based on independently generated copies of $\delta_1^*, \ldots, \delta_n^*, \delta_1^{**}, \ldots, \delta_n^{**}$. Then, the authors suggest to employ those b, which minimize these conditional expectations of $CV^*(b)$ and $CV^{**}(b)$ with respect to $b > 0$ as the empirical bandwidths $\hat{b}_1$ and $\hat{b}_2$ instead of the previously obtained $\hat{b}^*$ and $\hat{b}^{**}$. The bandwidth that we are actually interested in must be $\hat{b}_0$, heuristically speaking. In [28], it is shown that the cross-validation bandwidths $\hat{b}^*$ and $\hat{b}^{**}$ minimizes the asymptotic MISE (with respect to the weight function p). For $\hat{b}_0$, the authors suggest to take

$$\hat{b}_0 = \hat{b}_1^2 / \hat{b}_2$$

as an empirically accessible bandwidth for estimator (3.5) in the deconvolution regression model (3.2). This definition is motivated by a linear back-extrapolation procedure, where the relation $\log(\hat{b}_0) - \log(\hat{b}_1) \approx \log(\hat{b}_1) - \log(\hat{b}_2)$ is applied. The authors show adaptivity of $\hat{b}_0$ within the framework of the deconvolution kernel regression estimators (3.5), that is, the ratio of $\hat{b}_0$ and the optimal b_0, which minimizes the asymptotic MISE of estimator (3.5) weighted with the function p, under small error asymptotics, that is, the scaling parameter or the variance of the δ_j tends to zero. Numerical simulations indicate that this type of bandwidth selection works well in practice for errors-in-variable problems.

Now we focus on the question whether the convergence rates attained by the estimator (3.5) are optimal under the given conditions. Remember that we succeeded in showing rate optimality in the density deconvolution context in Theorems 2.13 and 2.14.

For that purpose, we have to construct two different admitted distributions of the vector-valued random variables (W_j, Y_j). We imagine that those distributions compete to be the true regression function m. Therefore, those distributions, which are rather close to each other with respect to some information distance while their corresponding regression functions at x are far away from each other, are particularly useful. As $W_j = X_j + \delta_j$, let us construct two possible distributions for the unobserved (X_j, Y_j), since, then, the distribution of the (W_j, Y_j) can be derived from there. Consider

$$f_{(X,Y),\theta}(x,y) = f_1(x)f_2(y) + \theta\, C\Delta_X(x)\Delta_Y(y), \tag{3.15}$$

with $\theta \in \{0,1\}$, the supersmooth Cauchy density f_1 as defined in (2.51), $f_2(y) = (1/2)\cdot\exp(-|y|)$, $\Delta_X(x) = a_n^{-\beta}\cos(2a_n x)f_0(a_n x)$, $\Delta_Y(y) = (1/4)\cdot \operatorname{sign}(y)\exp(-|y|)$ and f_0 as in (2.24), as two bivariate densities competing to be the true joint density of (X_j, Y_j). The constant $C > 0$ is still to be chosen. Also, the sequence $(a_n)_n \uparrow \infty$ will be specified later.

First, we have to verify that $f_{(X,Y),\theta}(x,y)$ is a density function for $\theta \in \{0,1\}$. For $\theta = 0$, we realize that both f_1 and f_2 are univariate densities; hence, it follows rather easily that $f_{(X,Y),0}(x,y)$ is nonnegative and integrates to one. If $\theta = 1$, we use the inequalities $f_2(y) \geq 2|\Delta_Y(y)|$ and

$$\begin{aligned} |\Delta_X(x)| &\leq \text{const}\cdot\big|(1-\cos(a_n x))/(\pi a_n^2 x^2)\big| \leq \text{const}/\big(1+a_n^2x^2\big) \\ &\leq \text{const}\cdot\big(1+x^2\big)^{-1} = c\cdot f_1(x), \end{aligned} \tag{3.16}$$

for all $x, y \in \mathbb{R}$ and some constant $c > 0$ to show that $f_{(X,Y),\theta}(x,y) \geq 0$ can be guaranteed when the constant $C > 0$ is chosen sufficiently small. We have employed the result

$$\sup_{x\in\mathbb{R}} f_0(x)/f_1(x) \leq \text{const},$$

which can be shown by elementary analysis. As Δ_Y integrates to zero due to its symmetry with respect to $(0,0)$, we have shown that $f_{(X,Y),1}(x,y)$ is a bivariate density function.

Next, condition (3.14) and the membership of the corresponding pair (m, f_X) in $\mathcal{P}$ have to be shown. We fix that

$$\begin{aligned} f_X(x) = f_{X,\theta}(x) &= \int f_{(X,Y),\theta}(x,y)\,\mathrm{d}y \\ &= f_1(x)\cdot \underbrace{\int f_2(y)\,\mathrm{d}y}_{=1} + \theta C \Delta_X(x)\cdot \underbrace{\int \Delta_Y(y)\,\mathrm{d}y}_{=0} \\ &= f_1(x) \end{aligned}$$

for both $\theta \in \{0,1\}$ by using a symmetry argument for Δ_Y. The regression function is calculated by

$$\begin{aligned} m(x) = m_\theta(x) = E(Y_1 \mid X_1 = x) &= \int y f_{(X,Y),\theta}(x,y)\,\mathrm{d}y / f_X(x) \\ &= f_1^{-1}(x)\cdot\Big(f_1(x)\cdot \underbrace{\int y f_2(y)\,\mathrm{d}y}_{=0} + \theta C\Delta_X(x)\cdot \int y\Delta_Y(y)\,\mathrm{d}y\Big) \\ &= \theta\cdot 2C\Delta_X(x)\cdot \int_0^\infty y\Delta_Y(y)\,\mathrm{d}y / f_1(x) \\ &= \theta\cdot C'\,\Delta_X(x)/f_1(x), \end{aligned}$$

with some constant $C' > 0$, again using symmetry of f_2 and Δ_Y appropriately. Then, we study the conditional expectation of Y_1^2 given $X_j = x$, thus

$$\begin{aligned} E\big(Y_1^2 \mid X_j = x\big) &= \int y^2 f_{(X,Y),\theta}(x,y)\,\mathrm{d}y / f_X(x) \\ &= f_1^{-1}(x)\cdot\Big(f_1(x)\cdot\int y^2 f_2(y)\,\mathrm{d}y + \theta C\Delta_X(x)\cdot \underbrace{\int y^2\Delta_Y(y)\,\mathrm{d}y}_{=0}\Big) \\ &\leq C'', \end{aligned}$$

for some positive constant C'' for all $x \in \mathbb{R}$. We have used the symmetry of Δ_Y and the exponential decay of Δ_Y and f_2 so that the existence of their absolute second moment is ensured. Combining the above inequality with the boundedness of $f_X = f_1$, we derive that both τ^2 and $m^2 f_X$ are bounded uniformly in n on the whole real line. Hence, condition (3.14) is satisfied if C_3 is large enough.

Concerning the conditions contained in the constraint $(m, f_X) \in \mathcal{P}$, we conventionally put $x = 0$; here, x denotes the point where $m(x)$ shall be estimated, although this proof is extendable to general $x \in \mathbb{R}$. We have already verified that $\|m^2 f_X\|_\infty, \|f_X\|_\infty \leq C_1$ for C_1 large enough but independent of n. We have

$$p = p_\theta = m\cdot f_X = \theta\cdot C'\,\Delta_X\,,$$

so that integrability of p on the whole real line follows as well. Furthermore, we have

$$p^{\mathrm{ft}}(t) = \theta \cdot C' \Delta_X^{\mathrm{ft}}(t)$$
$$= \theta \cdot C' \cdot \Big[\frac{1}{2} a_n^{-\beta-1} f_0^{\mathrm{ft}}\Big(\frac{t+2a_n}{a_n}\Big) + \frac{1}{2} a_n^{-\beta-1} f_0^{\mathrm{ft}}\Big(\frac{t-2a_n}{a_n}\Big)\Big],$$

where Lemma A.1(e) and (f) have been used. We realize that p^{ft} is compactly supported and, hence, integrable on the whole of $\mathbb{R}$. Therefore, we recall that $f_0^{\mathrm{ft}}(t) = (1-|t|)_+$. Also, we have $f_X^{\mathrm{ft}}(t) = f_1^{\mathrm{ft}}(t) = \exp(-|t|)$ so that $f_X^{\mathrm{ft}} \in L_1(\mathbb{R})$ can be fixed as well.

We have $f_X(0) = f_1(0) = 1/\pi > 0$ so that the validity of the Hölder conditions remains to be checked. We easily recognize that f_1 is differentiable infinitely often on the whole real line. Therefore, (2.29) is satisfied for any $\beta \geq 1$ and $\delta > 0$ if C is large enough. With respect to p we derive that

$$p^{(\lfloor\beta\rfloor)}(x) = \theta \cdot C' a_n^{-\beta} a_n^{\lfloor\beta\rfloor} \vartheta(a_n x),$$

where $\vartheta(x) = \dfrac{\mathrm{d}^{\lfloor\beta\rfloor}}{\mathrm{d}x^{\lfloor\beta\rfloor}}\big(\cos(2x) f_0(x)\big)$. Note that ϑ is continuously differentiable on the whole of $\mathbb{R}$. By induction, we may derive that f_0 along with each of its derivatives are bounded on the whole real line so that ϑ and ϑ' are bounded functions, leading to

$$\big|p^{(\lfloor\beta\rfloor)}(y) - p^{(\lfloor\beta\rfloor)}(\tilde{y})\big| \leq O\big(a_n^{-\beta+\lfloor\beta\rfloor}\big) \cdot |a_n y - a_n \tilde{y}|^{\beta-\lfloor\beta\rfloor} \leq \mathrm{const} \cdot |y-\tilde{y}|^{\beta-\lfloor\beta\rfloor},$$

with a uniform constant. Hence, we have verified that $(m, f_X) \in \mathcal{P}$ for both $\theta \in \{0,1\}$, n sufficiently large, any $(a_n)_n \uparrow \infty$ when C in (2.29), C_1 are sufficiently large, C_2 is sufficiently small.

As the validity of the given conditions is evident, we are now going to prove optimality of the convergence rates. Assume an arbitrary estimator $\hat{m}(x)$ of $m(x)$, for $x=0$, based on the observed data $(W_1,Y_1),\ldots,(W_n,Y_n)$. Therefore, any map $\hat{m} : \mathbb{R} \times (\mathbb{R}^2)^n \to \mathbb{R}$ is eligible as the general regression estimator. Then, we insert the data, that is, $\hat{m}(x) = \hat{m}\big(x;(W_1,Y_1),\ldots,(W_n,Y_n)\big)$. We define the set

$$G(m,\epsilon) = \big\{z \in (\mathbb{R}^2)^n : \big|\hat{m}(0;z) - m(0)\big|^2 \geq \epsilon\big\},$$

for any admitted regression function m and $\epsilon > 0$. Furthermore, the following set is introduced,

$$D_{m,\tilde{m}} = \big\{z \in (\mathbb{R}^2)^n : \big|\hat{m}(0;z) - m(0)\big|^2 \leq \big|\hat{m}(0;z) - \tilde{m}(0)\big|^2\big\},$$

for two competing admitted regression functions m and $\tilde{m}$. Therefore, $D_{m,\tilde{m}}$ contains all values of the data where m is preferable to $\tilde{m}$ when deciding between those two functions with respect to their value at 0 based on the estimator $\hat{m}$.

We realize that $D_{m,\tilde{m}} \subseteq G(\tilde{m},\epsilon)$ whenever $\epsilon < |m(0)-\tilde{m}(0)|^2/4$ because, under the assumed existence of some $z \in (\mathbb{R}^2)^n$ with $|\hat{m}(0;z)-\tilde{m}(0)| < \sqrt{\epsilon} < |m(0)-\tilde{m}(0)|/2$ and $|\hat{m}(0;z)-m(0)| \leq |\hat{m}(0;z)-\tilde{m}(0)| < \sqrt{\epsilon}$, the triangle inequality implies that

$$|m(0)-\tilde{m}(0)| \leq \big|\hat{m}(0;z)-\tilde{m}(0)\big| + \big|\hat{m}(0;z)-m(0)\big| \leq 2\sqrt{\epsilon} < |m(0)-\tilde{m}(0)|,$$

leading to contradiction.

On the other hand, we have $(\mathbb{R}^2)^n \backslash D_{m,\tilde{m}} \subseteq G(m,\epsilon)$ for all $\epsilon < |m(0)-\tilde{m}(0)|^2/4$, because assuming the existence of some $z \in (\mathbb{R}^2)^n$ with $\big|\hat{m}(0;z)-m(0)\big| > \big|\hat{m}(0;z)-\tilde{m}(0)\big|$ and $\big|\hat{m}(0;z)-\tilde{m}(0)\big| < \big|\hat{m}(0;z)-m(0)\big| < \sqrt{\epsilon}$ leads to the conclusion

$$|m(0)-\tilde{m}(0)| \leq \big|\hat{m}(0;z)-\tilde{m}(0)\big| + \big|\hat{m}(0;z)-m(0)\big| \leq 2\sqrt{\epsilon} < |m(0)-\tilde{m}(0)|,$$

and, hence, we have contradiction, too.

Assuming that the estimator $\hat{m}$ achieves the rate $(\epsilon_n)_n$ in a weak, pointwise, uniform sense, we have

$$\lim_{c\to\infty}\Big(\limsup_{n\to\infty}\sup_{(m,f_X)\in\mathcal{P}} P\big[\big((W_1,Y_1),\ldots,(W_n,Y_n)\big) \in G(m,c\epsilon_n)\big]\Big) = 0,$$

by definition. Hence,

$$\lim_{c\to\infty}\Big(\limsup_{n\to\infty} P_0\big[\big((W_1,Y_1),\ldots,(W_n,Y_n)\big) \in G(m_0,c\epsilon_n)\big] \\ + P_1\big[\big((W_1,Y_1),\ldots,(W_n,Y_n)\big) \in G(m_1,c\epsilon_n)\big]\Big) = 0, \tag{3.17}$$

where P_θ, $\theta \in \{0,1\}$ denotes probability under the assumption that $f_{(X;Y);\theta}$, given by (3.15), as the true density of the (X_j,Y_j); analogously, m_θ denotes the corresponding regression function. If, for some fixed but arbitrary $c>0$, we have

$$4c\epsilon_n < |m_0(0)-m_1(0)|^2 \tag{3.18}$$

for any integer $n>0$ sufficiently large, the inclusions derived above give us

$$\begin{aligned}
&P_0\big[\big((W_1,Y_1),\ldots,(W_n,Y_n)\big) \in G(m_0,c\epsilon_n)\big] \\
&\qquad + P_1\big[\big((W_1,Y_1),\ldots,(W_n,Y_n)\big) \in G(m_1,c\epsilon_n)\big] \\
&\geq P_0\big[\big((W_1,Y_1),\ldots,(W_n,Y_n)\big) \in (\mathbb{R}^2)^n\backslash D_{m_0,m_1}\big] \\
&\qquad + P_1\big[\big((W_1,Y_1),\ldots,(W_n,Y_n)\big) \in D_{m_0,m_1}\big] \\
&= 1 - P_0\big[\big((W_1,Y_1),\ldots,(W_n,Y_n)\big) \in D_{m_0,m_1}\big] \\
&\qquad + P_1\big[\big((W_1,Y_1),\ldots,(W_n,Y_n)\big) \in D_{m_0,m_1}\big] \\
&\geq 1 - \sup_{D\in\mathfrak{B}((\mathbb{R}^2)^n)} \Big|P_0\big[\big((W_1,Y_1),\ldots,(W_n,Y_n)\big) \in D\big] \\
&\qquad - P_1\big[\big((W_1,Y_1),\ldots,(W_n,Y_n)\big) \in D\big]\Big|,
\end{aligned} \tag{3.19}$$

where $\mathfrak{B}((\mathbb{R}^2)^n)$ denotes the collection of all Lebesgue-measurable subsets of $(\mathbb{R}^2)^n$. Writing

$$P_\theta(D) = P_\theta\big[\big((W_1,Y_1),\ldots,(W_n,Y_n)\big) \in D\big],$$

we define the probability measure P_θ on $(\mathbb{R}^2)^n$. The total variation distance (TV) between the two measures P_0 and P_1 is defined by

$$\mathrm{TV}(P_0,P_1) = \sup_{D\in\mathfrak{B}((\mathbb{R}^2)^n)} \big|P_0(D) - P_1(D)\big|.$$

Therefore, (3.19) is equivalent with $1 - \mathrm{TV}(P_0,P_1)$. The TV can equivalently be represented by the $L_1(\mathbb{R})$-distance between the corresponding Lebesgue densities by the following lemma.

Lemma 3.4. (Scheffé's lemma) *Assume two probability measures P, Q on $\mathbb{R}^d$ for some $d \geq 1$ having the Lebesgue densities p and q, respectively. Then, we have*

$$\|p-q\|_1 = 2\, TV(P,Q).$$

Proof. We consider, for some arbitrary Lebesgue measurable $B \subseteq \mathbb{R}^d$, that

$$\begin{aligned}
\|p-q\|_1 &= \int \big|p(x) - q(x)\big|\, \mathrm{d}x \\
&\geq \int_B \big(p(x) - q(x)\big)\, \mathrm{d}x + \int_{\mathbb{R}^d\setminus B} \big(q(x) - p(x)\big)\, \mathrm{d}x \\
&= \int_B p(x)\, \mathrm{d}x - \int_B q(x)\, \mathrm{d}x + 1 - \int_B q(x)\, \mathrm{d}x - 1 + \int_B p(x)\, \mathrm{d}x \\
&= 2\Big(\int_B p(x)\, \mathrm{d}x - \int_B q(x)\, \mathrm{d}x\Big). \qquad (3.20)
\end{aligned}$$

By exchanging p and q, the left side of the above inequality stays as it is, while the right side changes its sign. Therefore, we may conclude that

$$\|p-q\|_1 \geq 2\Big|\int_B p(x)\, \mathrm{d}x - \int_B q(x)\, \mathrm{d}x\Big| = 2\big|P(B) - Q(B)\big|.$$

As this inequality holds for any measurable $B \subseteq \mathbb{R}^d$, it follows that

$$\|p-q\|_1 \geq 2\,\mathrm{TV}(P,Q).$$

As we want to prove that the above inequality holds as an equality, it remains to be shown that the reverse inequality is valid, too. For that purpose, we introduce the set

$$A = \big\{x \in \mathbb{R}^d \,:\, p(x) > q(x)\big\}.$$

As both p and q are densities and, hence, Lebesgue-measurable functions, the set $A \subseteq \mathbb{R}^d$ is Lebesgue-measurable. When putting $B = A$, the inequality (3.20) becomes an equality. Then,

$$\|p-q\|_1 = 2\Big(\int_A p(x)\,\mathrm{d}x - \int_A q(x)\,\mathrm{d}x\Big) \le 2\,\mathrm{TV}(P,Q)$$

proves the lemma. ■

The Lebesgue density p_θ of the measure P_θ is identical with the density of the sample $\big((W_1,Y_1),\ldots,(W_n,Y_n)\big)$. As the data (W_j,Y_j), $j=1,\ldots,n$, are i.i.d., we have

$$p_\theta\big((w_1,y_1),\ldots,(w_n,y_n)\big) = \prod_{j=1}^n f_{(W,Y);\theta}(w_j,y_j),$$

when $f_{(W,Y);\theta}$ denotes the density of (W_1,Y_1). Applying Lemma 3.4 gives us

$$\begin{aligned}
1-\mathrm{TV}(P_0,P_1) &= 1-\frac{1}{2}\|p_0-p_1\|_1\\
&= 1-\frac{1}{2}\int \underbrace{\max\{p_0(x),p_1(x)\}}_{\le p_0(x)+p_1(x)}\,\mathrm{d}x + \frac{1}{2}\int \min\{p_0(x),p_1(x)\}\,\mathrm{d}x\\
&\ge \frac{1}{2}\int \min\{p_0(x),p_1(x)\}\,\mathrm{d}x\\
&\ge \frac{1}{4}\Big(\int \sqrt{p_0(x)}\sqrt{p_1(x)}\,\mathrm{d}x\Big)^2\\
&\ge \frac{1}{4}\Big(\int\cdots\int \prod_{j=1}^n \sqrt{f_{(W,Y);0}(w_j,y_j)}\sqrt{f_{(W,Y);1}(w_j,y_j)}\,\mathrm{d}w_1\mathrm{d}y_1\cdots\mathrm{d}w_n\mathrm{d}y_n\Big)^2\\
&\ge \frac{1}{4}\Big(\iint \sqrt{f_{(W,Y);0}(w,y)}\sqrt{f_{(W,Y);1}(w,y)}\,\mathrm{d}w\mathrm{d}y\Big)^{2n},
\end{aligned}$$

where we have used LeCam' inequality (Lemma 2.11) along with the fact that the densities p_0 and p_1 integrate to one. For simplicity, we have written $x=\big((w_1,y_1),\ldots,(w_n,y_n)\big)$. In the sequel, H denotes the Hellinger distance as defined in (2.45). By (2.46) and (2.47), we conclude that

$$\begin{aligned}
1-\mathrm{TV}(P_0,P_1) &= \frac{1}{4}\Big(1-\frac{1}{2}H^2(f_{(W,Y);0},f_{(W,Y);1})\Big)^{2n}\\
&\ge \frac{1}{4}\Big(1-\frac{1}{2}\chi^2(f_{(W,Y);0},f_{(W,Y);1})\Big)^{2n}\\
&\ge \text{const} > 0
\end{aligned}$$

for all n sufficiently large, if the χ^2-distance converges to zero at the rate $1/n$. That follows by elementary analysis.

Summarizing, (3.17) is violated whenever (3.18) holds true and

$$\chi^2(f_{(W,Y);0},f_{(W,Y);1}) = O(1/n), \tag{3.21}$$

where $O(\cdots)$ does not depend on the parameter c, which is assumed to tend to infinity, afterwards. Thus, let us derive an upper bound on that χ^2-distance. Note that

$$
\begin{aligned}
P\big[W_1 \in A, Y_1 \in B\big] &= P\big[X_1 + \delta_1 \in A, Y_1 \in B\big] \\
&= \int_{a\in A, b\in B} \mathrm{d}P\big[X_1 + \delta_1 = a, Y_1 = b\big] \\
&= \int\int_{a\in A, b\in B} \mathrm{d}P\big[X_1 = a - s, Y_1 = b\big] g(s)\,\mathrm{d}s \\
&= \int_{a\in A, b\in B} \int f_{(X,Y);\theta}(a-s, b) g(s)\,\mathrm{d}s\mathrm{d}a\mathrm{d}b,
\end{aligned}
$$

by Fubini's theorem with $f_{(X,Y);\theta}$ as specified in (3.15). That gives us the specific convolution relation between the observation density of (W_1, Y_1) and the unobserved joint density of (X_1, Y_1), more concretely

$$
f_{(W,Y);\theta}(w,y) = \int f_{(X,Y);\theta}(w-s, y) g(s)\,\mathrm{d}s.
$$

We have

$$
\begin{aligned}
&\chi^2(f_{(W,Y);0}, f_{(W,Y);1}) \\
&= \iint f^{-1}_{(W,Y);0}(w,y)\big(f_{(W,Y);1}(w,y) - f_{(W,Y);0}(w,y)\big)^2\,\mathrm{d}w\mathrm{d}y \\
&= \iint \Big(\int f_2(y) f_1(w-s) g(s)\,\mathrm{d}s\Big)^{-1}\big(f_{(W,Y);1}(w,y) - f_{(W,Y);0}(w,y)\big)^2\,\mathrm{d}w\mathrm{d}y \\
&= \iint \big[f_1 * g\big]^{-1}(w) f_2^{-1}(y)\big(f_{(W,Y);1}(w,y) - f_{(W,Y);0}(w,y)\big)^2\,\mathrm{d}w\mathrm{d}y \\
&= C^2 \int f_2^{-1}(y)\Delta_Y^2(y)\,\mathrm{d}y \cdot \int \big[f_1 * g\big]^{-1}(w)\big|\big[\Delta_X * g\big](w)\big|^2\,\mathrm{d}w \\
&\le O\big(a_n^{-2\beta}\big) \cdot \int \big[f_1 * g\big]^{-1}(w)\big|\big[\cos(2a_n\cdot) f_0(a_n\cdot) * g\big](w)\big|^2\,\mathrm{d}w
\end{aligned}
$$

The following steps are very similar to those used in the proof of Theorem 2.13. As in (2.55), we conclude that

$$
\big[f_1 * g\big](w) \ge \text{const} \cdot (1+w^2)^{-1}
$$

holds true for all $w \in \mathbb{R}$. Using the corresponding Fourier-analytic tools as in (2.57), we may derive that

$$
\begin{aligned}
&\chi^2(f_{(W,Y);0}, f_{(W,Y);1}) \\
&\le O\big(a_n^{-2\beta}\big) \cdot \int \Big\{\big|\big[\cos(2a_n\cdot) f_0(a_n\cdot) * g\big]^{\mathrm{ft}}(t)\big|^2 \\
&\qquad\qquad + \big|\big(\big[\cos(2a_n\cdot) f_0(a_n\cdot) * g\big]^{\mathrm{ft}}\big)'(t)\big|^2\Big\}\,\mathrm{d}t \\
&\le O\big(a_n^{-2\beta-2}\big) \cdot \int \Big\{\big|f_0^{\mathrm{ft}}(t/a_n - 2) + f_0^{\mathrm{ft}}(t/a_n + 2)\big|^2 \big|g^{\mathrm{ft}}(t)\big|^2 \\
&\qquad + a_n^{-2}\Big\{\big|(f_0^{\mathrm{ft}})'(t/a_n - 2) + (f_0^{\mathrm{ft}})'(t/a_n + 2)\big|^2 \big|g^{\mathrm{ft}}(t)\big|^2\Big\} \\
&\qquad + \Big\{\big|f_0^{\mathrm{ft}}(t/a_n - 2) + f_0^{\mathrm{ft}}(t/a_n + 2)\big|^2 \big|(g^{\mathrm{ft}})'(t)\big|^2\Big\}\,\mathrm{d}t,
\end{aligned}
$$

where we have used Lemma A.1(e) and (f). We recall that f_0^{ft} is supported on $[-1,1]$; hence, the supports of $f_0^{\mathrm{ft}}(\cdot/a_n-2)$ and $f_0^{\mathrm{ft}}(\cdot/a_n+2)$ are equal to $[a_n, 3a_n]$ and $[-3a_n, -a_n]$, respectively. Therefore, we conclude that

$$\chi^2(f_{(W,Y);0}, f_{(W,Y);1}) = O\big(a_n^{-2\beta-1}\big) \cdot \max\big\{\big|g^{\mathrm{ft}}(t)\big|^2 + \big|(g^{\mathrm{ft}})'(t)\big|^2 : |t| \in [a_n, 3a_n]\big\}.$$

We impose the same conditions on g as in Theorem 2.13(a) and (b), respectively. That gives us

$$\chi^2(f_{(W,Y);0}, f_{(W,Y);1}) = \begin{cases} O\big(a_n^{-2\beta-2\alpha-1}\big), & \text{in the setting of (a)}, \\ O\big(a_n^{-2\beta-1}\big)\exp\big(-d\cdot a_n^{\gamma}\big), & \text{in the setting of (b)}, \end{cases}$$

for some constant $d>0$.

Choosing $a_n \asymp n^{1/(2\beta+2\alpha+1)}$ and $a_n = c_a \cdot \big(\log n\big)^{1/\gamma}$, with a constant $c_a>0$ sufficiently large, the condition (3.21) is verified.

With respect to condition (3.18), we derive that

$$\begin{aligned} \big|m_0(0)-m_1(0)\big|^2 &= (C')^2 \Delta_X^2(0)/f_1^2(0) \\ &\geq \mathrm{const}\cdot a_n^{-2\beta}\cos^2(2a_n\cdot 0) f_0^2(a_n\cdot 0) \\ &\geq \mathrm{const}\cdot a_n^{-2\beta}. \end{aligned}$$

Hence, for any rate $(\epsilon_n)_n$ which converges to zero faster than $a_n^{-2\beta}$, that is, $\epsilon_n/a_n^{-2\beta}$ tends to zero as $n\to\infty$, (3.18) is satisfied for n sufficiently large for any fixed but arbitrary $c>0$ so that no estimator can achieve faster rates than $a_n^{-2\beta}$ in the weak sense, given the above definition of $(a_n)_n$. As already mentioned, our proof can be extended to the estimation of $m(x)$ at any $x\in\mathbb{R}$ rather than $x=0$.

Therefore, the following theorem has been proven.

Theorem 3.5. *In the errors-in-variables regression model (3.2), assume that condition (3.14) holds. Further, assume that $\beta\geq 1$, C_1, C_3 sufficiently large and $C_2>0$ sufficiently small. Then, there exists no estimator $\hat{m}(x)$ of the regression function $m(x)$ based on the i.i.d. data $(W_1,Y_1),\ldots,(W_n,Y_n)$ generated by model (3.2), which satisfies*

$$\lim_{c\to\infty}\Big(\limsup_{n\to\infty}\sup_{(m,f_X)\in\mathcal{P}} P\big[\big|\hat{m}(x)-m(x)\big|^2 \geq c\epsilon_n\big]\Big) = 0,$$

if
(a) g is ordinary smooth (see (2.31)) and satisfies $\big|(g^{\mathrm{ft}})'(t)\big| \leq const\cdot|t|^{-\alpha}$, in addition, and $\epsilon_n = o\big(n^{-2\beta/(2\beta+2\alpha+1)}\big)$;
or
(b) g is supersmooth (see (2.32)) and satisfies $\big|(g^{\mathrm{ft}})'(t)\big| \leq const\cdot\exp(-D'|t|^{\gamma})$ for some $D'>0$, in addition, and $\epsilon_n = o\big((\log n)^{-2\beta/\gamma}\big)$.

Therefore, we have derived optimality for the same convergence rates as in the density deconvolution problem.

3.4 Berkson Regression

There is another errors-in-variables problem in nonparametric regression, which is closely related to the model (3.2), but not identical. In literature, it is usually referred to as the Berkson regression model. It was first mentioned in the paper of [1], which had been published before conventional nonparametric techniques such as kernel smoothing were introduced. The main difference between the Berkson model and (3.2) concerns the fact that, in the Berkson context, the covariate is affected by additive noise after it was measured.

Mathematically, the Berkson model may be formalized by the statistical experiment where we observe the i.i.d. data $(X_1, Y_1), \ldots, (X_n, Y_n)$, where

$$Y_j = m(X_j + \delta_j) + \varepsilon_j \,, \qquad j = 1, \ldots, n. \tag{3.22}$$

The random variables δ_j, which describe the covariate noise, are i.i.d.; we may assume that $\delta_j, X_j, \varepsilon_j$ are independent. The ε_j represent the regression error as usual. As the regression error shall be solely stochastic without any biasing effects, we may assume that the ε_j are centered and satisfy $E\varepsilon_j = 0$. Therefore, the degenerated case where all $\delta_j \equiv 0$, almost surely, corresponds to the standard nonparametric regression model.

Our goal is to estimate the regression function m.

Most approaches to the Berkson problem use parametric or semiparametric models, which includes the assumption of known shape constraints on the regression function m, which allows only for finitely many real-valued parameters. For instance, polynomial regression is considered in [71]. Other papers on the Berkson problem are given by [6] or [110]. The first nonparametric approach to the Berkson problem – to our best knowledge – is given by [32]. A recent work of [13] combines the model (3.2) and the Berkson problem by considering a regression dataset, which are affected by both Berkson noise and so-called classical error in the sense of (3.2).

To underline the practical applicability of the Berkson model, we mention the following real data problem, which is described in [32]. The data are taken from a study of the US Department of Agriculture in 1978. Observations drawn by three methods of estimating the areas growing specific crops are given. The first method is aerial photography; the corresponding measurement of n areas are denoted by $W_1, \ldots, W_n$. The second and the third methods are satellite imagery and personal interviews of the farmers, where the measurements of the same areas are reported by $Y_1, \ldots, Y_n$ and $X_1, \ldots, X_n$, respectively. The second method enjoys the best accuracy, and the third method is least precise. The first method is between the others. Now, one is interested in the prediction of the observation Y_j based on the measurement W_j. Thus, the investigation of some functional relation between those quantities is required. If the observations $(W_1, Y_1), \ldots, (W_n, Y_n)$ are directly accessible, we have a usual nonparametric regression problem. On the other hand, if only the data $(X_1, Y_1), \ldots, (X_n, Y_n)$ are available, while we are still interested in the link between W_j and Y_j, we have the Berkson regression problem (3.22) as the authors suggest modeling of W_j by $W_j = X_j + \delta_j$; hence, a noisy version of X_j. Other applications can be found in the book of [16].

To understand why (3.22) is a model where deconvolution is applicable, let us consider the conditional expectation of Y_j given $X_j = x$,

$$\begin{aligned} E\big(Y_j \mid X_j = x\big) &= E\big(m(X_j + \delta_j) \mid X_j = x\big) + E\big(\varepsilon_j \mid X_j = x\big) \\ &= Em(x + \delta_j) + E\varepsilon_j \\ &= \int m(x+s) g^{-}(s)\, \mathrm{d}s \\ &= \int m(x-s) g(s)\, \mathrm{d}s \\ &= \big[m * g\big](x), \end{aligned} \tag{3.23}$$

where $g^{-}(x) = g(-x)$, and g^{-} represents the density of the δ_j, or equivalently, g denotes the density of the random variables $-\delta_j$. Therefore, as in (3.2), it seems reasonable to assume that the density g is known at this stage, so that we can concentrate on the deconvolution step. We notice that the empirically accessible function $m_{(X,Y)}(x) = E\big(Y_j \mid X_j = x\big)$, which is, in fact, the regression function of the direct observations, is equal to the convolution of the desired regression function m and the density g. Again, we are in the general framework of statistical deconvolution problems as described in Chap. 1.

To use Fourier techniques, the regression function m must be assumed to be integrable or at least squared-integrable. Otherwise, the Fourier transform of m is not well-defined. There, we realize a major difference between Berkson regression and density deconvolution, where the target function is a density and, hence, integrable by definition in the latter setting. On the other hand, a general regression function need not be integrable on the whole real line. In the previous section where the classical errors-in-variables regression problem is addressed, we need not assume the integrability of the regression function itself but of its product with the design density. This condition is significantly weaker as any bounded regression function is admitted. However, in problem (3.2), we deconvolved the functions $p = m \cdot f_X$ and f_X, while, in the Berkson setting, the regression function m itself shall be deconvolved from the error density g. For those reasons, it seems convenient to impose the condition that g is compactly supported as in [32]. Further, this condition motivates us to use a discrete-transform model as described in the following subsection.

3.4.1 Discrete-Transform Approach

We discuss a Fourier series approach rather than Fourier integration methods as have been considered so far. Our strategy leans on the fact that any $L_2(\mathbb{R})$-function f, which is supported on some compact interval $[a,b]$, may be represented by its Fourier series (see (A.9) in the appendix). Therefore, whenever the regression function m is assumed to vanish outside the interval $[a,b]$, we have

$$m = \frac{1}{b-a} \sum_k \exp\Big(-\frac{2\pi i k}{b-a}\cdot\Big) m^{\text{ft}}\big(2\pi k/(b-a)\big)\chi_{[a,b]},$$

where the sum is to be taken over all integers k, and the infinite sum must be understood as an $L_2(\mathbb{R})$-limit. This implies that the function m may uniquely be reconstructed from its Fourier coefficients

$$m_k = m^{\text{ft}}\big(2\pi k/(b-a)\big).$$

Hence, the Fourier transform $m^{\text{ft}}(t)$ need not be known for all $t \in \mathbb{R}$, but knowledge at the discrete grid points $t_k = 2\pi k/(b-a)$, integer k, is sufficient. The fact that the function m is uniquely reconstructable by the Fourier coefficients $m^{\text{ft}}\big(2\pi k/(b-a)\big)$ is a pay-back for the condition saying that m is supported on $[a,b]$. Therefore, our strategy is estimating the m_k, which are countably many, in fact, and, then, replacing the m_k by their estimators $\hat{m}_k$ in order to estimate the function m.

From (3.23), we learn that

$$\int \exp(itx) E\big(Y_j \mid X_j = x\big)\, dx = m^{\text{ft}}(t)\, g^{\text{ft}}(t) \tag{3.24}$$

for all $t \in \mathbb{R}$, when applying Lemma A.1(b). Thus, as we observe $(X_j, Y_j), j = 1, \ldots, n$, we shall be able to produce an empirical version of the left side of the equation (3.24); that empirical version is denoted by $\hat{\psi}_h(t)$. Then, we define

$$\hat{h}_k = \hat{\psi}_h\big(2\pi k/(b-a)\big),$$

for any integer k. Then, by (3.24), it is reasonable to suggest

$$\hat{m}_k = \hat{h}_k / g^{\text{ft}}\big(2\pi k/(b-a)\big),$$

as the estimators for the coefficients m_k. The final estimator for m is defined by

$$\hat{m}(x) = \frac{1}{b-a} \sum_{|k| \le K_n} \exp\Big(-\frac{2\pi i k}{b-a} x\Big) \hat{m}_k \chi_{[a,b]}(x), \tag{3.25}$$

with some integer K_n. Hence, the infinite sum of the Fourier series representation has been reduced to a finite sum. In fact, this is required to ensure well-definiteness of the estimator. By that truncation of the Fourier coefficients, we avoid an explosion of the stochastic error by too many estimated coefficients. The band-limiting parameter K_n corresponds to the reciprocal of the bandwidth in kernel regularization methods.

The discrete transform approach may also be used in density deconvolution. Such an approach is given by [59]. In that setting, the empirical characteristic function $\hat{\psi}_h(t) = n^{-1} \sum_{j=1}^n \exp(itY_j)$ is employed, and $\hat{h}_k = \hat{\psi}_h\big(2\pi k/(b-a)\big)$. The authors use a sine–cosine series rather than the complex Fourier expansion as in (A.9), see the appendix, in particular (A.13) and (A.14), for this transformation and the resulting deconvolution methods.

In the Berkson regression context, finding an appropriate empirical version $\hat{\psi}_h(t)$ is more challenging. First, let us assume that the design density f_X is known. Then, we suggest

$$\hat{\psi}_h(t) = \frac{1}{n}\sum_{j=1}^{n} Y_j \exp(itX_j)/\max\{f_X(X_j), \rho_n(t)\},$$

with a ridge function $\rho_n(t) \geq 0$ to be selected, as an empirical version for $m^{\mathrm{ft}}(t)$. Indeed, by using some conditional expectations, we have

$$\begin{aligned}
E\hat{\psi}_h(t) &= \frac{1}{n}\sum_{j=1}^{n} E\big(Y_j \exp(itX_j)/\max\{f_X(X_j), \rho_n(t)\}\big) \\
&= E\big(E(Y_1 \mid X_1)\exp(itX_1)/\max\{f_X(X_1), \rho_n(t)\}\big) \\
&= \int E(Y_1 \mid X_1 = x)\exp(itx)\frac{1}{\max\{f_X(x), \rho_n(t)\}} f_X(x)\,\mathrm{d}x \\
&= \int Em(x+\delta_1)\exp(itx)\frac{f_X(x)}{\max\{f_X(x), \rho_n(t)\}}\,\mathrm{d}x \\
&= \int \big[m * g\big](x)\exp(itx)\frac{f_X(x)}{\max\{f_X(x), \rho_n(t)\}}\,\mathrm{d}x. \qquad (3.26)
\end{aligned}$$

Therefore, in view of the expectation, ρ_n should be very close to zero. In the extreme case $\rho_n \equiv 0$, we would have unbiasedness of the estimator, that is, $E\hat{\psi}_h(t) = m^{\mathrm{ft}}(t)g^{\mathrm{ft}}(t)$, according to Lemma A.1(b). Furthermore, the variance is bounded above as follows.

$$\begin{aligned}
&\operatorname{var}\hat{\psi}_h(t) \\
&\leq \frac{1}{n^2}\sum_{j=1}^{n} E\big|Y_j \exp(itX_j)/\max\{f_X(X_j), \rho_n(t)\}\big|^2 \\
&= \frac{1}{n}\int E(Y_1^2 \mid X_1 = x)\big[\max\{f_X(x), \rho_n(t)\}\big]^{-2} f_X(x)\chi_{(0,\infty)}(f_X(x))\,\mathrm{d}x \\
&\leq \frac{1}{n}\int \Big(\int m^2(x+u)g(-u)\,\mathrm{d}u + E\varepsilon_1^2\Big)\min\big\{f_X^{-1}(x), f_X(x)\rho_n^{-2}(t)\big\} \\
&\qquad\qquad \cdot \chi_{(0,\infty)}(f_X(x))\,\mathrm{d}x \\
&\leq \frac{1}{n}\int \big(E\varepsilon_1^2 + \big[m^2 * g\big](x)\big)\min\big\{f_X^{-1}(x), f_X(x)\rho_n^{-2}(t)\big\}\chi_{(0,\infty)}(f_X(x))\,\mathrm{d}x, \qquad (3.27)
\end{aligned}$$

where $\lambda(A)$ denotes the Lebesgue measure of some set A, and we write supp f_X for the support of the function f_X. We realize that the variance requires the reverse property of ρ_n compared to the bias term; more concretely, $\rho_n(t) > 0$ shall not be too small.

Therefore, we choose the most convenient condition where f_X is supported on some interval $[-R, R]$, $R > 0$, but

$$f_X(x) \geq c_X > 0\,, \quad \text{for all } x \in \text{supp}\,(m^2 * g).$$

The validity of that assumption implies that both m and g are compactly supported. In particular, when m is supported on some $[a, b]$ and g is supported on some $[-d, d]$, we have to assume that $R \geq \max\{|b|, |a|\} + d$. Those conditions are satisfied, for example, when f_X is the uniform density on $[-R, R]$. Then, the ridge function $\rho_n(t)$ becomes redundant, so we may select $\rho_n(t) \equiv 0$, leading to

$$E\hat{\psi}_h(t) = \int [m * g](x) \exp(itx)\,\mathrm{d}x = m^{\mathrm{ft}}(t) g^{\mathrm{ft}}(t),$$

according to (3.26) and Lemma A.1(b). Also, (3.27) gives us

$$\operatorname{var} \hat{\psi}_h(t) \leq 2Rn^{-1}c_x^{-1}\left(\|m\|_\infty^2 + E\varepsilon_1^2\right).$$

Assuming that none of the Fourier coefficients $g^{\mathrm{ft}}(2\pi k/(b-a))$, integer k, vanishes, we insert the specific form of $\hat{h}_k = \hat{\psi}_h(2\pi k/(b-a))$ into the generic estimator type (3.25) so that we define the estimator for $m(x)$ by

$$\begin{aligned} \hat{m}(x) = & \frac{1}{b-a} \sum_{|k| \leq K_n} \exp\left(-\frac{2\pi \mathrm{i} k}{b-a} x\right) \\ & \cdot \frac{1}{n} \sum_{j=1}^{n} Y_j \exp\left(\mathrm{i}2\pi k X_j/(b-a)\right) / \left[f_X(X_j)\, g^{\mathrm{ft}}\left(2\pi k/(b-a)\right)\right] \chi_{[a,b]}(x), \end{aligned} \tag{3.28}$$

with an integer-valued parameter sequence $(K_n)_n$ still to be determined. The previous results lead to the following proposition, which claims some upper bound on the MISE of estimator (3.28).

Proposition 3.6. *Consider the Berkson regression model (3.22) under the assumptions that the design density f_X is known, and the functions m, g, f_X are supported on the intervals $[a, b]$, $[-d, d]$, $[-R, R]$, respectively, with $R > \max\{|a|, |b|\} + d\}$,*

$$\begin{aligned} & f_X(x) \geq c_X > 0\,, \quad \textit{for all } x \in \textit{supp}\,(m^2 * g), \\ & E\varepsilon_1^2 \leq C_\varepsilon < \infty, \|m\|_\infty^2 \leq C_m\,, \\ & g^{\mathrm{ft}}\left(2\pi k/(b-a)\right) \neq 0, \qquad \textit{for all integer } k. \end{aligned}$$

Applying estimator $\hat{m}$ as defined in (3.28), we derive that

$$\begin{aligned} E\|\hat{m} - m\|_2^2 \leq & \frac{2R(C_m + C_\varepsilon)}{n(b-a)c_X} \cdot \sum_{|k| \leq K_n} \left|g^{\mathrm{ft}}\left(2\pi k/(b-a)\right)\right|^{-2} \\ & + \frac{1}{b-a} \sum_{|k| > K_n} \left|m^{\mathrm{ft}}\left(2\pi k/(b-a)\right)\right|^2. \end{aligned}$$

Proof. We apply the discrete version of Parseval's identity stated in (A.10), which is derived from Theorem A.7, along with the representation (A.9) with respect to m, to obtain that

$$E\|\hat{m}-m\|_2^2 = \frac{1}{b-a}\sum_{|k|\leq K_n} E\big|\hat{m}_k - m^{\mathrm{ft}}\big(2\pi k/(b-a)\big)\big|^2 + \frac{1}{b-a}\sum_{|k|>K_n} \big|m^{\mathrm{ft}}\big(2\pi k/(b-a)\big)\big|^2 .$$

Using the usual bias-variance decomposition and the bounds in (3.26) and (3.27), we derive that

$$\begin{aligned} E\big|\hat{m}_k - m^{\mathrm{ft}}\big(2\pi k/(b-a)\big)\big|^2 &= \operatorname{var}\hat{m}_k + \big|E\hat{m}_k - m^{\mathrm{ft}}\big(2\pi k/(b-a)\big)\big|^2 \\ &\leq n^{-1}2Rc_X^{-1}\,(C_m + C_\varepsilon)/\big|g^{\mathrm{ft}}\big(2\pi k/(b-a)\big)\big|^2 . \end{aligned}$$

This bound on the coefficient estimators gives us the upper bound on the MISE as stated. ■

3.4.2 Convergence Rates

To derive convergence rates we have to impose smoothness constraints on the regression function m as we did for density deconvolution in Sect. 2.4.3 and for classical errors-in-variables regression in Subsection 3.3.2. As we consider the MISE (mean integrated squared error), Sobolev conditions seem appropriate. However, we use the discrete Sobolev constraints designed for compactly supported functions as given in (A.12). Compared to the standard Sobolev conditions (2.30) where the summation over the coefficients is replaced by integration, the previous type of Sobolev assumptions have the advantage that the support boundaries are exempted from smoothness constraints; under some circumstances (in particular, smooth periodic continuability), jump discontinuities of m at the endpoints are allowed by (A.12) for any $\beta > 0$. Please see the appendix, Sect. A.3, for more detailed information. Thus, we impose

$$\sum_k \big|m^{\mathrm{ft}}\big(2\pi k/(b-a)\big)\big|^2\big(1+|k|^{2\beta}\big) \leq C \tag{3.29}$$

for some constant $C > 0$ and the smoothness degree $\beta > 0$ of m, where the sum is to be taken over all integers k. Therefore, the constant factor $1/(b-a)$ may be included in the constant C. All those functions m, which are square integrable, supported on $[a,b]$, uniformly bounded by C on the whole real line and satisfy (3.29) are collected in the Sobolev class $\mathcal{F}^D_{C,\beta}$.

With respect to the error density, we also need some conditions, especially addressing the decay of the sequence of the Fourier coefficients. In density deconvolution and classical errors-in-variables regression problems, we considered ordinary smooth and supersmooth error densities, see (2.31) and

(2.32). In our setting, however, it seems redundant to impose a positive lower bound on $|g^{\mathrm{ft}}(t)|$ for all $t \in \mathbb{R}$, since only the Fourier coefficients, that is, $g^{\mathrm{ft}}(t)$ for the discrete points $t = 2\pi k/(b-a)$, integer k, are used in the construction of estimator (3.28), leaning on the result that the function m is uniquely determined by its Fourier coefficients. Therefore, it is sufficient to restrict the assumption of a lower bound to those grid points. We define a discrete version of ordinary smooth error densities by

$$c_g/\big(1+|k|^{2\alpha}\big) \leq \big|g^{\mathrm{ft}}\big(2\pi k/(b-a)\big)\big|^2 \leq C_g/\big(1+|k|^{2\alpha}\big) \tag{3.30}$$

for all integer k and some constants $c_g, C_g > 0$ and smoothness parameter $\alpha > 0$. Also, there is an analogue for supersmooth densities in the discrete setting, and it is

$$c_g \exp\big(-d_g|k|^\gamma\big) \leq \big|g^{\mathrm{ft}}\big(2\pi k/(b-a)\big)\big| \leq C_g \exp\big(-D_g|k|^\gamma\big) \tag{3.31}$$

for all integer k and some constants $C_g, c_g, D_g, d_g > 0$. The parameter $\gamma > 0$ measures the smoothness of g.

Obviously, as no assumption about the behavior of $g^{\mathrm{ft}}(t)$ between those grid points is required, our consideration includes densities g whose Fourier transforms have some isolated zeros at specific points. As long as those zeros do not coincide with one of the grid points $2\pi k/(b-a)$, integer k, the conditions (3.30) and (3.31) can still be satisfied in some cases. This is particularly important because compactly supported densities are likely to show oscillatory behavior in the Fourier domain. This remark is also given in [32]. On the other hand, the case where $g^{\mathrm{ft}}\big(2\pi k/(b-a)\big)$ is zero for some k is discussed in [94]. In the latter work, the analyticity of m^{ft}, that is, the approximability of m^{ft} at the zeros of g^{ft} by its Taylor series, which is indeed guaranteed by the compact support of m, is used to estimate $m^{\mathrm{ft}}(t)$ at those zeros by empirical information acquired in small neighborhoods of the zeros. Then, the effect caused by the zeros of g^{ft} can be cleaned up up to deterioration of the convergence rates by a logarithmic factor. Nevertheless, in the current book, we keep the assumption of the more elementary conditions (3.30) or (3.31).

Then, the following result is implied.

Proposition 3.7. *Consider the Berkson regression model (3.22) under the assumptions that the design density f_X is known, and*

the functions g, f_X are supported on the intervals $[-d,d], [-R,R]$, respectively,
$f_X(x) \geq c_X > 0$, *for all* $x \in supp\big(m^2 * g\big)$,
$E\varepsilon_1^2 \leq C_\varepsilon < \infty$.

We apply estimator $\hat{m}$ as defined in (3.28). Then,
(a) for error densities g satisfying (3.30), we select the parameter $K_n \asymp n^{1/(2\beta+2\alpha+1)}$ and obtain that

$$\sup_{m\in\mathcal{F}_{C,\beta}^D} E\|\hat{m}-m\|_2^2 = O\big(n^{-2\beta/(2\beta+2\alpha+1)}\big).$$

(b) for error densities g satisfying (3.31), we select the parameter $K_n = (C_k \log n)^{1/\gamma}$ with $C_K \in \big(0, 1/(2d_g)\big)$ and obtain that

$$\sup_{m\in\mathcal{F}_{C,\beta}^D} E\|\hat{m}-m\|_2^2 = O\big((\log n)^{-2\beta/\gamma}\big).$$

Proof. The proof is based on Proposition 3.6, from which we learn that

$$\begin{aligned}
\sup_{m\in\mathcal{F}_{C,\beta}^D} E\|\hat{m}-m\|_2^2 &= O\big(1/n\big)\cdot \sum_{|k|\le K_n} \big|g^{\mathrm{ft}}\big(2\pi k/(b-a)\big)\big|^{-2} \\
&\qquad + O(1)\cdot \sup_{m\in\mathcal{F}_{C,\beta}^D} \sum_{|k|>K_n} \big|m^{\mathrm{ft}}\big(2\pi k/(b-a)\big)\big|^2 \\
&= O\big(1/n\big)\cdot \sum_{|k|\le K_n} \big|g^{\mathrm{ft}}\big(2\pi k/(b-a)\big)\big|^{-2} \\
&\qquad + O(1)\cdot \sup_{m\in\mathcal{F}_{C,\beta}^D} \sum_{|k|>K_n} \big|m^{\mathrm{ft}}\big(2\pi k/(b-a)\big)\big|^2 (1+|k|^{2\beta})\underbrace{(1+|k|^{2\beta})^{-1}}_{=O\left(K_n^{-2\beta}\right)} \\
&= O\big(1/n\big)\cdot \sum_{|k|\le K_n} \big|g^{\mathrm{ft}}\big(2\pi k/(b-a)\big)\big|^{-2} + O\big(K_n^{-2\beta}\big),
\end{aligned}$$

when applying the membership of m in $\mathcal{F}_{C,\beta}^D$. In case (a), it follows that

$$\sup_{m\in\mathcal{F}_{C,\beta}^D} E\|\hat{m}-m\|_2^2 = O\big(n^{-1}K_n^{1+2\alpha}\big) + O\big(K_n^{-2\beta}\big),$$

so that the convergence rates are optimized by the given selection of the sequence $(K_n)_n$. In case (b), we have

$$\sup_{m\in\mathcal{F}_{C,\beta}^D} E\|\hat{m}-m\|_2^2 = O\big(n^{-1}K_n \exp\big(2d_g K_n^\gamma\big)\big) + O\big(K_n^{-2\beta}\big).$$

Again, inserting the selection of K_n leads to the desired rates. ∎

We realize that the same rates occur as in density deconvolution (Theorem 2.9) and classical errors-in-variables regression (Theorem 3.3). The rates in part (a) of Proposition 3.7 were first established by [32] in the Berkson regression context. Also, the selection of parameter K_n, which corresponds to one by the bandwidth b, is comparable. Again, the case of ordinary smooth error densities in Proposition 3.7(a) suffers from the fact that the stated selection of K_n is nonadaptive since it contains the smoothness degree β in its definition. The discussion of data-driven selection of K_n is deferred as we have to focus on an even more severe problem first.

The construction of the estimator (3.28) and its theoretical properties such as Proposition 3.7 require full knowledge of the design density f_X, what seems unrealistic in most practical applications. Therefore, estimator (3.28) must be viewed as an oracle estimator. This limited applicability caused us to give Proposition 3.7 as a proposition rather than a theorem.

However, the design density f_X is empirically accessible by the direct data $X_1, \ldots, X_n$ given in the standard Berkson model (3.22). Therefore, it seems doable to mimic the term $1/f_X(X_j)$ by an estimator in (3.28). The quantity

$$D_j(X_j) = 2(n-1) \min\{|X_k - X_j| : k \neq j\}$$

may be proposed as the empirical version of $1/f_X(X_j)$, which shall replace $1/\max\{f_X(X_j), \rho_n(t)\}$ in the estimator (3.28). Such a technique has been roughly suggested in [58]; however, it has not been studied in the framework of that paper.

To avoid the assumption of a known design density f_X, we change the model into the fixed design case. This means that we no longer assume that the covariates $X_1, \ldots, X_n$ are observations of some i.i.d. random variables; but they are just real numbers. For a discussion on the fixed design model in general regression problems, see Sect. 3.1. Also, we assume that the design points are sufficiently dense in $[-R, R]$. In particular, we impose the condition

$$\sum_{j=1}^{n} \left|X_{(j)} - X_{(j+1)}\right|^2 + (-R - X_{(1)})^2 + (R - X_{(n)})^2 \leq C_D/n \tag{3.32}$$

for all integer n and some constant $C_D > 0$, where $X_{(1)} < \cdots < X_{(n)}$ denote the covariates $X_1, \ldots, X_n$ with ascending order. Note that this condition is satisfied by, for example, the equidistant constellation of the n design points on the interval $[-R, R]$.

For simplicity, we assume that $a = -\pi$ and $b = \pi$. Then, (3.24) may be written as the following sum over disjoint intervals

$$\begin{aligned} m^{\mathrm{ft}}(t) g^{\mathrm{ft}}(t) = \int_{\min\{-\pi-d, X_{(1)}\}}^{X_{(1)}} & \exp(itx) \big[m * g\big](x) \, \mathrm{d}x \\ & + \sum_{j=1}^{n-1} \int_{X_{(j)}}^{X_{(j+1)}} \exp(itx) \big[m * g\big](x) \, \mathrm{d}x \\ & + \int_{X_{(n)}}^{\max\{X_{(n)}, \pi+d\}} \exp(itx) \big[m * g\big](x) \, \mathrm{d}x. \end{aligned}$$

We recall that $m * g$ is supported on $[-\pi - d, \pi + d]$. Then, consider

$$\hat{\psi}_h(t) = \sum_{j=1}^{n-1} Y_{(j)} \int_{X_{(j)}}^{X_{(j+1)}} \exp(itx) \, \mathrm{d}x \tag{3.33}$$

as an empirical version of $m^{\mathrm{ft}}(t)g^{\mathrm{ft}}(t)$, where $Y_{(j)}$ denotes Y_k, which is observed at the covariate $X_{(j)}$. With respect to the expectation of $\hat{\psi}_h(t)$ while appreciating that the $X_1,\ldots,X_n$ are deterministic under the current assumptions, we obtain that

$$\begin{aligned} E\hat{\psi}_h(t) &= \sum_{j=1}^{n-1} EY_{(j)} \int_{X_{(j)}}^{X_{(j+1)}} \exp(itx)\,\mathrm{d}x \\ &= \sum_{j=1}^{n-1} \big[m*g\big](X_{(j)}) \int_{X_{(j)}}^{X_{(j+1)}} \exp(itx)\,\mathrm{d}x. \end{aligned}$$

From there, we conclude that

$$\begin{aligned} &\Big|E\hat{\psi}_h(t) - m^{\mathrm{ft}}(t)g^{\mathrm{ft}}(t)\Big| \\ &\le \Big|\int_{\min\{-\pi-d,X_{(1)}\}}^{X_{(1)}} \exp(itx)\big[m*g\big](x)\,\mathrm{d}x\Big| \\ &+ \sum_{j=1}^{n-1}\Big|\big[m*g\big](X_{(j)}) \int_{X_{(j)}}^{X_{(j+1)}} \exp(itx)\,\mathrm{d}x - \int_{X_{(j)}}^{X_{(j+1)}} \big[m*g\big](x)\exp(itx)\,\mathrm{d}x\Big| \\ &\qquad + \Big|\int_{X_{(n)}}^{\max\{X_{(n)},\pi+d\}} \exp(itx)\big[m*g\big](x)\,\mathrm{d}x\Big|, \end{aligned}$$

so that

$$\begin{aligned} &\Big|E\hat{\psi}_h(t) - m^{\mathrm{ft}}(t)g^{\mathrm{ft}}(t)\Big| \\ &\le \|m*g\|_\infty \max\{X_{(1)}+\pi+d,0\} + \|m*g\|_\infty \max\{-X_{(n)}+\pi+d,\,0\} \\ &+ \sum_{j=1}^{n-1}(X_{(j+1)}-X_{(j)}) \\ &\qquad \cdot \sup\Big\{\Big|\big[m*g\big](X_{(j)}) - \big[m*g\big](x)\Big| \,:\, x\in[X_{(j)},X_{(j+1)}]\Big\}. \end{aligned}$$

Let us assume that $m*g$ is continuously differentiable on the whole real line, and its derivative is uniformly bounded. Then, it follows that

$$\begin{aligned} &\big|E\hat{\psi}_h(t) - m^{\mathrm{ft}}(t)g^{\mathrm{ft}}(t)\big| \\ &\le \|m*g\|_\infty \max\{X_{(1)}+\pi+d,0\} + \|m*g\|_\infty \max\{-X_{(n)}+\pi+d,\,0\} \\ &\qquad + \sum_{j=1}^{n-1}(X_{(j+1)}-X_{(j)})^2\,\big\|\big[m*g\big]'\big\|_\infty, \end{aligned} \tag{3.34}$$

holds true. Equation (3.34) gives us an upper bound on the proximity of the expectation $\hat{\psi}_h(t)$ and the desired quantity $m^{\mathrm{ft}}(t)g^{\mathrm{ft}}(t)$.

Furthermore, we consider the variance of $\hat{\psi}_h(t)$. We may use standard arguments for decomposing the variance of independent random variables.

$$\begin{aligned}
\operatorname{var} \hat{\psi}_h(t) &= \operatorname{var} \sum_{j=1}^{n-1} Y_{(j)} \int_{X_{(j)}}^{X_{(j+1)}} \exp(itx)\, dx \\
&= \sum_{j=1}^{n-1} \Big| \int_{X_{(j)}}^{X_{(j+1)}} \exp(itx)\, dx \Big|^2 \operatorname{var} Y_{(j)} \\
&\leq \sum_{j=1}^{n-1} \big(X_{(j+1)} - X_{(j)}\big)^2 E|Y_{(j)}|^2 \\
&\leq \sum_{j=1}^{n-1} \big(X_{(j+1)} - X_{(j)}\big)^2 \big(\|m * g\|_\infty^2 + E\varepsilon_1^2\big).
\end{aligned}$$

Combining this result with (3.34), we obtain, by the usual decomposition of the mean squared error into the variance and the bias term, that

$$\begin{aligned}
& E\big|\hat{\psi}_h(t) - m^{\mathrm{ft}}(t) g^{\mathrm{ft}}(t)\big|^2 \\
& \leq \sum_{j=1}^{n-1} \big(X_{(j+1)} - X_{(j)}\big)^2 \big(\|m * g\|_\infty^2 + E\varepsilon_1^2\big) \\
& + \Big(\|m * g\|_\infty \max\{X_{(1)} + \pi + d, 0\} + \|m * g\|_\infty \max\{-X_{(n)} + \pi + d\,,\, 0\} \\
& \qquad\qquad + \sum_{j=1}^{n-1} \big(X_{(j+1)} - X_{(j)}\big)^2 \|[m * g]'\|_\infty\Big)^2 .
\end{aligned} \tag{3.35}$$

To estimate the regression function m, we apply the Fourier series approach again as introduced in (3.25) as a general scheme. Since we have chosen the boundaries $a = -\pi$ and $b = \pi$ specifically, we suggest to define

$$\hat{m}_k = \hat{\psi}_h(k)/g^{\mathrm{ft}}(k),$$

with $\hat{\psi}_h$ as in (3.33). Thus, we propose the estimator

$$\hat{m}_1(x) = \frac{1}{2\pi} \sum_{|k| \leq K_n} \exp\big(-ikx\big) \hat{m}_k \chi_{[-\pi,\pi]}(x)\,. \tag{3.36}$$

Unlike the previously suggested estimator (3.28), in fact, estimator (3.36) does not require knowledge of the design density, which does not exist in the assumed fixed design setting.

As a new condition, we have assumed that the derivative of $m * g$ exists and is uniformly bounded on the whole real line. We realize that

$$\big|[m*g]'(x)\big| = \Big|\int m'(x-y)g(y)\,\mathrm{d}y\Big| \le \int \big|m'(x-y)\big|g(y)\,\mathrm{d}y \le \|m'\|_\infty$$

as the density g integrates to one. Therefore, the new condition follows whenever we assume uniform boundedness of m' on the whole of $\mathbb{R}$ by an appropriate constant. This sufficient condition is included in the definition of the modified class for the regression function; we define $\mathcal{F}_{C,\beta}^{D,'}$ as the collection of all regression functions m which are contained in $\mathcal{F}_{C,\beta}^{D}$ as occurring in Proposition 3.7, and which satisfy $\|m'\|_\infty \le C$; also, we put $a = -\pi$ and $b = \pi$. Equipped with those results, we are able to give the following proposition.

Proposition 3.8. *Consider the Berkson regression model (3.22) in the fixed design case, that is, $X_1, \ldots, X_n$ are deterministic real numbers contained in some interval $[-R, R]$, which satisfy (3.32) with respect to $[-R, R], R > 0$. Furthermore, we assume that*

the functions m, g are supported on the intervals $[-\pi, \pi]$ and $[-d, d]$, respectively,
we have $R > \pi + d$,
$E\varepsilon_1^2 \le C_\varepsilon < \infty$.

We apply estimator $\hat{m}_1$ as defined in (3.36). Then,
(a) for error densities g satisfying (3.30), we select the parameter $K_n \asymp n^{1/(2\beta+2\alpha+1)}$ and obtain that

$$\sup_{m\in\mathcal{F}_{C,\beta}^{D,'}} E\|\hat{m}_1 - m\|_2^2 = O\big(n^{-2\beta/(2\beta+2\alpha+1)}\big).$$

(b) for error densities g satisfying (3.31), we select the parameter $K_n = \big(C_k \log n\big)^{1/\gamma}$ with $C_K \in \big(0, 1/(2d_g)\big)$ and obtain that

$$\sup_{m\in\mathcal{F}_{C,\beta}^{D,'}} E\|\hat{m}_1 - m\|_2^2 = O\big((\log n)^{-2\beta/\gamma}\big).$$

Proof. Let us consider the MISE under the fixed design assumption. Writing $m_k = m^{\mathrm{ft}}(k)$, we may apply the discrete version of Parseval's identity (see Theorem A.7 in the appendix) to obtain

$$\sup_{m\in\mathcal{F}_{C,\beta}^{D,'}} E\|\hat{m}_1 - m\|_2^2 = \sup_{m\in\mathcal{F}_{C,\beta}^{D,'}} \Big(\frac{1}{2\pi}\sum_{|k|\le K_n} E\big|\hat{m}_k - m_k\big|^2 + \frac{1}{2\pi}\sum_{|k|>K_n} |m_k|^2\Big)$$
$$\le \sup_{m\in\mathcal{F}_{C,\beta}^{D,'}} \frac{1}{2\pi}\sum_{|k|\le K_n} E\big|\hat{m}_k - m_k\big|^2 + O\big(K_n^{-2\beta}\big)$$

$$\leq \sup_{m\in\mathcal{F}_{C,\beta}^{D,\prime}} \frac{1}{2\pi} \sum_{|k|\leq K_n} \big|g^{\mathrm{ft}}(k)\big|^{-2} E\big|\hat{m}_k - m^{\mathrm{ft}}(k)g^{\mathrm{ft}}(k)\big|^2 + O\big(K_n^{-2\beta}\big)$$

$$\leq O\big(K_n\big)\cdot\Big(\sup_{k\in\mathbb{Z},|k|\leq K_n} \big|g^{\mathrm{ft}}(k)\big|^{-2}\Big)\cdot\Big(\sum_{j=1}^{n-1}\big(X_{(j+1)}-X_{(j)}\big)^2 + \Big[\sum_{j=1}^{n-1}\big(X_{(j+1)}-X_{(j)}\big)^2\Big]^2$$

$$+\big(\max\{X_{(1)}+\pi+d,0\}\big)^2 + \big(\max\{-X_{(n)}+\pi+d\,,\,0\}\big)^2\Big) + O\big(K_n^{-2\beta}\big)\,,$$

where we have used (3.35) in the last step. Then, we apply condition (3.32) to obtain the upper bound

$$O\big(n^{-1}K_n\big)\cdot\Big(\sup_{k\in\mathbb{Z},|k|\leq K_n}\big|g^{\mathrm{ft}}(k)\big|^{-2}\Big) + O\big(K_n^{-2\beta}\big)$$

uniformly on the MISE. The specific selection of K_n leads to the convergence rates as stated. ■

We realize that the convergence rates could be kept from Proposition 3.7 in the fixed design case where we can remove the condition of known f_X, which is a serious problem in practice. In the sequel, we concentrate on the question whether Proposition 3.8 is extendable to the random design case in the setting of Proposition 3.7. We adopt estimator $\hat{m}_1$ as defined in (3.36). At least, it is applicable under random covariate information.

Mathematically, as the random variables X_j, δ_j, and ε_j are all independent, the findings under fixed design in Proposition 3.8 may be considered as conditional calculations, given the data $X_1,\ldots,X_n$. In particular, we may derive the unconditional MISE by the equation

$$E\big\|\hat{m}_1 - m\big\|_2^2 = E\,E\big(\big\|\hat{m}_1 - m\big\|_2^2 \mid X_1,\ldots,X_n\big).$$

The inner conditional expectation in the above equation was subject of our studies in the proof of Proposition 3.8. However, we are not guaranteed that condition (3.32) still holds almost surely. Nevertheless, as this condition is only used in the last step of the proof of Proposition 3.8, we may copy the previous steps and apply the outer expectation so that we derive that

$$\sup_{m\in\mathcal{F}_{C,\beta}^{D,\prime}} E\|\hat{m}_1 - m\|_2^2$$

$$\leq O\big(K_n\big)\cdot\Big(\sup_{k\in\mathbb{Z},|k|\leq K_n}\big|g^{\mathrm{ft}}(k)\big|^{-2}\Big)\cdot\Big(\sum_{j=1}^{n-1} E\big(X_{(j+1)}-X_{(j)}\big)^2 + E\Big[\sum_{j=1}^{n-1}\big(X_{(j+1)}-X_{(j)}\big)^2\Big]^2 + E\big(\max\{X_{(1)}+\pi+d,0\}\big)^2$$

$$+ E\big(\max\{-X_{(n)}+\pi+d\,,\,0\}\big)^2\Big) + O\big(K_n^{-2\beta}\big). \quad (3.37)$$

We may use the fact that

$$\sum_{j=1}^{n-1}\big(X_{(j+1)}-X_{(j)}\big)^2 \le 2R\cdot\sum_{j=1}^{n-1}\big(X_{(j+1)}-X_{(j)}\big) \le 4R^2$$

holds almost surely to derive that

$$E\Big(\sum_{j=1}^{n-1}\big(X_{(j+1)}-X_{(j)}\big)^2\Big)^2 \le O(1)\cdot E\sum_{j=1}^{n-1}\big(X_{(j+1)}-X_{(j)}\big)^2.$$

Therefore, the investigation of the second moment of $\sum_{j=1}^{n-1}\big(X_{(j+1)}-X_{(j)}\big)^2$ reduces to that of the first absolute moment.

To analyze the expectation $E\big(X_{(j+1)}-X_{(j)}\big)^2$, we consider that

$$\sum_{j=1}^{n-1}\big(X_{(j+1)}-X_{(j)}\big)^2 = \sum_{j=1}^{n}\big[\min\{X_k : X_k > X_j\}-X_j\big]^2,$$

holds almost surely. If the set $\{X_k : X_k > X_j\}$ is empty, we conventionally put $\min\{X_k : X_k > X_j\} = X_j$. Then, the validity of the above equality can be verified by permutation of the addends. Thus,

$$\begin{aligned}
E\sum_{j=1}^{n-1}\big(X_{(j+1)}-X_{(j)}\big)^2 &= \sum_{j=1}^{n}E\big[\min\{X_k : X_k > X_j\}-X_j\big]^2\\
&= \sum_{j=1}^{n}\int_{s>0}P\big[\big(\min\{X_k : X_k > X_j\}-X_j\big)^2 > s\big]\,\mathrm{d}s\\
&= \sum_{j=1}^{n}\int_{s>0}P\big[\min\{X_k : X_k > X_j\} > X_j+\sqrt{s}\,\big]\,\mathrm{d}s\\
&= \sum_{j=1}^{n}\int_{s>0}E\,P\big[\min\{X_k : X_k > X_j\} > X_j+\sqrt{s}\mid X_j\big]\,\mathrm{d}s\\
&= \sum_{j=1}^{n}\int_{s=0}^{4R^2}\int_{x=-R}^{R}P\big[\min\{X_k : X_k > x\} > x+\sqrt{s}\mid X_j = x\big]f_X(x)\,\mathrm{d}x\mathrm{d}s.
\end{aligned}$$

Those restrictions of the integration domain are justified by the inclusion of the support of f_X in $[-R,R]$. We realize that $\min\{X_k : X_k > x\} > x+\sqrt{s}$, under the condition $X_j = x$, is equivalent to the assertion that none of the X_k is contained in $(x, x+\sqrt{s}\,]$; at least one of the X_k except X_j lies in $(x+\sqrt{s},\infty)$ and the all other X_k are contained in $(-\infty, x]$. By an elementary result from probability theory, we recognize a multinomial distribution so that

$$
\begin{aligned}
& P\big[\min\{X_k : X_k > x\} > x + \sqrt{s} \mid X_j = x\big] \\
&= \sum_{k=1}^{n-1} \frac{(n-1)!}{k!(n-1-k)!} \Big(\int_{-\infty}^{x} f_X(y)\,\mathrm{d}y\Big)^{n-1-k} \Big(\int_{x+\sqrt{s}}^{\infty} f_X(y)\,\mathrm{d}y\Big)^{k} \\
&= \Big(1 - \int_{x}^{x+\sqrt{s}} f_X(y)\,\mathrm{d}y\Big)^{n-1} - \Big(\int_{-\infty}^{x} f_X(y)\,\mathrm{d}y\Big)^{n-1}.
\end{aligned}
$$

Then, the above inequality may be continued by

$$
\begin{aligned}
& E\sum_{j=1}^{n-1} \big(X_{(j+1)} - X_{(j)}\big)^2 \\
&\quad \le n \int_{s=0}^{4R^2} \int_{x=-R}^{R} \Big(1 - \int_{x}^{x+\sqrt{s}} f_X(y)\,\mathrm{d}y\Big)^{n-1} f_X(x)\,\mathrm{d}x\mathrm{d}s.
\end{aligned}
$$

Then, the above inequality is continued by Fubini's theorem and the definition $S_n = c(\log n)^2/n^2$ with some constant $c > 0$,

$$
\begin{aligned}
& \le n \int_{x=-R+S_n}^{R-S_n} \int_{s=0}^{4R^2} \Big(1 - \int_{x}^{x+\sqrt{s}} f_X(y)\,\mathrm{d}y\Big)^{n-1} f_X(x)\,\mathrm{d}s\mathrm{d}x \\
& \quad + n \int_{x\in[-R,R]\backslash[-R+S_n,R-S_n]} \int_{s=0}^{4R^2} \Big(1 - \int_{x}^{x+\sqrt{s}} f_X(y)\,\mathrm{d}y\Big)^{n-1} f_X(x)\,\mathrm{d}s\mathrm{d}x \\
& \le n \int_{x=-R+S_n}^{R-S_n} \int_{s=0}^{4R^2} \Big(1 - \int_{x}^{x+\sqrt{s}} f_X(y)\,\mathrm{d}y\Big)^{n-1} f_X(x)\,\mathrm{d}s\mathrm{d}x \,+\, O\big((\log n)^2/n\big) \\
& \le n \int_{x=-R+S_n}^{R-S_n} \int_{s=0}^{S_n} \max\big\{1 - c_X\sqrt{s}, 0\big\}^{n-1} f_X(x)\,\mathrm{d}s\mathrm{d}x \\
& \quad + n \int_{x=-R+S_n}^{R-S_n} \int_{s=S_n}^{4R^2} \max\big\{1 - c_X\sqrt{S_n}, 0\big\}^{n-1} f_X(x)\,\mathrm{d}s\mathrm{d}x \,+\, O\big((\log n)^2/n\big) \\
& \le O\big((\log n)^2/n\big) \,+\, o(n^{-1}) \,=\, O\big((\log n)^2/n\big),
\end{aligned}
$$

when choosing the constant $c > 0$ sufficiently large in the definition of S_n. This upper bound is uniform in f_X. We have used the conditions assumed in Proposition 3.7.

In view of (3.37), the term $E(\max\{X_{(1)}+\pi+d\,,\,0\})^2$ remains to be studied. We have $X_{(1)} = \min\{X_1,\ldots,X_n\}$. Consider that

$$
\begin{aligned}
E\big(\max\{X_{(1)} + \pi + d\,,\,0\}\big)^2 &= \int_{s=0}^{\infty} P\big[\min\{X_1,\ldots,X_n\} > \sqrt{s} - \pi - d\big]\,\mathrm{d}s \\
= \int_{s=0}^{\infty} \big(P\big[X_1 > \sqrt{s} - \pi - d\big]\big)^n\,\mathrm{d}s &\le \int_{s=0}^{\infty} \big(P\big[X_1 > \sqrt{s} - R\big]\big)^n\,\mathrm{d}s
\end{aligned}
$$

$$= \int_{s=0}^{\infty} \Big(1 - \int_{x=-R}^{\sqrt{s}-R} f_X(x)\,\mathrm{d}x\Big)^n \mathrm{d}s \leq \int_{s=0}^{S} \big(1 - c_X\sqrt{s}\big)^n \mathrm{d}s + o(n^{-1})$$
$$\leq \frac{1}{c_X(n+1)}\Big(1 - \big(1 - c_X\sqrt{S}\big)^{n+1}\Big) + o(n^{-1}) = O(1/n)\,,$$

for some fixed $S > 0$ sufficiently small. The term $E\big(\max\{-X_{(n)} + \pi + d\,, 0\}\big)^2$ can be treated analogously. Again, the bounds are uniform in f_X.

Therefore, we come to the following conclusions: for ordinary smooth error densities (3.30), we observe some slight loss in the speed of convergence by the factor $O\big((\log n)^2\big)$ for estimator (3.36) in the random design case with unknown design density f_X. In the paper of [32], the authors claim that the convergence rates of Proposition 3.7 can be kept even without assuming knowledge of f_X by appropriate estimation of $m * g$ by local linear smoothing. Obviously, their procedure requires the use of a second bandwidth parameter (in addition to the smoothing parameter K_n) for the linear smoother, whose appropriate selection is not an easy task. On the other hand, the estimator $\hat{m}_1$ as defined in (3.36) does not require any additional smoothing, while the logarithmic factor must be seen as a disadvantage. However, there might be some potential to reduce the power of the logarithm or even to completely remove this disturbing factor. We leave this question open. However, no deterioration of the convergence rates of the estimator (3.36) occurs in the supersmooth error and random design case (Proposition 3.7(b)) because it can be shown that the additional logarithmic factor affects only the variance term, which converges to zero with an algebraic rate and, hence, is dominated by the logarithmic bias term.

Furthermore, we realize the problem that the optimal selection of parameter K_n in the ordinary smooth error case (Proposition 3.8(a)) requires knowledge of the commonly unknown smoothness degree β of the regression function m. Therefore, we shall consider some data-driven procedure to select this parameter suitably; as we did in Sect. 2.5 in density deconvolution. In standard error-free nonparametric regression, which is included in our framework by assuming that $\delta_j \equiv 0$ almost surely, the usual cross-validation (CV) function is equal to

$$\mathrm{CV}(K) = \sum_{j=1}^{n} \big|Y_j - \hat{m}_{-j}(X_j; K)\big|^2\,,$$

defined for all integers K, where $\hat{m}_{-j}(X_j; K)$ denotes a regression estimator $\hat{m}(x)$ for $x = X_j$ when putting the smoothing parameter K_n equal to K, and the observation (X_j, Y_j) is omitted to avoid stochastic dependence between the estimator and its argument $x{=}X_j$. Therefore, estimator $\hat{m}_{-j}(\cdot; K)$ is based on the sample $(X_1, Y_1), \ldots, (X_{j-1}, Y_{j-1}), (X_{j+1}, Y_{j+1}), \ldots, (X_n, Y_n)$ only. Then, the parameter $K_n = \hat{K}_n$ is defined by

$$\hat{K}_n = \arg\min_K \mathrm{CV}(K).$$

For literature on CV in standard nonparametric regression, see, for example, [111].

In the Berkson setting, however, we face the problem that the appropriate CV-function is equal to

$$\mathrm{CV}(K) = \sum_{j=1}^{n} \big|Y_j - \hat{m}_{1;-j}(X_j + \delta_j; K)\big|^2,$$

with $\hat{m}_1$ as defined in (3.36), where the random variables δ_j are not available. In [32], the authors mention that problem and suggest to replace the unknown quantity $\hat{m}_{1;-j}(X_j + \delta_j; K)$ by the accessible conditional expectation

$$E\big(\hat{m}_{1;-j}(X_j + \delta_j; K) \mid X_1, \ldots, X_n\big),$$

where that expectation is calculable by integration where the density $-g$ of the δ_j is known. The numerical simulations in [32] indicate that smoothing parameter selection based on this modified CV-function shows satisfactory performance, although a rigorous asymptotic theory for the Berkson problem is missing so far.

An optimality proof of the rates in Proposition 3.7(a) is given in [32].

We also mention that it seems impossible to relax the condition saying that the design density f_X is bounded away from zero on the support of $h = m * g$ without losing speed of convergence since, otherwise, the function h to be deconvolved is hardly accessible at those points where f_X is low. In particular, this affects the situation where m or g are not compactly supported. In the paper of [13], such problems are discussed. The authors suggest kernel methods for noncompactly supported regression functions; however, they mention that the rates from compactly supported models cannot be maintained.

A problem, which is closely related to the Berkson regression setting, is the problem of *inverse regression*. In that latter model, we observe the i.i.d. data $(X_1, Y_1), \ldots, (X_n, Y_n)$ where

$$Y_j = \big[g * f\big](X_j) + \varepsilon_j,$$

where the regression error ε_j and the design variables X_j are independent. There exists also a fixed design version where the covariates $X_1, \ldots, X_n$ are assumed to be deterministic. In a multivariate context, this problem is also related to image deblurring, which will be studied in the following chapter.

4

Image and Signal Reconstruction

4.1 Discrete Observation Scheme and Blind Deconvolution

Problems of reconstructing images or signals, which are affected by random noise, may also be viewed as problems in the field of nonparametric statistics. Then, one may consider a noisy image or signal as an empirical observation, where the true image or signal is the quantity that we are interested in.

Such problems of signal recovery can be classified by the dimensionality of the underlying signal. Conventional images are usually observed on a plane so that the dimension d is two. In this case, we have a bivariate problem. Models in which the dimension of the image is $d > 1$ are called multivariate problems; spatial problems, that is, $d = 3$, also have their applications; while, of course, the most elementary – and also useful – situation is the univariate problem $d = 1$.

In the current section, we assume that the image is observed at a finite number of points only. We classify the following model of the true images and the corresponding observation scheme:

(PGI) Pixel grid image. Such an image consists of finitely many grid points only. To describe the model mathematically, we denote the grid points by $x_1, x_2, \ldots \in \mathbb{R}^d$. In many applications, they may be assumed to be located in the d-dimensional cube $[-1,1]^d$. At each grid point, we observe the brightness or the intensity of the image, which may be represented by a real-valued random variable Y_j, j. Apparently, Y_j may be written as the true (but unobserved) intensity $y_j \in \mathbb{R}$ of the image at the grid point x_j, contaminated by some i.i.d. random variable ε_j, which is centered by the condition $E\varepsilon_j = 0$, that is, $Y_j = y_j + \varepsilon_j$. Our goal is to detect the true values $y_1, y_2, \ldots$ of the image. Pixel grid image occur in electronic imaging and scanning, in particular. Of course, any computer is able to save only finitely many objects.

Note that the model (PGI) described above may be seen as a regression problem. Pixel grid images only allow for countably many parameters, which

A. Meister, *Deconvolution Problems in Nonparametric Statistics*, Lecture Notes in Statistics 193, DOI 10.1007/978-3-540-87557-4_4,

determines the image uniquely. However, the pixels are assumed to be very dense; we will make this condition precise later.

Model (PGI) is not classified as a deconvolution problem. In this setting, the only perturbance of the true image is caused by random noise, which contaminates the intensity of any grid point independently and additively. However, as another kind of corrupting effects, a pointspread phenomenon may occur; it is also often referred to as a blurring effect. This means that a point in the true image is not mapped to a point in the recorded image; but the intensity mass, which is concentrated in one point of the true image, spreads so that all other grid points are also influenced by the true brightness of point x_j.

Therefore, we choose the linear model where the intensity of the blurred image Y_j at the grid point x_j is the sum of the specific contributions of the true brightness at each grid point, more precisely, we have

$$Y_i = \sum_j g_{i,j} y_j, \quad \text{integer } i,$$

where the sum is to be taken over all integers j. At this stage, our model is purely deterministic and does not suffer from any random contamination. The weight coefficients $g_{i,j}$ describe the impact of how the true brightness y_j at the grid point x_j influences the recorded brightness Y_j at x_i. Since the total intensity should be preserved by pointspread effects, we assume the scaling condition

$$\sum_j g_{i,j} = 1,$$

for all integer i. The function g is called the blurring function or the pointspread function.

Throughout this chapter, we assume that the shape of the pointspread effects do not depend on the specific grid point; but are spatially homogeneous. Hence, the vector $x_i - x_j$ shall determine the weight $g_{i,j}$ uniquely. Therefore, we may postulate the existence of a function $g : \mathbb{R}^d \to \mathbb{R}$ so that the representation

$$g_{i,j} = g(x_i - x_j)$$

is justified. Consequently, we may write

$$Y_j = \sum_j g(x_i - x_j) y_j. \tag{4.1}$$

On the other hand, Y_j and y_j may also be represented as $Y(x_j)$ and $y(x_j)$, respectively, where Y and y are functions mapping $\mathbb{R}^d$ to $\mathbb{R}$. Note that all values of Y and y except those taken at the grid points; and all values of g except those taken at the points $x_i - x_j$, for integer i, j, are meaningless and may be chosen arbitrarily. To satisfy the scaling condition, we assume that

$$\sum_j g(x_i - x_j) = 1, \tag{4.2}$$

for all grid points x_i. Then, (4.1) is a discrete convolution equation.

In practical applications, we shall appreciate that the quality of a recorded image may suffer from both blurring and random noise effects. Therefore, the considered model in discrete image deconvolution, for example, (PGI) under additional blurring, may be interpreted as the statistical experiment where the data (x_j, Y_j), integer j, with

$$Y(x_j) = \sum_k g(x_j - x_k)y(x_k) + \varepsilon_j\,, \tag{4.3}$$

are observed. That model is discussed and used in, for example, [11] or [60]. Also, see the book of [108] for a comprehensive overview on discrete image deblurring and related problems.

In the sequel, we study the framework and the methodology for the model (4.3), which has been suggested by [60]. In their work, the authors take into account the realistic condition that the pointspread function g might be unknown. Such models are often referred to as *blind deconvolution*. In literature, g is sometimes described by a parametric model, that is, the framework for g allows only finitely many parameters to be unknown, see, for example, [12]. Other classes of papers assume that the true image y has known support and the background is uniform. In that setting, the pointspread function g is estimable from observation drawn from the uniform background; and only weak nonparametric conditions on g are required, see, for example, [79] and [128] to give a brief survey on some other approaches to blind deconvolution.

Following the framework of [60], we assume equidistant location of the grid points and choose the grid $\mathbb{Z}^d$ as the domain of the blurring function g. Hence, (4.3) becomes

$$Y(j) = \sum_k g(j-k)y(k) + \varepsilon_j, \tag{4.4}$$

where the sum is to be taken over all $k \in \mathbb{Z}^d$, and $j \in \mathbb{Z}^d$. Furthermore, the approach

$$g(j) = n^{-d} C_d \tilde{g}(j/n) \tag{4.5}$$

is used with the renormalizing factor

$$C_d = \Big(n^{-d} \sum_{j \in \mathbb{Z}^d} \tilde{g}(j/n)\Big)^{-1},$$

and a (Lebesgue) density function g being defined on the domain $\mathbb{R}^d$. The definition of the latter factor guarantees that the blurring process is intensity-preserving, that is,

$$\sum_{j \in \mathbb{Z}^d} g(j) = 1.$$

Then, n^{-1}, with a large integer n, describes the pixel width; therefore, n measures the denseness of the grid. Hence, good estimation results may be expected when n is very large. As $n \to \infty$, we have $C_d \to 1$ so that the scaling constant becomes less essential for large n. That setting describes the approximation of a continuous signal by a pixel grid image.

We assume that $\tilde{g}$ is supported on the d-dimensional ball around the origin with the radius λ_n/n. This assumption covers the realistic condition that the blurring effects are restricted to some bounded neighborhood around the true point whose light intensity spreads; and the support of g is included in a ball R_n with the origin as its center and the radius $\lambda_n > 0$. Also, we assume that the true image y is supported on a known bounded set S.

To solve the convolution equation (4.4), Fourier techniques seem appropriate again. We assume that n/λ_n is bounded, and that S is included in a ball around the origin of the radius $\leq O(n)$. Then, the blurred and noisy image Y vanishes outside the set T, defined by

$$T = \{j + k : j \in R, k \in S\}.$$

A more specific version of the Fourier transform is introduced, which we denote by grid-Fourier transform. It is defined by

$$\xi^{\mathrm{Ft}}(t) = \sum_{j \in \mathbb{Z}^d} \xi(j) \exp\big(\mathrm{i}t \cdot j\big), \quad t \in \mathbb{R}^d, \tag{4.6}$$

for any function ξ defined on $\mathbb{Z}^d$, where the notation ξ^{Ft} helps to distinguish between the grid-Fourier transform and the previously used Fourier transform ξ^{ft} as has been introduced in Chap. 1. The product $j \cdot t$ is to be understood as the standard inner product in $\mathbb{R}^d$ – also known as the dot-product. We can show that the grid-Fourier transform satisfies the central deconvolution formula (Lemma A.1(b) and Lemma A.5) in the discrete-grid interpretation of convolution. That means

$$\begin{aligned}
Y^{\mathrm{Ft}}(t) &= \sum_j \exp(\mathrm{i}j \cdot t) Y(j) \\
&= \sum_j \exp(\mathrm{i}j \cdot t) \sum_k g(j-k) y(k) \\
&= \sum_k y(k) \exp(\mathrm{i}k \cdot t) \sum_j \exp\big(\mathrm{i}(j-k) \cdot t\big) g(j-k) \\
&= \sum_k y(k) \exp(\mathrm{i}k \cdot t) \sum_l \exp\big(\mathrm{i}l \cdot t\big) g(l) \\
&= y^{\mathrm{Ft}}(t) \cdot g^{\mathrm{Ft}}(t).
\end{aligned} \tag{4.7}$$

Again, all indefinite sums shall be taken over the whole of $\mathbb{Z}^d$. We have used that the set $\mathbb{Z}^d$ is invariant with respect to an additive shift in $\mathbb{Z}^d$, that is, $\{x + y : y \in \mathbb{Z}^d\} = \mathbb{Z}^d$ for all $x \in \mathbb{R}^d$. Therefore, in the Fourier domain, the grid-Fourier transform of the true image y^{Ft} can be reconstructed from Y^{Ft} by simple division by $g^{\mathrm{Ft}}(t)$. Again, at this stage, we have to assume that the blurring function g is known. It corresponds to the error density g in density deconvolution.

Of course, the sum over all $j \in \mathbb{Z}^d$ occurring in the definition of the grid-Fourier transform may be reduced to the sum taken over those $j \in \mathbb{Z}^d$, which are contained in the support of the function to be transformed. For instance, we have

$$g^{\mathrm{Ft}}(t) = \sum_{j \in R} g(j) \exp(\mathrm{i} j \cdot t),$$

$$y^{\mathrm{Ft}}(t) = \sum_{j \in S} y(j) \exp(\mathrm{i} j \cdot t),$$

$$Y^{\mathrm{Ft}}(t) = \sum_{j \in T} Y(j) \exp(\mathrm{i} j \cdot t).$$

The deconvolution equality (4.7) may be applied to our model equation (4.4), so that we have

$$Y^{\mathrm{Ft}}(t) = g^{\mathrm{Ft}}(t) y^{\mathrm{Ft}}(t) + \varepsilon^{\mathrm{Ft}}(t) \tag{4.8}$$

for all $t \in \mathbb{R}^d$, where we define

$$\varepsilon^{\mathrm{Ft}}(t) = \sum_{j \in T} \varepsilon_j \exp(\mathrm{i} j \cdot t).$$

Note that the restricted summation to the finite set T is justified by the assumption that Y is supported on T. This guarantees that only finitely many i.i.d. random variables $\varepsilon_1, \varepsilon_2, \ldots$ are involved. By an abuse of the notation, we denote that ε_k, $k \in \mathbb{Z}$, which occurs at the grid point $x_k = j \in \mathbb{Z}^d$, by ε_j.

Equation (4.8) underlines that estimating the pointspread function g from a known true image y, on the one hand, and estimating the true image y from a known pointspread function g, on the other hand, are generally the same statistical problem. Differences may occur only due to different assumptions on the shape of g and y. Hall and Qiu [60] mention that the blurring function g is usually assumed to be smooth, while the true image y is usually allowed to have some jump discontinuities coming from the frontiers of some recorded objects.

The strategy suggested by [60] is as follows: First, one shall estimate the pointspread function g of an underlying measurement system – in the present context, we may interpret the measurement system as a camera – by taking a picture of a known test image or pattern. A rather simple image is recommended such as a square-block test pattern. Thus, with respect to this experiment, the true image y' is known and the outcome image Y' is observed. By (4.8), the blurring function g is accessible from known y' and Y'. Assuming that the pointspread function g is a characteristic of the camera, which does not change in time, the information about g, which is acquired by the first experiment, can be used in further experiments. Now, assume we want to reconstruct a true real-life image y from a blurred and noisy observation Y, that is, a picture taken by the camera, which has been tested before. We may insert the empirical version of the pointspread function g to estimate the true real image y based on Y.

According to (4.8), it seems reasonable to consider

$$\hat{g}^{\mathrm{Ft}}(t) = (Y')^{\mathrm{Ft}}(t)(y')^{\mathrm{Ft}}(-t)|(y')^{\mathrm{Ft}}(t)|^r / \max\{|(y')^{\mathrm{Ft}}(t)|, \rho(t)\}^{r+2} \tag{4.9}$$

as an empirical version of $g^{\mathrm{Ft}}(t)$. The function $\rho(t) > 0$ is a so-called ridge function. The estimator (4.9) is the analogue of the ridge-parameter estimator as used in density deconvolution, see Sect. 2.2.3. Furthermore, some parameter $r \geq 0$ is introduced. That regularization is used to avoid division by zero when we are not guaranteed that $(y')^{\mathrm{Ft}}$ is nonvanishing. In fact, we are dealing with compactly supported blurring functions y' where one frequently faces the problem of periodic zeros of $(y')^{\mathrm{Ft}}$. Note that if $|(y')^{\mathrm{Ft}}(t)| > \rho(t)$, we have the elementary empirical version of g^{Ft}, namely

$$\hat{g}^{\mathrm{Ft}}(t) = (Y')^{\mathrm{Ft}}(t)/(y')^{\mathrm{Ft}}(t).$$

Otherwise, $(y')^{\mathrm{Ft}}$ is close to zero and the regularization becomes efficient.

The problem of reconstructing g from its grid-Fourier transform g^{Ft} can be solved by considering the multivariate Fourier series of g^{Ft}, see, for example, Sect. A.4 of the appendix along with the representation (A.20). We have

$$\begin{aligned}(2\pi)^{-d}\int_A \exp(-\mathrm{i}t\cdot j) g^{\mathrm{Ft}}(t)\,\mathrm{d}t &= \sum_k g(k)(2\pi)^{-d}\int_A \exp(\mathrm{i}t\cdot k)\exp(-\mathrm{i}t\cdot j)\,\mathrm{d}t \\ &= \sum_k g(k)\delta_{k,j} = g(j),\end{aligned}$$

hence, a simple way of Fourier inversion is possible in this setting, where $A = [-\pi,\pi]^d$. Then, the estimator for the pointspread function g, which is proposed by [60], is defined by

$$\begin{aligned}\hat{g}(j) = (2\pi)^{-d}\int_A &\exp(-\mathrm{i}j\cdot t)(Y')^{\mathrm{Ft}}(t)(y')^{\mathrm{Ft}}(-t)|(y')^{\mathrm{Ft}}(t)|^r \\ &/ \max\{|(y')^{\mathrm{Ft}}(t)|, \rho(t)\}^{r+2}\,\mathrm{d}t. \end{aligned}\tag{4.10}$$

To study the quality of this estimator, [60] introduce the MSSE-criterion (Mean Summed Squared Error); it is defined by

$$\mathrm{MSSE}(\hat{g}, g) = E\sum_{j\in\mathbb{Z}^d}\big|\hat{g}(j) - g(j)\big|^2 = \sum_{j\in\mathbb{Z}^d} E\big|\hat{g}(j) - g(j)\big|^2.$$

As the domain of the true image is a discrete grid, the integral is replaced by a sum, compared to the MISE-criterion, which is used in density estimation. To give an upper bound on the MSSE, we may split the estimator (4.10) into a deterministic and a random noise part, that is,

$$\begin{aligned}\hat{g}(j) &= (2\pi)^{-d} \int_A \exp(-ij \cdot t) g^{\mathrm{ft}}(t) |(y')^{\mathrm{Ft}}(t)|^{r+2} \\ &\qquad\qquad / \max\{|(y')^{\mathrm{Ft}}(t)|, \rho(t)\}^{r+2}\, \mathrm{d}t \\ &+ (2\pi)^{-d} \int_A \exp(-ij \cdot t) \varepsilon^{\mathrm{Ft}}(t) (y')^{\mathrm{Ft}}(-t) |(y')^{\mathrm{Ft}}(t)|^{r} \\ &\qquad\qquad / \max\{|(y')^{\mathrm{Ft}}(t)|, \rho(t)\}^{r+2}\, \mathrm{d}t,\end{aligned}$$

by (4.8). Note that

$$E\varepsilon^{\mathrm{Ft}}(t) = \sum_{j \in T} E\varepsilon_j \exp(ij \cdot t) = 0$$

for all $t \in \mathbb{R}^d$. It follows from there that

$$\begin{aligned}E\hat{g}(j) = (2\pi)^{-d} \int_A \exp(-ij \cdot t) g^{\mathrm{ft}}(j) |(y')^{\mathrm{Ft}}(t)|^{r+2} \\ / \max\{|(y')^{\mathrm{Ft}}(t)|, \rho(t)\}^{r+2}\, \mathrm{d}t,\end{aligned}$$

so that

$$\begin{aligned}&\left| g(j) - E\hat{g}(j) \right|^2 \\ &= \left| (2\pi)^{-d} \int_A \exp(-ij \cdot t) g^{\mathrm{ft}}(t) \left(1 - \frac{\left|(y')^{\mathrm{Ft}}(t)\right|^{r+2}}{\max\{\left|(y')^{\mathrm{Ft}}(t)\right|, \rho(t)\}^{r+2}} \right) \mathrm{d}t \right|^2.\end{aligned}$$

With respect to the variance term, we use the result from probability theory that any deterministic shift of a random variable can be removed when calculating the variance. Hence, we derive that

$$\begin{aligned}\operatorname{var} \hat{g}(j) &= \operatorname{var}\Big((2\pi)^{-d} \int_A \exp(-ij \cdot t) \varepsilon^{\mathrm{Ft}}(t) (y')^{\mathrm{Ft}}(-t) |(y')^{\mathrm{Ft}}(t)|^{r} \\ &\qquad\qquad / \max\{|(y')^{\mathrm{Ft}}(t)|, \rho(t)\}^{r+2}\, \mathrm{d}t \Big) \\ &= \operatorname{var}\Big(\sum_{k \in T} \varepsilon_k\, (2\pi)^{-d} \int_A \exp\big(i(k-j) \cdot t\big) (y')^{\mathrm{Ft}}(-t) |(y')^{\mathrm{Ft}}(t)|^{r} \\ &\qquad\qquad / \max\{|(y')^{\mathrm{Ft}}(t)|, \rho(t)\}^{r+2}\, \mathrm{d}t \Big) \\ &\leq \sum_{k \in T} E\, \varepsilon_k^2 \Big| (2\pi)^{-d} \int_A \exp\big(i(k-j) \cdot t\big) (y')^{\mathrm{Ft}}(-t) |(y')^{\mathrm{Ft}}(t)|^{r} \\ &\qquad\qquad / \max\{|(y')^{\mathrm{Ft}}(t)|, \rho(t)\}^{r+2}\, \mathrm{d}t \Big|^2,\end{aligned}$$

using the independence of the ε_j. We may assume that the second moment of the ε_k exists; it is denoted by σ^2.

Combining those results, we may apply the following decomposition of the MSSE.

$$
\begin{aligned}
\mathrm{MSSE}(\hat{g},g) &= \sum_{j\in\mathbb{Z}} \mathrm{var}\,\hat{g}(j) + \big|E\hat{g}(j) - g(j)\big|^2 \\
&= \sigma^2 \sum_{k\in T}\sum_{j\in\mathbb{Z}} \Big|(2\pi)^{-d}\int_A \exp\big(\mathrm{i}(k-j)\cdot t\big)(y')^{\mathrm{Ft}}(-t)|(y')^{\mathrm{Ft}}(t)|^r \\
&\qquad\qquad / \max\{|(y')^{\mathrm{Ft}}(t)|,\rho(t)\}^{r+2}\,\mathrm{d}t\Big|^2 \\
&\quad + \sum_{j\in\mathbb{Z}} \Big|(2\pi)^{-d}\int_A \exp(-\mathrm{i}j\cdot t)g^{\mathrm{ft}}(t)\left(1 - \frac{\big|(y')^{\mathrm{Ft}}(t)\big|^{r+2}}{\max\{\big|(y')^{\mathrm{Ft}}(t)\big|,\rho(t)\}^{r+2}}\right)\mathrm{d}t\Big|^2 \\
&= \sigma^2\,(\#T)\sum_{j\in\mathbb{Z}} \Big|(2\pi)^{-d}\int_A \exp\big(-\mathrm{i}j\cdot t\big)(y')^{\mathrm{Ft}}(-t)|(y')^{\mathrm{Ft}}(t)|^r \\
&\qquad\qquad / \max\{|(y')^{\mathrm{Ft}}(t)|,\rho(t)\}^{r+2}\,\mathrm{d}t\Big|^2 \\
&\quad + (2\pi)^{-d}\int_A \big|g^{\mathrm{ft}}(t)\big|^2 \left(1 - \frac{\big|(y')^{\mathrm{Ft}}(t)\big|^{r+2}}{\max\{\big|(y')^{\mathrm{Ft}}(t)\big|,\rho(t)\}^{r+2}}\right)^2 \mathrm{d}t \\
&= (2\pi)^{-d}\sigma^2\,(\#T)\int_A |(y')^{\mathrm{Ft}}(t)|^{2+2r} / \max\{|(y')^{\mathrm{Ft}}(t)|,\rho(t)\}^{2r+4}\,\mathrm{d}t \\
&\quad + (2\pi)^{-d}\int_A \big|g^{\mathrm{ft}}(t)\big|^2 \left(1 - \frac{\big|(y')^{\mathrm{Ft}}(t)\big|^{r+2}}{\max\{\big|(y')^{\mathrm{Ft}}(t)\big|,\rho(t)\}^{r+2}}\right)^2 \mathrm{d}t,
\end{aligned} \tag{4.11}
$$

where we have applied the discrete multivariate version of Parseval's identity to both the bias term and the variance (see (A.21) in the Appendix Sect. A.4). We realize that the reduction of the summation domain from $\mathbb{Z}^d$ to the finite set T in the definition of $(Y')^{\mathrm{Ft}}$ is essential to obtain a finite variance term.

To consider the asymptotic MSSE, [60] propose that the denseness of the grid points tends to infinity, that is, $n\to\infty$, in view of (4.5). Then, (4.11) must also be adapted to the rescaled version. In particular, [60] define $g_n^{\mathrm{Ft}}(t) = g^{\mathrm{Ft}}(t/n)$, $(y')^{\mathrm{ft}}(t) = n^{-d}(y')^{\mathrm{ft}}(t/n)$, $\rho_n = n^{-d}\rho$, $A_n = [-n\pi,n\pi]^d$, and $\tau = n^{-d}\#T$, transferred to our notation. With respect to the asymptotics, they also consider n^dMSSE rather than the MSSE itself. In this setting, the authors derive upper bounds on the convergence rates of estimator (4.10), and study lower bounds with respect to any estimator under certain conditions. Surprisingly, the convergence rates do not depend on the smoothness of the pointspread function g in many cases; that effect is also observed for the ridge-parameter estimator (2.21) in the field of density deconvolution, see [58].

Now, as g is empirically accessible, we are able to reconstruct an image y from an observed blurred and noisy version Y. Hall and Qiu [60] suggest the plug-in estimator

$$\hat{y}(j) = \frac{1}{(2\pi)^d} \operatorname{Re} \int_A \frac{Y^{\mathrm{Ft}}(t)}{\hat{g}^{\mathrm{Ft}}(t)} \chi_{(\gamma,\infty)}(|\hat{g}^{\mathrm{Ft}}(t)|) \exp(-ij \cdot t)\, dt,$$

with $\hat{g}^{\mathrm{Ft}}(t)$ as in (4.9) and a threshold parameter $\gamma > 0$. Those techniques are called Wiener filter methods. As an alternative estimator, the authors propose the ridge-parameter approach

$$\hat{y}_2(j) = \frac{1}{(2\pi)^d} \operatorname{Re} \int_A \frac{Y^{\mathrm{Ft}}(t)}{\left|\hat{g}^{\mathrm{Ft}}(t)\right|^2 + \alpha|t|^\beta} \overline{\hat{g}^{\mathrm{Ft}}(t)} \exp(-ij \cdot t)\, dt,$$

with the parameters $\alpha, \beta > 0$ to be chosen suitably. Generally, to estimate y, one may use the same methods as for the previous estimation of g because of the symmetry of the problems as mentioned before; however, differences in the smoothness constraints shall be appreciated.

Finally, we mention that the PGI model can be generalized to continuous models. A possible extension of PGI is the experiment where the Lebesgue integral replaces the discrete sum in the definition of (4.3). In such images, the brightness is described by a function f mapping $\mathbb{R}^d$ to $\mathbb{R}$. Nevertheless, we do not observe the whole image; but only at particular points so that the model involving some grid points $x_1, x_2, \ldots \in \mathbb{R}^d$ is applicable again. The observation structure is given by the dataset $(x_1, Y_1), (x_2, Y_2), \ldots$, where

$$Y_j = \big[f * g\big](x_j) + \varepsilon_j$$

for all integer j, where the ε_j are again i.i.d. random variables with $E\varepsilon_j = 0$. This problem is also referred to as an inverse regression problem. The image to be reconstructed is continuous while the observation scheme is still discrete. We give a brief draft of how estimators can be constructed in this problem.

Our goal is to estimate the function f based on the given data (x_j, Y_j), $j = 1, \ldots, n$, and knowledge of g. For the sake of simplicity, we assume that f and $f * g$ are supported on the d-dimensional cube $[-\pi, \pi]^d$. Also, f and g are square integrable. Elementary rescaling arguments make the following results applicable to functions f with more general compact support. Also, we assume regular equidistant location of the grid points on $[-\pi, \pi]^d$. More concretely, we change the subscript of the grid point x_j into

$$x_j = \left(2\pi j_1/\lfloor n^{1/d} \rfloor - \pi, \ldots, 2\pi j_d/\lfloor n^{1/d} \rfloor - \pi\right)^t,$$

where $j_k \in \{1, \ldots, \lfloor n^{1/d} \rfloor\}$, for all $k \in \{1, \ldots, d\}$. Then, we may use an appropriate counting map from $\{1, \ldots, \lfloor n^{1/d} \rfloor^d\}$ to $\{(j_1, \ldots, j_d) : j_k = 1, \ldots, \lfloor n^{1/d} \rfloor, k = 1, \ldots, d\}$. In the current setting, $n - \lfloor n^{1/d} \rfloor^d$ of the data are left unused; however, at an asymptotic point of view, we may forgo those residual data. To simplify the notation, we write $x_{j_1,\ldots,j_d}$ instead of x_j.

It seems reasonable to apply a multivariate Fourier series approach as explained in the Appendix Sect. A.4. By (A.18), we may establish the representability

$$f(x) = (2\pi)^{-d} \sum_{k\in\mathbb{Z}^d} \exp(-\mathrm{i}x \cdot k) f^{\mathrm{ft}}(k) \cdot \chi_{[-\pi,\pi]^d}(x),$$

where the infinite sum represents an $L_2(\mathbb{R}^d)$-limit. Considering the typical deconvolution formula in its suitable version (A.15), we are motivated to construct an empirical version of

$$\begin{aligned}(f*g)^{\mathrm{ft}}(t) &= \int \cdots \int \exp\Big(\mathrm{i}\sum_{j=1}^d t_j y_j\Big) \big[f*g\big](y_1,\ldots,y_d)\, \mathrm{d}y_1 \cdots \mathrm{d}y_d \\ &\approx \sum_{j_1=0}^{\lfloor n^{1/d}\rfloor -1} \cdots \sum_{j_d=0}^{\lfloor n^{1/d}\rfloor -1} \int_{y_1=2\pi j_1/\lfloor n^{1/d}\rfloor -\pi}^{2\pi(j_1+1)/\lfloor n^{1/d}\rfloor -\pi} \cdots \int_{y_d=2\pi j_d/\lfloor n^{1/d}\rfloor -\pi}^{2\pi(j_d+1)/\lfloor n^{1/d}\rfloor -\pi} \exp\Big(\mathrm{i}\sum_{j=1}^d t_j y_j\Big) \\ &\qquad \mathrm{d}y_1 \cdots \mathrm{d}y_d \,\big[f*g\big](x_{j_1,\ldots,j_d}),\end{aligned} \tag{4.12}$$

where the latter approximation is based on smoothness of the function $f*g$, which has to be assumed. On the other hand, we derive that

$$\begin{aligned}&E \sum_{j_1=0}^{\lfloor n^{1/d}\rfloor -1} \cdots \sum_{j_d=0}^{\lfloor n^{1/d}\rfloor -1} \int_{y_1=2\pi j_1/\lfloor n^{1/d}\rfloor -\pi}^{2\pi(j_1+1)/\lfloor n^{1/d}\rfloor -\pi} \cdots \int_{y_d=2\pi j_d/\lfloor n^{1/d}\rfloor -\pi}^{2\pi(j_d+1)/\lfloor n^{1/d}\rfloor -\pi} \exp\Big(\mathrm{i}\sum_{j=1}^d t_j y_j\Big) \\ &\qquad \mathrm{d}y_1 \cdots \mathrm{d}y_d \, Y_{j_1,\ldots,j_d} \\ &= \sum_{j_1=0}^{\lfloor n^{1/d}\rfloor -1} \cdots \sum_{j_d=0}^{\lfloor n^{1/d}\rfloor -1} \int_{y_1=2\pi j_1/\lfloor n^{1/d}\rfloor -\pi}^{2\pi(j_1+1)/\lfloor n^{1/d}\rfloor -\pi} \cdots \int_{y_d=2\pi j_d/\lfloor n^{1/d}\rfloor -\pi}^{2\pi(j_d+1)/\lfloor n^{1/d}\rfloor -\pi} \exp\Big(\mathrm{i}\sum_{j=1}^d t_j y_j\Big) \\ &\qquad \mathrm{d}y_1 \cdots \mathrm{d}y_d \,\big[f*g\big](x_{j_1,\ldots,j_d}),\end{aligned}$$

where $Y_{j_1,\ldots,j_d}$ denotes that Y_j, which is observed at $x_j = x_{j_1,\ldots,j_d}$. Obviously, we shall consider

$$\begin{aligned}\hat{\psi}(t) &= \sum_{j_1=0}^{\lfloor n^{1/d}\rfloor -1} \cdots \sum_{j_d=0}^{\lfloor n^{1/d}\rfloor -1} \int_{y_1=2\pi j_1/\lfloor n^{1/d}\rfloor -\pi}^{2\pi(j_1+1)/\lfloor n^{1/d}\rfloor -\pi} \cdots \int_{y_d=2\pi j_d/\lfloor n^{1/d}\rfloor -\pi}^{2\pi(j_d+1)/\lfloor n^{1/d}\rfloor -\pi} \exp\Big(\mathrm{i}\sum_{j=1}^d t_j y_j\Big) \\ &\qquad \mathrm{d}y_1 \cdots \mathrm{d}y_d \, Y_{j_1,\ldots,j_d}\,,\end{aligned}$$

as an estimator of $(f*g)^{\mathrm{ft}}(t)$ for any $t \in \mathbb{R}^d$. To quantify the accuracy of the approximation in (4.12), one has to impose some smoothness constraints on

$f * g$. In higher-dimensional problems, one may derive higher-order approximations, compared to (4.12), which shall be more precise. Nevertheless, for $d \leq 2$, the above estimator $\hat{\psi}(t)$ seems appropriate. Anyway, we may derive an empirical version for the coefficients $\left[f * g\right]^{\text{ft}}(k) = f^{\text{ft}}(k) g^{\text{ft}}(k)$ by $\hat{\psi}(k)$. To construct a deconvolution estimator for f, we may apply the general discrete scheme as introduced in Sect. 3.4.1.

A widely-used generalization to continuous image deblurring where the observation structure is continuous, too, is studied in the following section.

4.2 White Noise Model

Assume a one-dimensional image or signal, which is represented by a function $f : \mathbb{R} \to \mathbb{R}$. Then, $f(x)$ may be interpreted as the intensity or brightness of the image at the point $x \in \mathbb{R}$. The signal is observed over the whole real line; therefore, its brightness is not concentrated at finitely many discrete points, unlike PGI as discussed in the previous section.

Again, we assume that the image suffers from random noise effects. At this stage, we assume that an image f supported on $[0,1]$ is observed at the equidistant points $x_1 < \cdots < x_n$ with $x_j = j/n$, $j = 1, \ldots, n$. Then, we observe the brightness $f(j/n)$, however, corrupted by a random variable ε_j. Then, the observations may be modeled as the dataset $Z_1, \ldots, Z_n$ with $Z_j = f(j/n) + \varepsilon_j$, where the ε_j are i.i.d. random variables. Equivalently to the observation of the data $Z_1, \ldots, Z_n$, we may observe the simple function

$$Y_n(x) = \frac{1}{n} \sum_{j=1}^{n} Z_j \cdot \chi_{[0,x]}(x_j) = \frac{1}{n} \sum_{j=1}^{n} f(j/n) \cdot \chi_{[0,x]}(x_j) + \frac{1}{n} \sum_{j=1}^{n} \varepsilon_j \cdot \chi_{[0,x]}(x_j) \tag{4.13}$$

for all $x \in [0,1]$. This can be easily seen by the fact that data $Z_1, \ldots, Z_m$ are uniquely reconstructable from the step function Y_m, and vice versa. For large n and continuous f, we may apply the approximation of the deterministic term

$$\frac{1}{n} \sum_{j=1}^{n} f(j/n) \cdot \chi_{[0,x]}(x_j) \approx \int_0^x f(y)\,\mathrm{d}y.$$

The second term in (4.13) may approximately be viewed as $n^{-1/2} W(x)$, for large n, where W denotes a standard Wiener process. Intuitively, this can be justified by a central limit theorem in a functional sense; note that the $\varepsilon_1, \ldots, \varepsilon_n$ are i.i.d. A rigorous theoretical study of the equivalence of discrete and continuous observation schemes is given by, for example, [5] for nonparametric regression and [104] for density estimation.

All those aspects motivate us to introduce the so-called white noise model, where one observes the stochastic process $Y(x), x \in \mathbb{R}$, defined by

$$Y(x) - Y(0) = \int_0^x f(t)\,\mathrm{d}t + n^{-1/2} \cdot \big(W(x) - W(0)\big),$$

where our goal is still to estimate the function f. Note that this model differs significantly from all other models, which have been introduced in this book so far. Now, we observe a curve over the whole of $\mathbb{R}$. However, as the noise level of the observation tends to zero – consider the scaling factor $n^{-1/2}$ in the above equation – we shall be able to reconstruct f from the observed and noisy image Y consistently.

We recall the definition of the standard Wiener process. A stochastic process $W(x), x \in \mathbb{R}$, is called a Wiener process if

1. $W(0) = 0$
2. $W(x)$ is continuous on $\mathbb{R}$ almost surely
3. The increments $W(x_1) - W(x_2), W(x_2) - W(x_3), \ldots, W(x_{n-1}) - W(x_n)$ are independent for all integer n and all real numbers $x_1 < \cdots < x_n$.
4. The increments $W(y) - W(x)$ are normally distributed with the mean 0 and the variance $y - x$, for all x, y with $y > x$.

We give the book of [116] as a reference. Lots of literature can be found on Wiener processes and their broad fields of application. Apart from their importance in the field of signal recovery, they are a basic foundation of nowadays financial mathematics.

Again, we consider image blurring models where the image f is not only affected by random noise (as modeled by the Wiener process) but also suffers from pointspread effects. We assume that the pointspread effects can be described by a linear operator. In particular, when assuming spatial homogeneity of the pointspread effects, that is, every point of the true signal is spread under the same rule, this operator can be interpreted as a convolution operator. Therefore, we consider the model, under which the stochastic process $Y(x)$, $x \in \mathbb{R}$ is observed, where Y is defined by

$$Y(x) - Y(0) = \int_0^x [f * g](t)\,\mathrm{d}t + n^{-1/2}\big(W(x) - W(0)\big) \tag{4.14}$$

for all $x \in \mathbb{R}$. Our goal is the reconstruction of the true signal f. In particular, we assume that g is a density, which may be seen as the blurring density or the pointspread function as studied in the discrete context for PGI in the previous section. The assertion that g integrates to one can be justified whenever the pointspread effect is intensity-preserving.

Such models as (4.14) are often referred to as white noise models. They have received considerable attention in nonparametric statistics in general, see, for example, [38, 52, 53, 78]. The latter two papers are concerned with the reconstruction of a boundary curve in the multivariate version of the model (4.14).

To derive a differential version of (4.14), we recall the concept of Ito integrals. Therein, we consider the random variable

$$S_n(f) = \sum_{j=0}^{n-1} f(x_{j,n})\big(W(x_{j+1,n}) - W(x_{j,n})\big), \tag{4.15}$$

with $a = x_{0,n} < \cdots < x_{n,n} = b$ for some interval $I = [a, b]$ and some function $f : \mathbb{R} \to \mathbb{R}$. Obviously, we may define the integral

$$I_{a,b}(f) = \int_a^b f(x)\,\mathrm{d}W(x), \tag{4.16}$$

with its formal symbol $\mathrm{d}W$ by $S_n(f)$ for the simple functions

$$f(x) = \sum_{j=0}^{n-1} f_j \cdot \chi_{[x_{j,n}, x_{j+1,n})}(x), \tag{4.17}$$

for $x \in [a, b)$ and some real-valued coefficients $f_0, \ldots, f_{n-1}$. We derive that

$$\begin{aligned}
E\big|S_n(f)\big|^2 &= E\Big|\sum_{j=0}^{n-1} f(x_{j,n})\big(W(x_{j+1,n}) - W(x_{j,n})\big)\Big|^2 \\
&= \Big|\sum_{j=0}^{n-1} f(x_{j,n})\underbrace{E\big(W(x_{j+1,n}) - W(x_{j,n})\big)}_{=0}\Big|^2 \\
&\qquad + \sum_{j=0}^{n-1} f^2(x_{j,n})\mathrm{var}\big(W(x_{j+1,n}) - W(x_{j,n})\big) \\
&= \sum_{j=0}^{n-1} f^2(x_{j,n})\big(x_{j+1,n} - x_{j,n}\big)
\end{aligned} \tag{4.18}$$

for any function $f : \mathbb{R} \to \mathbb{R}$, where we have used the unbiasedness and the independence of the increments of W as postulated in the definition of the Wiener process. In particular, for the functions with the shape (4.17), (4.18) is equivalent to the formula

$$E\Big|\int_a^b f(x)\,\mathrm{d}W(x)\Big|^2 = \int_a^b \big|f(x)\big|^2\,\mathrm{d}x, \tag{4.19}$$

which is also known as the *Ito isometry*.

This equality can be extended to more general functions. We assume that $f : \mathbb{R} \to \mathbb{R}$ is continuous and bounded on its restriction to the interval $[a, b]$. Define the approximating simple functions

$$f_n(x) = \sum_{j=0}^{n-1} f\big(a + (b-a)j/n\big)\,\chi_{[a+(b-a)j/n, a+(b-a)(j+1)/n)}(x)$$

for large integer n. By dominated convergence, we derive that

$$\int_a^b \big|f_n(x) - f(x)\big|^2\,\mathrm{d}x \stackrel{n\to\infty}{\longrightarrow} 0.$$

This means that the function sequence $(f_n)_n$, restricted to the support $[a, b]$, converges with respect to the $L_2([a, b])$-norm. Therefore, $(f_n)_n$ is a Cauchy sequence with respect to the $L_2([a, b])$-norm. Note that the theory of the $L_p(\mathbb{R})$-spaces as defined in Sect. 2.2.1 can be extended to integrals with respect to some probability measure instead of the Lebesgue measure. We define the space

$$L_2^P = \{X \text{ random variable} : E|X|^2 = \int |X(\omega)|^2 dP(\omega) < \infty\}.$$

One can show that L_2^P is a Hilbert space, equipped with the norm $\|X\|_{P,2} = (E|X|^2)^{1/2}$. Then, according to (4.19), the stochastic process

$$\left(\int_a^b f_n(x)\, \mathrm{d}W(x)\right)_n$$

is a Cauchy sequence with respect to the norm $\|\cdot\|_{P,2}$. Because of the completeness of the space L_2^P, we may establish that the above sequence of integrals converges in this space L_2^P; the corresponding $\|\cdot\|_{P,2}$-limit is then defined as the integral $I_{a,b}(f)$ as introduced in (4.16) for simple functions. Then, the Ito isometry (4.19) also remains valid for general continuous and bounded functions on their restriction to $[a, b]$, where the condition of boundedness can be omitted as f is continuous on its restricted compact domain. Also, the convergence with respect to the norm $\|\cdot\|_{P,2}$ implies convergence of the expected values of $\int_a^b f_n(x)\, \mathrm{d}W(x)$ to the expectation of $\int_a^b f(x)\, \mathrm{d}W(x)$, which can be shown, in fact, by rather simple application of Jensen's inequality. Obviously, we have

$$\begin{aligned} E\int_a^b f_n(x)\, \mathrm{d}W(x) &= \sum_{j=0}^{n-1} f(a+(b-a)j/n) E\big(W(a+(b-a)(j+1)/n) \\ &\qquad -W(a+(b-a)j/n)\big) \\ &= 0 \end{aligned}$$

due to the unbiasedness of the increments of the Wiener process W, so that we conclude that

$$E\int_a^b f(x)\, \mathrm{d}W(x) = 0. \tag{4.20}$$

The integral $I_{a,b}(f)$ is called an *Ito integral.*

We mention that f may also be approximated by more general simple functions, which are locally constant on some disjoint and covering intervals $[x_{0,n}, x_{1,n}], \ldots, [x_{n-1,n}, x_{n,n}]$ as long as $\max(x_{j+1,n} - x_{j,n}) \to 0$ as $n \to \infty$.

By similar arguments the concept of the integral (4.19), along with the validity of the Ito isometry (4.19), may be extended to general $L_2(\mathbb{R})$-functions f. In particular, the denseness of all bounded, continuous, compactly supported functions in $L_2(\mathbb{R})$ is needed along with some truncation techniques. Then, one may put $a = -\infty$ and $b = \infty$. However, for our purpose, the above definition of the Ito integral for continuous, bounded, compactly supported functions f is sufficient. We mention that the Ito integral must not be defined as a Lebesgue–Stieltjes integral of each path $W(x, \omega)$ viewed as a deterministic function over $x \in \mathbb{R}$ for a fixed $\omega \in \Omega$, denoting the underlying probability space. That is due to the fact that a standard Wiener process does not have finite absolute variation almost surely. For further theory on stochastic integration and Ito calculus, see, for example, the books of [106] and [74].

Equipped with this theoretical background, we may define an equivalent version of the statistical model (4.14), where the process $Y(x)$ is represented by differentials. Still, we observe the process $Y(x)$, which satisfies the stochastic differential equation

$$\mathrm{d}Y(x) = \big[f * g\big](x)\,\mathrm{d}x + n^{-1/2}\,\mathrm{d}W(x), \text{ for all } x \in \mathbb{R} \text{ and } Y(0) = 0 \quad (4.21)$$

in its formal representation, where we stipulate that $f \in L_2(\mathbb{R})$ and g is a density contained in $L_2(\mathbb{R})$. Equivalently, this statistical experiment can be described by the observation of the random variables

$$\begin{aligned} Z(\varphi) &= \int \varphi(x)\,\mathrm{d}Y(x) \\ &= \int \varphi(x)[f * g](x)\,\mathrm{d}x + n^{-1/2} \int \varphi(x)\,\mathrm{d}W(x) \end{aligned}$$

for all bounded, continuous, and compactly supported functions $\varphi : \mathbb{R} \to \mathbb{R}$, which are dense in $L_2(\mathbb{R})$.

Now let us consider how to estimate the function f in the model (4.14) based on the observation of the stochastic process $Y(x), x \in \mathbb{R}$. We assume that the target function f and the pointspread function g are supported on the intervals $[-\pi, \pi]$ and $[-d, d]$, for some $d > 0$, respectively. From there, we are guaranteed that the support of $f * g$ is included in the interval $[-\pi - d, \pi + d]$. Apparently, observing the restriction of $Y(x)$ to $x \in [-\pi - d, \pi + d]$ is sufficient for the function f because none of the $Y(x)$, for $|x| > \pi + d$, contains any information about f.

Since a deconvolution step is required, it seems convenient to use Fourier methods again. We define the empirically fully accessible quantity

$$\hat{\psi}_h(t) = \int_{x=-\pi-d}^{\pi+d} \exp(itx)\,\mathrm{d}Y(x).$$

In practice, $\hat{\psi}_h(t)$ may be computed approximately by the function

$$\hat{\psi}_{m,h}(t) = \sum_{j=0}^{m-1} \exp(itx_j)\big(Y(x_{j+1}) - Y(x_j)\big),$$

where $x_j = (2\pi+2d)j/m-\pi-d$ for large m. Whenever the blurred signal $f*g$ is continuous on its restriction to $[-\pi-d,\pi+d]$, we may fix convergence of $\hat\psi_{m,h}(t)$ to $\hat\psi_h(t)$ for any t with respect to the $\big(E|\cdot|^2\big)^{1/2}$-distance as $m\to\infty$. This follows from the elementary analytic integration theory with respect to the integral $\int_{-\pi-d}^{\pi+d}\exp(itx)\big[f*g\big](x)\,dx$ and the Ito theory with respect to the stochastic part $\int_{-\pi-d}^{\pi+d}\exp(itx)\,dW(x)$.

According to (4.21), we obtain that

$$\begin{aligned}\hat\psi_h(t) &= \int_{x=-\pi-d}^{\pi+d}\exp(itx)\big[f*g\big](x)\,dx + n^{-1/2}\int_{x=-\pi-d}^{\pi+d}\exp(itx)\,dW(x)\\ &= f^{\mathrm{ft}}(t)g^{\mathrm{ft}}(t) + n^{-1/2}\int_{x=-\pi-d}^{\pi+d}\exp(itx)\,dW(x),\end{aligned}$$

where the deconvolution formula (Lemma A.5) has been employed. Therefore, for large n, it seems reasonable to consider

$$\hat\psi_f(t) = \hat\psi_h(t)/g^{\mathrm{ft}}(t)$$

as an empirical version of $f^{\mathrm{ft}}(t)$. As we have assumed that f is supported on $[-\pi,\pi]$, application of the Fourier series approach is obvious to apply Fourier inversion. Inspired by Theorem A.7, we suggest to use the estimator

$$\hat f(x) = \frac{1}{2\pi}\sum_{|j|\le K_n}\exp(-ixj)\hat\psi_f(j)\cdot\chi_{[-\pi,\pi]}(x)\,, \tag{4.22}$$

with some integer-valued smoothing sequence $(K_n)_n$. Again, by formula (A.9), extension to those f with more general but compact support is possible.

The following proposition gives us an equivalent representation of the MISE of the estimator (4.22).

Proposition 4.1. *Consider the statistical model (4.21), where $f\in L_2(\mathbb{R})$ and the density g are supported on $[-\pi,\pi]$ and $[-d,d]$, respectively. Assume that the function $f*g$ is continuous on its restriction to $[-\pi-d,\pi+d]$ and that $g^{\mathrm{ft}}(k)\neq 0$ for all integer k. Then, estimator (4.22) satisfies*

$$E\|\hat f - f\|_2^2 = \frac{\pi+d}{\pi n}\sum_{|j|\le K_n}\big|g^{\mathrm{ft}}(j)\big|^{-2} + \frac{1}{2\pi}\sum_{|j|>K_n}\big|f^{\mathrm{ft}}(j)\big|^2.$$

Proof. The discrete version of Parseval's identity (Theorem A.7) gives us

$$E\|\hat f - f\|_2^2 = \frac{1}{2\pi}\sum_{|j|\le K_n}E\big|\hat\psi_f(j) - f^{\mathrm{ft}}(j)\big|^2 + \frac{1}{2\pi}\sum_{|j|>K_n}\big|f^{\mathrm{ft}}(j)\big|^2.$$

The equation (4.20) allows us to derive that

$$E\hat{\psi}_h(t) = f^{\text{ft}}(t) g^{\text{ft}}(t)$$

holds true for all $t \in \mathbb{R}$. Note that the function $\exp(it\cdot)$ is continuous and bounded on the interval $[-\pi - d, \pi + d]$. Thus,

$$E\hat{\psi}_f(t) = f^{\text{ft}}(t).$$

When considering the variance of estimator $\hat{\psi}_f(t)$, we obtain that

$$\begin{aligned}
\operatorname{var} \hat{\psi}_f(t) &= n^{-1} \big|g^{\text{ft}}(t)\big|^{-2} \operatorname{var}\Big(\int_{x=-\pi-d}^{\pi+d} \exp(itx)\, \mathrm{d}W(x)\Big) \\
&= n^{-1} \big|g^{\text{ft}}(t)\big|^{-2} E\Big| \int_{x=-\pi-d}^{\pi+d} \exp(itx)\, \mathrm{d}W(x)\Big|^2 \\
&= n^{-1} \big|g^{\text{ft}}(t)\big|^{-2} \int_{x=-\pi-d}^{\pi+d} \big|\exp(itx)\big|^2\, \mathrm{d}x \\
&= 2(\pi + d)/\big(n\big|g^{\text{ft}}(t)\big|^2\big),
\end{aligned}$$

where we have used that an additive deterministic shift of a random variable may be removed when considering the variance, and the Ito isometry (4.19) for continuous and bounded functions on $[-\pi - d, \pi + d]$. Then, putting $t = j$, the assertion of the proposition has been established. ■

Proposition 4.1 allows us to derive the convergence rates of the MISE of the estimator (4.22) as in many other deconvolution problems. With respect to the pointspread function, we impose the discrete version of ordinary smooth and supersmooth densities, see (3.30) and (3.31) for $a = -\pi$ and $b = \pi$. They are advantageous, compared to the continuous versions (2.31) and (2.32), as we apply the Fourier series approach again. With respect to the signal function f, we assume that f is supported on $[-\pi, \pi]$ and satisfies the following smoothness constraints,

$$\sum_{j \in \mathbb{Z}} \big|f^{\text{ft}}(j)\big|^2 (1 + |j|^{2\beta}) \le C,$$

inspired by the inequality (A.12) from the appendix, for some $\beta > 1/2$. All those f are collected in the function class $\mathcal{F}^w_{\beta,C}$. Then, we may give the following theorem.

Theorem 4.2. *Consider the statistical model (4.21), where the blurring density g is supported on $[-d, d]$. Assume that $f * g$ is continuous on its restriction to $[-\pi - d, \pi + d]$ for all $f \in \mathcal{F}^w_{\beta,C}$. Then,*

(a) whenever g satisfies (3.30) with $a = -\pi$ and $b = \pi$, then select $K_n \asymp n^{1/(2\beta+2\alpha+1)}$ so that estimator (4.22) satisfies

$$\sup_{f \in \mathcal{F}^w_{\beta,C}} E\|\hat{f} - f\|_2^2 = O\big(n^{-2\beta/(2\beta+2\alpha+1)}\big).$$

(b) whenever g satisfies (3.31) with $a = -\pi$ and $b = \pi$, then select $K_n = \big(C_k \log n\big)^{1/\gamma}$ with $C_K \in \big(0, 1/(2d_g)\big)$, so that estimator (4.22) satisfies

$$\sup_{f \in \mathcal{F}^w_{\beta,C}} E\|\hat{f} - f\|_2^2 = O\big((\log n)^{-2\beta/\gamma}\big).$$

Considering Proposition 4.1, Theorem 4.2 can be proven analogously to the Theorems 3.3 and 3.7, for instance. For minimax theory and adaptivity in white noise topics, see, for example, the book of [78].

Also, white noise models can be generalized to the multivariate setting, where W denotes the d-dimensional Wiener process then. For instance, see [52] for boundary curve estimation from a d-variate white noise model.

4.3 Circular Model and Boxcar Deconvolution

In this section, the white noise model (4.21) is modified so that circular deconvolution problems can be considered. The general idea is that the data are not recorded on a plane or a line; but rather on a torus or circle. In the field of medical imaging, this situation occurs when the intensity of some rays is recorded on a torus-shaped screen, since it may be impossible to create rays with parallel direction.

Let us consider the univariate setting of circular deconvolution. As motivated by the above application, let us consider the problem where the original function $\big[f * g\big](x)$ in the model (4.21) is not observed on the real line with respect to x but on a circle curve of the radius 1. Then, however, we realize self-overlapping of the function. This means that the blurred intensities $\big[f * g\big](x + 2\pi k)$, for all integer k, overlap each other at the same point x of the circle. After all, any point of the circle can uniquely be represented by an angle $x \in \big[-\pi, \pi\big)$. Therefore, we may change the model (4.21) so that we assume to observe the stochastic process $Y(x), x \in \big[-\pi, \pi\big)$, driven by the stochastic differential equation

$$\mathrm{d}Y(x) = \sum_{j \in \mathbb{Z}} \big[f * g\big](x + 2\pi j)\,\mathrm{d}x + n^{1/2}\,\mathrm{d}W(x),\ x \in \big[-\pi, \pi\big). \tag{4.23}$$

We assume that the true and desired signal $f \in L_2(\mathbb{R})$ is supported on $\big[-\pi, \pi\big]$, while the blurring density $g \in L_2(\mathbb{R})$ may well have noncompact support. Despite, the blurred version of the image is supported on $\big[-\pi, \pi\big)$. This situation is possible only in the circular deconvolution model. Circular (or periodic) deconvolution models have been studied in, for example, [73] or [72]. Circular models also occur in the field of density deconvolution, for example, [40]. Some related multivariate problems of spherical deconvolution or deconvolution on groups are studied in [121] or [129], for instance.

To get an intuition about how to construct an estimator of f in problem (4.23), we consider the Fourier transform of the circularly blurred function

$$\tilde{h}(x) = \sum_{j\in\mathbb{Z}} \big[f * g\big](x + 2\pi j) \cdot \chi_{[-\pi,\pi)}(x),\ x \in \mathbb{R}.$$

We have

$$\begin{aligned}
\tilde{h}^{\mathrm{ft}}(t) &= \sum_{j\in\mathbb{Z}} \int_{-\pi}^{\pi} \exp(\mathrm{i}tx)\big[f * g\big](x + 2\pi j)\,\mathrm{d}x \\
&= \sum_{j\in\mathbb{Z}} \exp(-\mathrm{i}t2\pi j) \int_{(2j-1)\pi}^{(2j+1)\pi} \exp(\mathrm{i}tx)\big[f * g\big](x)\,\mathrm{d}x
\end{aligned}$$

when using Lemma A.1(a) and (e). We realize that, for integer t, we have

$$\begin{aligned}
\tilde{h}^{\mathrm{ft}}(t) &= \sum_{j\in\mathbb{Z}} \int_{(2j-1)\pi}^{(2j+1)\pi} \exp(\mathrm{i}tx)\big[f * g\big](x)\,\mathrm{d}x = \int \exp(\mathrm{i}tx)\big[f * g\big](x)\,\mathrm{d}x \\
&= \big[f * g\big]^{\mathrm{ft}}(t) = f^{\mathrm{ft}}(t)g^{\mathrm{ft}}(t) \qquad (4.24)
\end{aligned}$$

by Lemma A.5. On the other hand, as f is supported on $[-\pi,\pi)$, knowledge of $f^{\mathrm{ft}}(t)$ for all integer t suffices to uniquely reconstruct the function f, see Sect. A.3 in the appendix where we have the Fourier series representation

$$f(x) = \frac{1}{2\pi}\sum_{j\in\mathbb{Z}} \exp(-\mathrm{i}xj) f^{\mathrm{ft}}(j) \cdot \chi_{[-\pi,\pi)}(x)$$

to be understood as an $L_2(\mathbb{R})$-limit, see Theorem A.7.

To identify the function $\tilde{h}$, we may use similar techniques as in the noncircular white noise model as studied in the previous section. In particular, we define the empirically accessible estimator of $f^{\mathrm{ft}}(t)$ by

$$\begin{aligned}
\tilde{\psi}_f(t) &= \int_{-\pi}^{\pi} \exp(\mathrm{i}tx)\,\mathrm{d}Y(x)/g^{\mathrm{ft}}(t) \\
&= \int_{-\pi}^{\pi} \exp(\mathrm{i}tx)\tilde{h}(x)\,\mathrm{d}x/g^{\mathrm{ft}}(t) + n^{-1/2}\int_{-\pi}^{\pi} \exp(\mathrm{i}tx)\,\mathrm{d}W(x)/g^{\mathrm{ft}}(t) \\
&= f^{\mathrm{ft}}(t) + n^{-1/2}\big[g^{\mathrm{ft}}(t)\big]^{-1} \int_{-\pi}^{\pi} \exp(\mathrm{i}tx)\,\mathrm{d}W(x),
\end{aligned}$$

where (4.24) has been applied. As motivated earlier, one has to apply a Fourier series approach by defining the final function estimator of f by

$$\hat{f}(x) = \frac{1}{2\pi} \sum_{|j|\le K_n} \exp(-\mathrm{i}xj)\tilde{\psi}_f(j) \cdot \chi_{[-\pi,\pi)}(x). \qquad (4.25)$$

The MISE of the estimator (4.25) can be represented in a very similar way as in Proposition 4.1. By the discrete version of Parseval's identity (see Theorem A.7), we have

$$\begin{aligned} E\|\hat{f}-f\|_2^2 &= \frac{1}{2\pi}\sum_{|j|\leq K_n} E\big|\tilde{\psi}_f(j)-f^{\mathrm{ft}}(j)\big|^2 + \frac{1}{2\pi}\sum_{|j|>K_n}\big|f^{\mathrm{ft}}(j)\big|^2 \\ &= \frac{1}{2\pi}\sum_{|j|\leq K_n} \operatorname{var}\tilde{\psi}_f(j) + \frac{1}{2\pi}\sum_{|j|>K_n}\big|f^{\mathrm{ft}}(j)\big|^2 \\ &= \frac{1}{2\pi n}\sum_{|j|\leq K_n}\big|g^{\mathrm{ft}}(j)\big|^{-2}\int_{-\pi}^{\pi}\big|\exp(\mathrm{i}jx)\big|^2\,\mathrm{d}x + \frac{1}{2\pi}\sum_{|j|>K_n}\big|f^{\mathrm{ft}}(j)\big|^2 \\ &= n^{-1}\sum_{|j|\leq K_n}\big|g^{\mathrm{ft}}(j)\big|^{-2} + \frac{1}{2\pi}\sum_{|j|>K_n}\big|f^{\mathrm{ft}}(j)\big|^2 \qquad (4.26) \end{aligned}$$

under application of (4.20) and (4.19). Then, by (4.26), we may extend Theorem 4.2 to the circular deconvolution model. We give

Theorem 4.3. *Consider the statistical model (4.23), where we assume that* $f \in \mathcal{F}^w_{\beta,C}$ *as in Theorem 4.2. The blurring density* g *is assumed to lie in* $L_2(\mathbb{R})$*. Then, for* $\beta > 1/2$*,*
(a) whenever g *satisfies (3.30) with* $a = -\pi$ *and* $b = \pi$*, then select* $K_n \asymp n^{1/(2\beta+2\alpha+1)}$ *so that estimator (4.25) satisfies*

$$\sup_{f\in\mathcal{F}^w_{\beta,C}} E\|\hat{f}-f\|_2^2 = O\big(n^{-2\beta/(2\beta+2\alpha+1)}\big)\,.$$

(b) whenever g *satisfies (3.31) with* $a = -\pi$ *and* $b = \pi$*, then select* $K_n = \big(C_k \log n\big)^{1/\gamma}$ *with* $C_K \in \big(0, 1/(2d_g)\big)$ *so that estimator (4.25) satisfies*

$$\sup_{f\in\mathcal{F}^w_{\beta,C}} E\|\hat{f}-f\|_2^2 = O\big((\log n)^{-2\beta/\gamma}\big).$$

Now we focus on the problem of proving optimality of the convergence rates in Theorem 4.3. The problem seems more challenging compared to density estimation or errors-in-variables regression as the observation consists of an entire function rather than finitely many data. Therefore, the general function estimator $\hat{f}$ of f shall be defined as a mapping from $C^0([-\pi,\pi])$ to $L_2(\mathbb{R})$, where $C^0([-\pi,\pi])$ denotes the set of all continuous function mapping $[-\pi,\pi]$ to $\mathbb{R}$. Then, we may insert the observed stochastic process $Y(x), x \in [-\pi,\pi]$, driven by the model (4.23), as the argument of $\hat{f}$, hence $\hat{f} = \hat{f}(Y)$. The estimator is defined as function-valued as we restrict our consideration to the MISE as an error criterion. For pointwise considerations, the generic estimator of f should be defined as a mapping from $C^0([-\pi,\pi]) \times [-\pi,\pi]$ to $\mathbb{R}$.

In the spirit of the lower bound proofs in the Sects. 2.4.6 and 3.3.2, we introduce the parameterized functions

$$f_\theta(x) = a_n \cdot \mathrm{Re} \sum_{j=K_n}^{2K_n} \theta_j \, \exp(\mathrm{i}jx) \cdot \chi_{[-\pi,\pi)}(x), \tag{4.27}$$

where $(a_n)_n \downarrow 0$ and the integer-valued sequence $(K_n)_n \uparrow \infty$ are still to be chosen; further, $\theta = (\theta_{K_n}, \ldots, \theta_{2K_n}) \in \{0,1\}^{(K_n+1)}$.

We have to check that all f_θ, for $\theta \in \{0,1\}^{(K_n+1)}$, are contained in $\mathcal{F}^w_{\beta,C}$. For that purpose, we recognize the Fourier coefficients of f_θ as

$$\begin{aligned} & f_\theta^{\mathrm{ft}}(k) \\ & = \frac{a_n}{2} \cdot \sum_{j=K_n}^{2K_n} \theta_j \underbrace{\int_{-\pi}^{\pi} \exp\big(\mathrm{i}(k-j)x\big)\, \mathrm{d}x}_{=2\pi\delta_{k,j}} + \frac{a_n}{2} \cdot \sum_{j=K_n}^{2K_n} \theta_j \underbrace{\int_{-\pi}^{\pi} \exp\big(\mathrm{i}(k+j)x\big)\, \mathrm{d}x}_{=2\pi\delta_{k,-j}} \\ & = \frac{1}{2}\pi \, a_n \, \theta_k \cdot \chi_{[K_n,2K_n]}\big(|k|\big). \end{aligned}$$

Hence, we realize that

$$\begin{aligned} \sum_k \big|f_\theta^{\mathrm{ft}}(k)\big|^2 (1+|k|^{2\beta}) & \le \frac{1}{4}\pi^2 a_n^2 \cdot \sum_{|k|=K_n}^{2K_n} \theta_{|k|}^2 (1+|k|^{2\beta}) \\ & \le O\big(a_n^2 K_n^{2\beta+1}\big), \end{aligned}$$

while appreciating that $|\theta_k| \le 1$ for all k, so that the selection $a_n = c_a \cdot K_n^{-\beta-1/2}$, with a sufficiently small constant $c_a > 0$, guarantees the membership of all f_θ in $\mathcal{F}^w_{\beta,C}$.

Then, consider the θ_j as i.i.d. random variables with $P\big[\theta_j = 0\big] = 1/2$, and assume we observe the stochastic process Y as defined in the model (4.23) when putting $f = f_\theta$. Without any loss of generality, we may assume that the estimator $\hat{f}$ is supported on $[-\pi,\pi]$ almost surely (as the target function f is also supported on that interval). Then, by the discrete version of Parseval's identity as given in Theorem A.7, we derive that

$$\begin{aligned} & \sup_{f \in \mathcal{F}^w_{\beta,C}} E\|\hat{f}(Y) - f\|_2^2 \ge E\, E_{f_\theta} \|\hat{f}(Y) - f_\theta\|_2^2 \\ & = \frac{1}{2\pi} \sum_{k\in\mathbb{Z}} E\, E_{f_\theta} \big|\hat{f}^{\mathrm{ft}}(Y;k) - f_\theta^{\mathrm{ft}}(k)\big|^2 \\ & \ge \frac{1}{2\pi} \sum_{k=K_n}^{2K_n} E\, E_{f_\theta} \big|\hat{f}^{\mathrm{ft}}(Y;k) - \pi a_n \theta_k/2\big|^2 \\ & = \frac{1}{4\pi} \sum_{k=K_n}^{2K_n} E\Big(E_{f_{\theta_{k,1}}} \big|\hat{f}^{\mathrm{ft}}(Y;k) - \pi a_n/2\big|^2 + E_{f_{\theta_{k,0}}} \big|\hat{f}^{\mathrm{ft}}(Y;k)\big|^2\Big) \end{aligned}$$

$$= \frac{1}{4\pi} \sum_{k=K_n}^{2K_n} E \int \left| \hat{f}^{\mathrm{ft}}\left(y_{f_{\theta_{k,0}}}(\omega); k\right)\right|^2 \mathrm{d}P(\omega)$$
$$+ \frac{1}{4\pi} \sum_{k=K_n}^{2K_n} E \int \left| \hat{f}^{\mathrm{ft}}\left(y_{f_{\theta_{k,1}}}(\omega); k\right) - \pi a_n/2\right|^2 \mathrm{d}P(\omega), \tag{4.28}$$

with the notation $\theta_{j,b} = \left(\theta_{K_n}, \ldots, \theta_{j-1}, b, \theta_{j+1}, \ldots, \theta_{2K_n}\right)$ and

$$y_f(\omega) = \int_0^{\cdot} \sum_{j\in\mathbb{Z}} \left[f * g\right](x + 2\pi j)\,\mathrm{d}x + n^{-1/2} W(\cdot, \omega),$$

viewed as a function on $[-\pi, \pi]$.

Now, we use *Girsanov's theorem* – see, for example, [80]. It says that an Ito process $Y(x), x \in [-\pi, \pi]$, driven by the stochastic differential equation

$$\mathrm{d}Y(x) = f(x)\,\mathrm{d}x + n^{-1/2}\,\mathrm{d}W(x)$$

with respect to some probability measure P has the same distribution as the Ito process, defined by

$$\mathrm{d}Y(x) = \tilde{f}(x)\,\mathrm{d}x + n^{-1/2}\,\mathrm{d}W(x)$$

under the probability measure Q where the corresponding Radon–Nikodym derivative is equal to

$$\frac{\mathrm{d}Q}{\mathrm{d}P} = \exp\Big(- n^{1/2} \int_{-\pi}^{\pi} (f(x) - \tilde{f}(x))\,\mathrm{d}W(x) - \frac{n}{2} \int_{-\pi}^{\pi} \left|f(x) - \tilde{f}(x)\right|^2 \mathrm{d}x\Big)$$

whenever $f, \tilde{f} \in L_2([-\pi, \pi])$. This is certainly not the most general assertion of Girsanov's theorem; however, it is sufficient for our purpose.

From there, we derive that

$$\int \left| \hat{f}^{\mathrm{ft}}\left(y_{f_{\theta_{k,1}}}(\omega); k\right) - \pi a_n/2\right|^2 \mathrm{d}P(\omega) = \int \left| \hat{f}^{\mathrm{ft}}\left(y_{f_{\theta_{k,0}}}(\omega); k\right) - \pi a_n/2\right|^2 \mathrm{d}Q(\omega)$$
$$= \int \left| \hat{f}^{\mathrm{ft}}\left(y_{f_{\theta_{k,0}}}(\omega); k\right) - \pi a_n/2\right|^2 \frac{\mathrm{d}Q}{\mathrm{d}P}(\omega)\mathrm{d}P(\omega),$$

where

$$\frac{\mathrm{d}Q}{\mathrm{d}P}(\omega) = \exp\Big(- n^{1/2} \int_{-\pi}^{\pi} \left(h_{\theta_{k,1}}(x) - h_{\theta_{k,0}}(x)\right)\mathrm{d}W(x)$$
$$- \frac{n}{2} \int_{-\pi}^{\pi} \left|h_{\theta_{k,1}}(x) - h_{\theta_{k,0}}(x)\right|^2 \mathrm{d}x\Big), \tag{4.29}$$

and $h_{\theta_{k,b}}(x) = \sum_{j\in\mathbb{Z}} \big[f_{\theta_{k,b}} * g\big](x+2\pi j)$. Therefore, (4.28) is bounded below by

$$\frac{1}{4\pi}\sum_{k=K_n}^{2K_n} E\int \Big(\big|\hat{f}^{\mathrm{ft}}\big(y_{f_{\theta_{k,0}}}(\omega);k\big)\big|^2 + \big|\hat{f}^{\mathrm{ft}}\big(y_{f_{\theta_{k,0}}}(\omega);k\big) - \pi a_n/2\big|^2\Big) \cdot \min\{(\mathrm{d}Q/\mathrm{d}P)(\omega),1\}\,\mathrm{d}P(\omega)$$

$$\geq \mathrm{const}\cdot a_n^2 \sum_{k=K_n}^{2K_n} E\int \min\{(\mathrm{d}Q/\mathrm{d}P)(\omega),1\}\,\mathrm{d}P(\omega). \tag{4.30}$$

For further consideration, we introduce the Kullback–Leibler distance between the probability measures P and Q, which is defined by

$$K(P,Q) = -\int \Big(\log\frac{\mathrm{d}Q}{\mathrm{d}P}(\omega)\Big)\,\mathrm{d}P(\omega).$$

As a useful result, we give the following lemma, which mainly goes back to [4].

Lemma 4.4. (Bretagnolle–Huber inequality) *Assume two probability measures P and Q, where any P-zero set is also a Q-zero set, so that the Radon–Nikodym derivative* $\mathrm{d}Q/\mathrm{d}P$ *exists. Then, we have*

$$\int \min\{\mathrm{d}Q/\mathrm{d}P,1\}\mathrm{d}P \geq \frac{1}{2}\exp\big(-K(P,Q)\big).$$

Proof. An obvious identity and elementary calculus give us

$$\begin{aligned}
-K(P,Q) &= \int \Big(\log\frac{\mathrm{d}Q}{\mathrm{d}P}\Big)\mathrm{d}P = \int \log\Big(\min\Big\{\frac{\mathrm{d}Q}{\mathrm{d}P},1\Big\}\cdot\max\Big\{\frac{\mathrm{d}Q}{\mathrm{d}P},1\Big\}\Big)\mathrm{d}P \\
&= \int \log\Big(\min\Big\{\frac{\mathrm{d}Q}{\mathrm{d}P},1\Big\}\Big)\mathrm{d}P + \int \log\Big(\underbrace{\max\Big\{\frac{\mathrm{d}Q}{\mathrm{d}P},1\Big\}}_{\leq 1+\mathrm{d}Q/\mathrm{d}P}\Big)\mathrm{d}P \\
&\leq \int \log\Big(\min\Big\{\frac{\mathrm{d}Q}{\mathrm{d}P},1\Big\}\Big)\mathrm{d}P + \int \log\big(1+\mathrm{d}Q/\mathrm{d}P\big)\mathrm{d}P.
\end{aligned}$$

As the log-function is concave (i.e., $(\mathrm{d}^2/\mathrm{d}x^2)\log x = -x^{-2} < 0$), we may apply Jensen's inequality so that

$$E\,\log X \leq \log(EX)$$

holds true for any positive-valued integrable random variable X. From there, we conclude that

$$\begin{aligned}
-K(P,Q) &\leq \log\int\Big(\min\Big\{\frac{\mathrm{d}Q}{\mathrm{d}P},1\Big\}\Big)\mathrm{d}P + \log\int\big(1+\mathrm{d}Q/\mathrm{d}P\big)\mathrm{d}P \\
&= \log\int\Big(\min\Big\{\frac{\mathrm{d}Q}{\mathrm{d}P},1\Big\}\Big)\mathrm{d}P + \log 2,
\end{aligned}$$

since P and Q are probability measures, implying that $\int \mathrm{d}P = \int \mathrm{d}Q = 1$. Then, the lemma can be shown by elementary calculus again. ■

By Lemma 4.4, we can impose the following lower bound on (4.30):

$$\begin{aligned}
&\text{const} \cdot a_n^2 \sum_{k=K_n}^{2K_n} E \exp\big(- K(P,Q)\big) \\
&\geq \text{const} \cdot a_n^2 \sum_{k=K_n}^{2K_n} E \exp\Big[E_P\Big(- n^{1/2} \int_{-\pi}^{\pi} \big(h_{\theta_{k,1}}(x) - h_{\theta_{k,0}}(x)\big)\, \mathrm{d}W(x) \\
&\qquad\qquad - \frac{n}{2}\int_{-\pi}^{\pi} \big|h_{\theta_{k,1}}(x) - h_{\theta_{k,0}}(x)\big|^2\, \mathrm{d}x\Big)\Big] \\
&= \text{const} \cdot a_n^2 \sum_{k=K_n}^{2K_n} E \exp\Big[- \frac{n}{2}\int_{-\pi}^{\pi} \big|h_{\theta_{k,1}}(x) - h_{\theta_{k,0}}(x)\big|^2\, \mathrm{d}x\Big]
\end{aligned}$$

under application of (4.29), where E_P denotes the expected value with respect to the probability measure P. As a convenient effect, the log contained in the Kullback–Leibler distance annuls the exp occurring in (4.29). Also, we have used (4.20) in the last step.

Then, we may establish $\text{const} \cdot a_n^2 K_n \asymp K_n^{-2\beta}$ as a uniform lower bound on the convergence rates of the MISE if we can verify that

$$\int_{-\pi}^{\pi} \big|h_{\theta_{k,1}}(x) - h_{\theta_{k,0}}(x)\big|^2\, \mathrm{d}x = O(1/n) \tag{4.31}$$

uniformly in all $|k| \in [K_n, 2K_n]$. As the functions $h_{\theta_{k,b}}$ are supported on $[-\pi, \pi]$, we may apply Theorem A.7 again so that the left side of (4.31) may be written as

$$\begin{aligned}
&\frac{1}{2\pi}\sum_{j\in\mathbb{Z}} \big|h^{\mathrm{ft}}_{\theta_{k,1}}(j) - h^{\mathrm{ft}}_{\theta_{k,0}}(j)\big|^2 = \frac{1}{2\pi}\sum_{j\in\mathbb{Z}} \big|f^{\mathrm{ft}}_{\theta_{k,1}}(j) - f^{\mathrm{ft}}_{\theta_{k,0}}(j)\big|^2 \big|g^{\mathrm{ft}}(j)\big|^2 \\
&= \frac{\pi}{2} a_n^2 \big|g^{\mathrm{ft}}(k)\big|^2 \leq \text{const} \cdot K_n^{-1-2\beta} \sup_{|l|\in[K_n,2K_n]} \big|g^{\mathrm{ft}}(l)\big|^2,
\end{aligned}$$

where we have used (4.24). Note that $\theta_{k,0}$ and $\theta_{k,1}$ coincide with respect to all components except the kth one. For ordinary smooth g satisfying (3.30), we have $\big|g^{\mathrm{ft}}(k)\big|^2 \asymp K_n^{-2\alpha}$, while for supersmooth g as in (3.31), the equation $\big|g^{\mathrm{ft}}(k)\big|^2 \leq \text{const} \cdot \exp\big(- 2D_g|K_n|^\gamma\big)$ holds true. Again, we put $a = -\pi$ and $b = \pi$. Hence, (4.31) is satisfied under the choices $K_n \asymp n^{1/(2\beta+2\alpha+1)}$ and $K_n = C_K \cdot \big(\log n\big)^{1/\gamma}$, respectively.

Summarizing, we have proven the following theorem, which establishes optimality of the convergence rates derived in Theorem 4.3.

Theorem 4.5. *Consider the setting and conditions of Theorem 4.3. Then, for* $\beta > 1/2$ *and an arbitrary estimator* $\hat{f}$ *of* f *based on the observation* $Y(x), x \in [-\pi, \pi]$ *from model (4.23),*

(a) whenever g satisfies (3.30) with $a = -\pi$ and $b = \pi$, we have

$$\sup_{f \in \mathcal{F}^w_{\beta,C}} E\|\hat{f} - f\|_2^2 \geq const \cdot n^{-2\beta/(2\beta+2\alpha+1)}$$

(b) whenever g satisfies (3.31) with $a = -\pi$ and $b = \pi$,

$$\sup_{f \in \mathcal{F}^w_{\beta,C}} E\|\hat{f} - f\|_2^2 = const \cdot (\log n)^{-2\beta/\gamma}.$$

We realize that no additional assumptions on the derivative of g^{ft} are required unlike in the Theorems 2.13 and 2.14. That is due to the fact that the minimax theory does not involve the χ^2-distance between the competing observation densities, but is based on their $L_2(\mathbb{R})$-distance.

As in density deconvolution (see Theorem 2.9(a)), we realize that the rate-optimal choice of the smoothing parameter K_n requires knowledge of the smoothness degree β of the target function f for ordinary smooth blurring densities, see Theorem 4.3(a). Therefore, data-driven selection of K_n should be considered as in Sect. 2.5 in the field of density estimation. However, we will consider a more general framework of fully data-driven estimators in statistical inverse problems and deconvolution, which is described in [18]. The authors refer to more general regularized deconvolution estimators. Transferred to our notation, those estimators can be defined by

$$\hat{f}(x) = \frac{1}{2\pi} \sum_{j \in \mathbb{Z}} w_j \exp(-ijx)\tilde{\psi}_f(j) \cdot \chi_{[-\pi,\pi]}(x), \tag{4.32}$$

with an appropriate sequence of weights w_j. The estimator (4.25) is included by choosing $w_j = \chi_{[0,K_n]}(j)$, which reduces the selection problem to the choice of the parameter K_n. Those simple cut-off estimators attain optimal convergence rates under common conditions, for example, Theorems 4.3 and 4.5. On the other hand, under more precise asymptotic studies, those estimators are less convenient, that is, sharp constants are not achieved by this simple choice of the weights. When allowing for all (deterministic) $w_j \geq 0$, we realize that

$$\begin{aligned} E\|\hat{f} - f\|_2^2 &= \frac{1}{2\pi} \sum_j E\left|w_j \tilde{\psi}_f(j) - f^{\mathrm{ft}}(j)\right|^2 \\ &= \frac{1}{2\pi} \sum_j \left(w_j^2 \operatorname{var} \tilde{\psi}_f(j) + \left|f^{\mathrm{ft}}(j) - w_j E\tilde{\psi}_f(j)\right|^2\right) \\ &= \frac{1}{2\pi} \sum_j \left(w_j^2 \left|g^{\mathrm{ft}}(j)\right|^{-2}(2\pi/n) + \left|f^{\mathrm{ft}}(j)\right|^2 |1 - w_j|^2\right) \end{aligned}$$

is minimized by the oracle selection

$$w_j = \left|h^{\mathrm{ft}}(j)\right|^2 / \left(2\pi/n + \left|h^{\mathrm{ft}}(j)\right|^2\right),$$

which is not applicable as it requires knowledge of $h^{\mathrm{ft}}(j)$. However, the coefficients w_j can be estimated since the $h^{\mathrm{ft}}(j) = f^{\mathrm{ft}}(j)g^{\mathrm{ft}}(j)$ are empirically accessible. Cavalier and Tsybakov [18] study the blockwise approach of [118] in deconvolution problems. Therein, we consider the following representation of the MISE of estimator (4.32):

$$E\|\hat{f} - f\|^2 = \frac{1}{2\pi}\Big\{\sum_{j\in\mathbb{Z}} w_j^2|\tilde{\psi}_f(j)|^2 - 2\mathrm{Re}\sum_{j\in\mathbb{Z}} w_j\tilde{\psi}_f(j)f^{\mathrm{ft}}(-j) + \sum_{j\in\mathbb{Z}}|f^{\mathrm{ft}}(j)|^2\Big\}.$$

The first term in the above equation is known; the third term is unknown but independent of the weights so that it can be neglected for the purpose of optimizing the weights. However, the mid term must be replaced by an estimator. A similar problem occurs in cross-validation in density deconvolution, see Sect. 2.5.1. Therefore, our goal is minimizing

$$S(w_1, w_2, \ldots) = \frac{1}{2\pi}\Big\{\sum_{j\in\mathbb{Z}} w_j^2|\tilde{\psi}_f(j)|^2 - 2\mathrm{Re}\sum_{j\in\mathbb{Z}} w_j\big(|\tilde{\psi}_f(j)|^2 - 2\pi/(n|g^{\mathrm{ft}}(j)|^2)\big)\Big\}.$$

We realize that

$$E\big|\tilde{\psi}_f(j)\big|^2 - 2\pi/(n|g^{\mathrm{ft}}(j)|^2) = \big|f^{\mathrm{ft}}(t)\big|^2,$$

which justifies the minimization of S instead of the MISE intuitively. In Stein's blockwise approach, one does not permit minimization of S over all $w_1, w_2, \ldots \in \mathbb{R}$, but the coefficients $g^{\mathrm{ft}}(j)w_j$ are assumed to be piecewise constant within the blocks B_k, $k = 1, \ldots, N$, where $B_k = \{n_k, \ldots, n_{k+1} - 1\}$ for a strictly increasing integer-valued sequence $(n_k)_k$. One can show that the function S is minimized by the selection

$$\hat{w}_j = \max\Big\{0, 1 - \frac{2\pi}{n}\sum_{l\in B_k}\big|g^{\mathrm{ft}}(l)\big|^{-2}\Big/\sum_{l\in B_k}\big|\tilde{\psi}_f(l)\big|^2\Big\} \tag{4.33}$$

if $j \in B_k$. Sharp adaptive asymptotics for such estimators under an additional penalty term are derived in [18].

The circular (or periodic) deconvolution model (4.23) with uniform blurring densities g is considered in the papers of [73] and [72]. Such topics are often referred to as *boxcar deconvolution.* Note that the Fourier transform of uniform densities has some periodic zeros so that the conditions (3.30) and (3.31) might not be satisfied, and some frequencies could be lost. If, indeed, $g^{\mathrm{ft}}(k) = 0$, for some integer k, there exists no consistent estimator of f in model (4.23). This can also be derived by the considerations used to prove Theorem 4.5. Note that the situation is different in noncircular deconvolution problems. There, one is able to construct consistent estimators of the target function f even if g^{ft} has some isolated zeros, while one has to accept deterioration of the convergence rates in some cases. As appropriate methods, we mention the ridge-parameter approach (see Sect. 2.2.3) as studied in [58], or local polynomial approximation around the zeros in the Fourier domain as investigated by [94]. The so-called ForWaRD algorithm, which uses specific wavelet techniques as introduced in [99], might address the boxcar deconvolution problem too, although the investigation of the convergence rates is restricted to standard error densities in that paper. Roughly speaking, in circular deconvolution models, only the $f^{\mathrm{ft}}(k)$, for integer k, are useful to reconstruct the function f, while in noncircular settings, we can use $f^{\mathrm{ft}}(t)$ for all $t \in \mathbb{R}$ so that an isolated zero of g^{ft} at some integer t can be compensated by information about $f^{\mathrm{ft}}(t)$ in small neighborhoods of those zeros.

However, the approaches of [73] and [72] focus on the cases where none of the $g^{\mathrm{ft}}(k)$, integer k, vanishes but the distance between the integer k and the zeros of g^{ft}, which are indeed periodic under the assumption of uniform blurring, may become arbitrarily small for large arguments of g^{ft}. Then the standard convergence rates of deconvolution (see, e.g., Theorem 4.3) can be kept up to some logarithmic loss in some cases. In particular, the ratio between the support of the blurring density g and the support of f must be irrational so that $g^{\mathrm{ft}}(\pi k)$ does not vanish for any integer k; note that the authors use the scaled Fourier series on the interval $[-1, 1]$. The authors use some results from number theory and diophantine approximation to develop the asymptotic theory for so-called badly approximable irrational numbers in order to classify that ratio between the support of g and f. In the paper of [75], this condition is relaxed.

A

Tools from Fourier Analysis

A.1 Fourier Transforms of $L_1(\mathbb{R})$-Functions

For any function $f \in L_1(\mathbb{R})$, we may define its Fourier transform f^{ft} by

$$f^{\mathrm{ft}}(t) = \int \exp(itx) f(x)\, \mathrm{d}x, \qquad t \in \mathbb{R}.$$

For the definition of the L_p-spaces, see Sect. 2.2.2. The following lemma summarizes some important properties

Lemma A.1. *Assume that $f, g \in L_1(\mathbb{R})$, $\lambda, \mu \in \mathbb{C}$. Then we have*

(a) linearity: $\big(\lambda f + \mu g\big)^{\mathrm{ft}} = \lambda f^{\mathrm{ft}} + \mu g^{\mathrm{ft}}$.

(b) convolution: $(f * g)^{\mathrm{ft}} = f^{\mathrm{ft}} g^{\mathrm{ft}}$.

(c) boundedness: $\sup_{t\in\mathbb{R}} |f^{\mathrm{ft}}(t)| \le \|f\|_1$.

(d) uniform continuity: $|f^{\mathrm{ft}}(t) - f^{\mathrm{ft}}(s)| \to 0$ *as* $|t-s| \to 0$.

(e) linear stretching: $\big[f(\cdot a + b)\big]^{\mathrm{ft}}(t) = a^{-1} \exp(-itb/a) f^{\mathrm{ft}}(t/a)$,

(f) Fourier shift: $\big[f(\cdot)\cos(a\cdot)\big]^{\mathrm{ft}}(t) = \frac{1}{2} f^{\mathrm{ft}}(t+a) + \frac{1}{2} f^{\mathrm{ft}}(t-a)$, *for all* $a \in \mathbb{R}$.

(g) symmetry: $f^{\mathrm{ft}}(-t) = \overline{f^{\mathrm{ft}}(t)}$ *for all* $t \in \mathbb{R}$ *and real-valued* f*; and* $f^{\mathrm{ft}}(-t) = f^{\mathrm{ft}}(t) \in \mathbb{R}$ *for all* $t \in \mathbb{R}$ *if, in addition,* f *is symmetric, that is,* $f(x) = f(-x)$ *for all* $x \in \mathbb{R}$.

Proof. (a) is a rather simple consequence of the linearity of integration; (c) follows by estimation when putting the absolute value inside the integral; (e) can be verified by a simple linear substitution in the integral occurring in the definition of the Fourier transform. Concerning (f), we use Euler's formula to show that

$$\begin{aligned}
&\int \exp(itx) f(x) \cos(ax)\,\mathrm{d}x \\
&= \frac{1}{2} \int f(x) \exp\big[\mathrm{i}x(t+a)\big] f(x)\,\mathrm{d}x + \frac{1}{2} \int f(x) \exp\big[\mathrm{i}x(t-a)\big] f(x)\,\mathrm{d}x \\
&= \frac{1}{2} f^{\mathrm{ft}}(t+a) + \frac{1}{2} f^{\mathrm{ft}}(t-a).
\end{aligned}$$

One can show (b) by substitution and Fubini's theorem. More concretely, we have

$$\begin{aligned}
(f * g)^{\mathrm{ft}}(t) &= \int \exp(itx) \int f(y) g(x-y)\,\mathrm{d}y\mathrm{d}x \\
&= \int \exp(ity) f(y) \int \exp(it(x-y))\, g(x-y)\,\mathrm{d}x\mathrm{d}y \\
&= \int \exp(ity) f(y)\,\mathrm{d}y \int \exp(ity) g(y)\,\mathrm{d}y = f^{\mathrm{ft}}(t) g^{\mathrm{ft}}(t).
\end{aligned}$$

To prove (d), consider that

$$\begin{aligned}
\big|f^{\mathrm{ft}}(t) - f^{\mathrm{ft}}(s)\big| &\leq \int \big|\exp(itx) - \exp(isx)\big| |f(x)|\,\mathrm{d}x \\
&= \int \big|\exp(\mathrm{i}(t-s)x) - 1\big| |f(x)|\,\mathrm{d}x.
\end{aligned}$$

In the above integral, the function to be integrated is bounded above by $2|f|$; furthermore, it converges pointwise to zero for any $x \in \mathbb{R}$ as $|t-s| \to 0$. Using the theorem of dominated convergence completes the proof of (d).

With respect to part (g), we consider that

$$f^{\mathrm{ft}}(-t) = \int \exp(-\mathrm{i}tx) f(x)\,\mathrm{d}x = \int \overline{\exp(itx)} f(x)\,\mathrm{d}x = \overline{f^{\mathrm{ft}}(t)}.$$

If, in addition, f is symmetric, we have

$$\begin{aligned}
f^{\mathrm{ft}}(t) &= \int_{-\infty}^{0} \exp(itx) f(x)\,\mathrm{d}x + \int_{0}^{\infty} \exp(itx) f(x)\,\mathrm{d}x \\
&= \int_{0}^{\infty} \exp(-\mathrm{i}tx) f(x)\,\mathrm{d}x + \int_{0}^{\infty} \exp(itx) f(x)\,\mathrm{d}x \\
&= 2 \int_{0}^{\infty} \cos(tx) f(x)\,\mathrm{d}x \in \mathbb{R},
\end{aligned}$$

for all t, where we notice by the above representation that $f^{\mathrm{ft}}(t)$ is symmetric due to the symmetry of the cos-function. ■

Therefore, Lemma A.1(c) says that the Fourier transform of any density function is bounded by 1. Also, we derive that

$$f^{\mathrm{ft}}(0) = \int \exp(\mathrm{i}x \cdot 0) f(x)\,\mathrm{d}x = \int f(x)\,\mathrm{d}x = 1, \tag{A.1}$$

for any density f.

The following theorem deals with the inversion of Fourier transforms. It is a modified version of Fourier's integration theorem.

Theorem A.2. *Assume that $f \in L_1(\mathbb{R})$ is bounded and continuous at some $x \in \mathbb{R}$; and, in addition, $f^{\mathrm{ft}} \in L_1(\mathbb{R})$. Then, we obtain*

$$f(x) = \frac{1}{2\pi} \int \exp(-\mathrm{i}tx) f^{\mathrm{ft}}(t)\,\mathrm{d}t.$$

Proof. We introduce the scaled normal distribution $\varphi_R(x) = \frac{1}{\sqrt{2\pi}R} \exp\big(-x^2/(2R^2)\big)$ whose Fourier transform is explicitly calculable and is well known from probability theory, we have

$$\varphi_R^{\mathrm{ft}}(t) = \exp\big(-(Rt)^2/2\big),$$

which is an $L_1(\mathbb{R})$-function. Now we consider that

$$I_R(x) = \int \sqrt{2\pi} R \varphi_R(t) \exp(-\mathrm{i}tx) f^{\mathrm{ft}}(t)\,\mathrm{d}t.$$

Because of the integrability of f^{ft}, the inequality $|\sqrt{2\pi} R \varphi_R(t)| \le 1$ and

$$\lim_{R\to\infty} \sqrt{2\pi} R \varphi_R(t) = 1$$

for any $t \in \mathbb{R}$, we conclude by dominated convergence that

$$I_R(x) \stackrel{R\to\infty}{\longrightarrow} \int \exp(-\mathrm{i}tx) f^{\mathrm{ft}}(t)\,\mathrm{d}t.$$

On the other hand, we derive by Fubini's theorem that

$$\begin{aligned}
I_R(x) &= \int \sqrt{2\pi} R \varphi_R(t) \exp(-\mathrm{i}tx) \int \exp(\mathrm{i}ty) f(y)\,\mathrm{d}y\mathrm{d}t \\
&= \int f(y) \int \sqrt{2\pi} R \varphi_R(t) \exp\big(\mathrm{i}t(y-x)\big)\,\mathrm{d}t\mathrm{d}y \\
&= \sqrt{2\pi} R \int f(y) \varphi_R^{\mathrm{ft}}(y-x)\,\mathrm{d}y \\
&= \sqrt{2\pi} \int f(z/R + x) \exp\big(-z^2/2\big)\,\mathrm{d}z.
\end{aligned}$$

Again, we may apply dominated convergence due to boundedness of f and its continuity at x. We have represented the limit of I_R, as $R \to \infty$, in two different ways and, hence, the results must coincide with each other:

$$\int \exp(-\mathrm{i}tx) f^{\mathrm{ft}}(t)\,\mathrm{d}t = f(x)\sqrt{2\pi}\int \exp\big(-z^2/2\big)\,\mathrm{d}z = 2\pi f(x),$$

as any normal density integrates to one, of course. Now the theorem follows. ■

We realize that the inverse Fourier transform nearly corresponds to the Fourier transform up to the different sign in its argument and the scaling factor $1/(2\pi)$. Nevertheless, if f is unbounded or f^{ft} is not integrable, the inverse Fourier transform does not necessarily exist.

We mention that there are also alternative definitions of the Fourier transform with respect to the sign of the argument and the scaling factors $1/\sqrt{2\pi}$ or $1/(2\pi)$, which may be included in the definition of the Fourier transform. Accordingly, Theorem A.2 has to be modified in that setting.

A.2 Fourier Transforms of $L_2(\mathbb{R})$-Functions

Now, we introduce the set $\mathcal{C}$ of all bounded and continuous functions in $L_1(\mathbb{R})$ with integrable Fourier transform and some element $f \in \mathcal{C}$; and further, some $g \in L_1(\mathbb{R})$. We easily derive that $\mathcal{C}$ is a linear space. Furthermore, using arguments as in the proof of Lemma 2.3, it follows that $\mathcal{C} \subseteq L_2(\mathbb{R})$. Using Theorem A.2 and Fubini's theorem, we derive that

$$\begin{aligned}
\int f(x)\overline{g(x)}\,\mathrm{d}x &= \frac{1}{2\pi}\int\int \exp(-\mathrm{i}tx) f^{\mathrm{ft}}(t)\,\mathrm{d}t\overline{g(x)}\,\mathrm{d}x \\
&= \frac{1}{2\pi}\int\int \exp(-\mathrm{i}tx)\overline{g(x)}\,\mathrm{d}x\, f^{\mathrm{ft}}(t)\,\mathrm{d}t \\
&= \frac{1}{2\pi}\int \overline{g}^{\mathrm{ft}}(-t) f^{\mathrm{ft}}(t)\,\mathrm{d}t \\
&= \frac{1}{2\pi}\int \overline{g^{\mathrm{ft}}(t)} f^{\mathrm{ft}}(t)\,\mathrm{d}t. \qquad \text{(A.2)}
\end{aligned}$$

Putting $g = f$ in (A.2), we realize that the Fourier transform of any function $f \in \mathcal{C}$ is contained in $L_2(\mathbb{R})$ and that

$$\|f\|_2^2 = \frac{1}{2\pi}\|f^{\mathrm{ft}}\|_2^2. \qquad \text{(A.3)}$$

Then, we use the normal density φ_R as defined in the proof of Theorem A.2. The following lemma allows us to approximate general $L_2(\mathbb{R})$-functions by functions in $\mathcal{C}$.

Lemma A.3. *The set $\mathcal{C}$ is dense in $L_2(\mathbb{R})$, that is, for any $f \in L_2(\mathbb{R})$, there exists a sequence $(f_n)_n$ with $f_n \in \mathcal{C}$, for all n, so that $\|f - f_n\|_2 \stackrel{n\to\infty}{\longrightarrow} 0$.*

Proof. We notice that $\varphi_R * f \in \mathcal{C}$ for any $f \in L_1(\mathbb{R})$ and $R > 0$ because the convolution of two $L_1(\mathbb{R})$-functions lies in $L_1(\mathbb{R})$; it is bounded by $\|f\|_1/(\sqrt{2\pi}R)$; it is continuous by dominated convergence due to the continuity of φ_R; further, by Lemma A.1(b), we have

$$\big|(\varphi_R * f)^{\mathrm{ft}}(t)\big| = \exp\big(-(Rt)^2/2\big)|f^{\mathrm{ft}}(t)| \le \exp\big(-(Rt)^2/2\big)\,\|f\|_1,$$

hence, integrability of $(\varphi_R * f)^{\mathrm{ft}}$ follows.

Now, we fix that the set of all bounded and compactly supported functions $f \in L_1(\mathbb{R})$ is dense in $L_2(\mathbb{R})$, that is, for any $f \in L_2(\mathbb{R})$, there exists a sequence $(f_n)_n$ consisting of bounded, compactly supported and integrable functions so that $\|f_n - f\|_2 \stackrel{n\to\infty}{\longrightarrow} 0$. That can be seen by putting $f_n(x) = \mathrm{med}\{-n, f(x), n\}\chi_{[-n,n]}(x)$. Further, by the definition of the Lebesgue measure, we know that any bounded, compactly supported, and integrable function f is approximable by monotonously increasing simple functions, that is, $f_n(x) = \sum_{k=1}^n f_{k,n}\,\chi_{A_{k,n}}(x)$ with some measurable sets $A_{k,n}$, in the pointwise sense. Considering $f = \max\{f,0\} - \max\{-f,0\}$, we may fix that, for any bounded, compactly supported, and integrable function f, we have a sequence $(f_n)_n$ of simple functions with $\|f_n - f\|_2 \stackrel{n\to\infty}{\longrightarrow} 0$. That approximability can be extended to the case where all $A_{k,n}$ are open and finite intervals. Finally, for an arbitrary $\varepsilon > 0$ and $f \in L_2(\mathbb{R})$, we derive that

$$\begin{aligned}
\|\varphi_R * f - f\|_2 \le {} & \Big\|\varphi_R * f - \varphi_R * \sum_{k=1}^n f_{k,n}\chi_{A_{k,n}}\Big\|_2 \\
& + \Big\|\Big(\varphi_R * \sum_{k=1}^n f_{k,n}\chi_{A_{k,n}}\Big) - \sum_{k=1}^n f_{k,n}\chi_{A_{k,n}}\Big\|_2 \\
& + \Big\|\sum_{k=1}^n f_{k,n}\chi_{A_{k,n}} - f\Big\|_2 \\
\le {} & 2\Big\|\sum_{k=1}^n f_{k,n}\chi_{A_{k,n}} - f\Big\|_2 \\
& + \sum_{k=1}^n |f_{k,n}|\Big(\int \big|[\varphi_R * \chi_{A_{k,n}}](x) - \chi_{A_{k,n}}(x)\big|^2\,\mathrm{d}x\Big)^{1/2} \\
\le {} & \varepsilon/2 + \sum_{k=1}^n |f_{k,n}|\Big(\int \Big[\int \varphi_1(y)\big|\chi_{A_{k,n}}(x - yR) - \chi_{A_{k,n}}(x)\big|\,\mathrm{d}y\Big]^2\,\mathrm{d}x\Big)^{1/2} \\
\le {} & \varepsilon/2 + \sum_{k=1}^n |f_{k,n}|\Big(\int \varphi_1(y)\underbrace{\int \big|\chi_{A_{k,n}}(x - yR) - \chi_{A_{k,n}}(x)\big|^2\,\mathrm{d}x}_{\le 2|y|R}\,\mathrm{d}y\Big)^{1/2},
\end{aligned}$$

where n, $A_{k,n}$, and $f_{k,n}$ depend on ε and f; but not on R. We have used the triangle inequality in $L_2(\mathbb{R})$ and Jensen's inequality, considering that φ_R, $R > 0$, are densities, along with Fubini's theorem. First, choosing n large enough and, then, $R > 0$ small enough, we realize that $\|\varphi_R * f - f\|_2^2$ can be bounded above by any $\varepsilon > 0$, which completes the proof. ■

Lemma A.3 provides the framework for extending the definition of the Fourier transform from $\mathcal{C} \subseteq L_1(\mathbb{R})$ to $L_2(\mathbb{R})$. Because of Lemma A.3, any $f \in L_2(\mathbb{R})$ has a corresponding sequence $(f_n)_n$ in $\mathcal{C}$, which converges to f in the $L_2(\mathbb{R})$-norm. Therefore, $(f_n)_n$ is a Cauchy sequence with respect to the $L_2(\mathbb{R})$-norm. It follows from (A.3) that

$$\|f_n - f_m\|_2^2 = \frac{1}{2\pi}\|f_n^{\mathrm{ft}} - f_m^{\mathrm{ft}}\|_2^2 \overset{m\to\infty}{\longrightarrow} 0, \quad \text{for all } n \in \mathbb{N}.$$

Hence, $(f_n^{\mathrm{ft}})_n$ is also a Cauchy sequence with respect to the $L_2(\mathbb{R})$-norm. Applying the result from functional analysis saying that $L_2(\mathbb{R})$ is complete, we conclude that $(f_n^{\mathrm{ft}})_n$ converges in $L_2(\mathbb{R})$; $L_2(\mathbb{R})$-limit then defines the Fourier transform of the function $f \in L_2(\mathbb{R})$. Note that (A.3) also guarantees uniqueness of that limit; otherwise, assume two sequences $(f_n)_n$ and $(\tilde{f}_n)_n$, which both approximate f; then, by (A.3), we have

$$\frac{1}{\sqrt{2\pi}}\|f_n^{\mathrm{ft}} - \tilde{f}_n^{\mathrm{ft}}\|_2 = \|f_n - \tilde{f}_n\|_2 \le \|f_n - f\|_2 + \|\tilde{f}_n - f\|_2 \overset{n\to\infty}{\longrightarrow} 0.$$

Therefore, $(f_n^{\mathrm{ft}})_n$ and $(\tilde{f}_n^{\mathrm{ft}})_n$ cannot converge to two different limits. Therefore, we also realize that, if $f \in L_1(\mathbb{R}) \cap L_2(\mathbb{R})$, the definitions of Fourier transforms for functions contained in $L_1(\mathbb{R})$ and $L_2(\mathbb{R})$ coincide with each other.

Note that the Fourier transform of any $L_2(\mathbb{R})$-function is also contained in $L_2(\mathbb{R})$. Further, by simple calculation and the Cauchy–Schwarz inequality, we have

$$\begin{aligned}&\Big|\langle f, g\rangle - \langle f_n, g_n\rangle\Big| \\ &\le \|g\|_2\|f - f_n\|_2 + \|g - g_n\|_2\|f\|_2 + \|f_n - f\|_2\|g_n - g\|_2 \overset{n\to\infty}{\longrightarrow} 0\end{aligned}$$

for all $f, g \in L_2(\mathbb{R})$, where $(f_n)_n$ and $(g_n)_n$ are two approximating sequences of f and g, respectively, in $\mathcal{C}$; and $\langle\cdot,\cdot\rangle$ denotes the $L_2(\mathbb{R})$-inner product. Therefore, the validity of (A.2) and, hence, (A.3) can be extended from $f \in \mathcal{F}, g \in L_1(\mathbb{R})$ to $f, g \in L_2(\mathbb{R})$. By such approximation arguments, many properties of the Fourier transform on $\mathcal{C}$ are extendable to the whole of $L_2(\mathbb{R})$. Considering Theorem A.2 again, it gives us

$$(f^{\mathrm{ft}})^{\mathrm{ft}}(x) = 2\pi f(-x), \quad \text{for all } x \in \mathbb{R},\ f \in \mathcal{C}.$$

Now, for any $f \in L_2(\mathbb{R})$, assume a sequence $(f_n)_n \subseteq \mathcal{C}$ with $\|f - f_n\|_2 \overset{n\to\infty}{\longrightarrow} 0$. Again, the existence of that sequence is guaranteed by Lemma A.3. By definition, we have $\|f^{\mathrm{ft}} - f_n^{\mathrm{ft}}\|_2 \overset{n\to\infty}{\longrightarrow} 0$. Further, by the extension of (A.3), we have

$$\|(f^{\text{ft}})^{\text{ft}} - (f_n^{\text{ft}})^{\text{ft}}\|_2 \overset{n\to\infty}{\longrightarrow} 0.$$

Therefore, we have

$$(f^{\text{ft}})^{\text{ft}}(x) = 2\pi f(-x), \quad \text{for all } x \in \mathbb{R},\ f \in L_2(\mathbb{R}). \tag{A.4}$$

Therefore, for any $f \in L_2(\mathbb{R})$, there exists some $g \in L_2(\mathbb{R})$ so that $g^{\text{ft}} = f$, namely $g = \frac{1}{2\pi} f^{\text{ft}}(-\cdot)$. Also, the solution for g is unique; since

$$g^{\text{ft}} = \tilde{g}^{\text{ft}} \Longrightarrow (g^{\text{ft}})^{\text{ft}} = (\tilde{g}^{\text{ft}})^{\text{ft}} \Longrightarrow g = \tilde{g},$$

for all $g, \tilde{g} \in L_2(\mathbb{R})$. Hence, the Fourier transform is an invertible mapping from $L_2(\mathbb{R})$ to $L_2(\mathbb{R})$. The main properties that we have derived are summarized in the following theorem.

Theorem A.4. *The Fourier transform on $L_2(\mathbb{R})$, defined by the unique continuation of the Fourier transform on $\mathcal{C}$, is a bijective mapping from $L_2(\mathbb{R})$ to $L_2(\mathbb{R})$. Its reverse mapping is equal to $f \mapsto \frac{1}{2\pi} f^{\text{ft}}(-\cdot)$. Further, we have*

$$\langle f, g\rangle = \frac{1}{2\pi}\langle f^{\text{ft}}, g^{\text{ft}}\rangle, \quad \textit{for all } f, g \in L_2(\mathbb{R}) \qquad \textit{(Plancherel's isometry)}$$

and

$$\|f\|_2^2 = \frac{1}{2\pi}\|f^{\text{ft}}\|_2^2, \quad \textit{for all } f \in L_2(\mathbb{R}). \qquad \textit{(Parseval's identity)}$$

To compare the Fourier transform on $L_1(\mathbb{R})$ to that on $L_2(\mathbb{R})$, we point out the difference that the Fourier transform does not map any $L_1(\mathbb{R})$-function to some $L_1(\mathbb{R})$-function. On the other hand, the Fourier transform of an $L_2(\mathbb{R})$-function is not necessarily continuous nor bounded, in general. Nevertheless, from Lemma A.1, we can keep the parts (a), (e), (f), and (g) for the $L_2(\mathbb{R})$-Fourier transforms where, in (g), "for all" must be replaced by "for almost all" in the Lebesgue sense; as well as a deconvolution result, given in the following lemma.

Lemma A.5. *We have*

$$(f * g)^{\text{ft}} = f^{\text{ft}} g^{\text{ft}}, \quad \textit{for all } f \in L_2(\mathbb{R}),\ g \in L_1(\mathbb{R}) \cap L_2(\mathbb{R}).$$

Proof. If, in addition, $f \in L_1(\mathbb{R}) \cap L_2(\mathbb{R})$, the assertion follows from Lemma A.1. Again, we use the fact that the intersection $L_1(\mathbb{R}) \cap L_2(\mathbb{R})$ is dense in $L_2(\mathbb{R})$, following from Lemma A.3. Then we may assume the existence of some sequence $(f_n)_n \subset L_1(\mathbb{R}) \cap L_2(\mathbb{R})$, which converges to f with respect to the $L_2(\mathbb{R})$-norm. By the boundedness of g^{ft}, which is ensured by Lemma A.1(c), and Parseval's identity, the right side of the equation is approximable by $f_n^{\text{ft}} g^{\text{ft}}$ with respect to $L_2(\mathbb{R})$-convergence. Concerning the left side, we consider that

$$\begin{aligned}\|(f_n * g)^{\mathrm{ft}} - (f * g)^{\mathrm{ft}}\|_2^2 &= 2\pi \|(f_n - f * g)\|_2^2 \\ &\leq 2\pi \int \Big| \int |f_n(x-y) - f(x-y)| |g(y)|^{1/2} |g(y)|^{1/2} \, \mathrm{d}y \Big|^2 \mathrm{d}x \\ &\leq 2\pi \int \int \big|f_n(x-y) - f(x-y)\big|^2 |g(y)| \, \mathrm{d}y \mathrm{d}x \, \|g\|_1 \\ &= 2\pi \, \|f_n - f\|_2^2 \cdot \|g\|_1^2 \overset{n\to\infty}{\longrightarrow} 0,\end{aligned}$$

by utilizing Parseval's identity, the Cauchy–Schwarz inequality, and Fubini's theorem. ∎

Also, we mention that Parseval's identity can be generalized to general spaces $L_p(\mathbb{R})$, $p > 1$. However, we then do not have equality any more. This generalization is called the *Hausdorff–Young inequality*; it says that

$$\|f^{\mathrm{ft}}\|_q \leq (2\pi)^{1/2-1/p} \|f\|_p, \tag{A.5}$$

for any $p \in [1,2]$ and $1/p + 1/q = 1$ and any $f \in L_p(\mathbb{R})$. Note that the domain of the Fourier transform can not only be extended to $L_2(\mathbb{R})$ but to any $L_q(\mathbb{R})$ with $q > 2$.

Now, we focus on smoothness and differentiability of f and on how it reflects in the Fourier transform. First, assume that $f \in L_1(\mathbb{R})$ is β-fold continuously differentiable on the whole real line, where all the derivatives $f^{(l)}(x)$, $l = 0, \ldots, \beta - 1$, converge to zero as $|x| \to \infty$. Then, we derive by $(l-1)$-times integration by parts that

$$\begin{aligned}(f^{(l)})^{\mathrm{ft}}(t) &= \int \exp(\mathrm{i}tx) f^{(l)}(x) \, \mathrm{d}x \\ &= \underbrace{\exp(\mathrm{i}tx) f^{(l-1)}(x) \Big|_{x=-\infty}^{\infty}}_{=0} - \mathrm{i}t \int \exp(\mathrm{i}tx) f^{(l-1)}(x) \, \mathrm{d}x \\ &\;\;\vdots \\ &= (-\mathrm{i}t)^l f^{\mathrm{ft}}(t),\end{aligned} \tag{A.6}$$

for all $l = 0, \ldots, \beta$. If, in addition, f and $f^{(l)}$ lie in $L_2(\mathbb{R})$, then we obtain by Parseval's identity that

$$\|f^{(l)}\|_2^2 = \frac{1}{2\pi} \int |t|^{2l} |f^{\mathrm{ft}}(t)|^2 \, \mathrm{d}t. \tag{A.7}$$

While the left side of the above equality is only well-defined for integer l, we may allow for any positive real-valued l on the right side. That provides a concept for extending the definition of differentiability and the derivatives to noninteger degrees.

An upper bound on the sum of the squared $L_2(\mathbb{R})$-norms of the derivatives $f^{(l)}$, $l = 0, \ldots, \beta$ may be assumed to formalize smoothness of the function f; this is used in mean integrated squared asymptotics, in particular. Note that

$$\sum_{l=0}^{\beta} \|f^{(l)}\|_2^2 = \frac{1}{2\pi} \int \sum_{l=0}^{\beta} |t|^{2l} |f^{\mathrm{ft}}(t)|^2 \, dt.$$

As any polynomial on the whole real line is bounded above by a constant times the sum of its terms in the smallest and the largest power, we have

$$\sum_{l=0}^{\beta} \|f^{(l)}\|_2^2 \leq \mathrm{const} \int (1 + |t|^{2l}) |f^{\mathrm{ft}}(t)|^2 \, dt.$$

On the other hand, the inequality

$$\sum_{l=0}^{\beta} \|f^{(l)}\|_2^2 \geq \frac{1}{2\pi} \cdot \int (1 + |t|^{2l}) |f^{\mathrm{ft}}(t)|^2 \, dt$$

holds true. Therefore, equivalently, we can impose smoothness conditions by

$$\int |f^{\mathrm{ft}}(t)|^2 (1 + |t|^{2\beta}) \, dt \leq C, \tag{A.8}$$

for some constants $C > 0$ and $\beta > 0$, where β is called the smoothness degree. As above, we realize that it is not problem to define condition (A.8) with noninteger smoothness degree. The assumption (A.8) is usually referred to as *Sobolev condition* on some function $f \in L_2(\mathbb{R})$.

Apart from the motivation of (A.8) by the first derivatives of f, we mention that a function need not be differentiable on the whole real line to satisfy (A.8) for $\beta = 1$, for instance. Just consider the triangle-shaped function

$$f(x) = \max\{1 - |x|, 0\},$$

which has the Fourier transform

$$f^{\mathrm{ft}}(t) = \frac{2(1 - \cos t)}{t^2},$$

which can be seen by rather simple calculation involving integration by parts. Apparently, $|f^{\mathrm{ft}}(t)|^2 \, |t|^2$ decays as $|t|^{-2}$ for $|t| \to \infty$ so that integrability of $|f^{\mathrm{ft}}(t)|^2 \, |t|^2$ can be verified. Therefore, the Sobolev condition (A.8) for $\beta = 1$ is also satisfied by some functions whose first derivatives exist only almost everywhere and are squared integrable on the whole real line. That motivates the concept of so-called *weak derivatives.*

More concretely, a function $f' \in L_2(\mathbb{R})$ is called the weak derivative of $f \in L_2(\mathbb{R})$ if

$$\langle f', g \rangle = -\langle f, g' \rangle,$$

holds true for all continuously differentiable $g \in L_1(\mathbb{R}) \cap L_2(\mathbb{R})$, with $g' \in L_1(\mathbb{R}) \cap L_2(\mathbb{R})$ and $\lim_{|x|\to\infty} g(x) = 0$. Therefore, g' may be interpreted as the classical derivative on the right side of the above equation, which is inspired by integration by parts, of course. Note that, in literature, there are several definitions of the weak derivative with respect to the conditions on g. However, as we see at the end, all those definitions are equivalent. In fact, integration by parts allows us to verify that formula for continuously differentiable square-integrable functions with square-integrable derivatives; and generalize the concept of differentiability. As the $L_2(\mathbb{R})$-inner product of some $f, g \in L_2(\mathbb{R})$, which uses the Lebesgue integral, ignores the values of f and g on sets having Lebesgue measure zero, those compactly supported and continuous functions f, whose derivatives exist up to finitely many sites and are bounded, are weakly differentiable, which affects the function $f(x) = \max\{(1-|x|), 0\}$. By definition, we may check that its weak derivative is equal to

$$f' = \chi_{(-1,0)} - \chi_{(0,1)}.$$

From there, f' is determined almost everywhere (in Lebesgue sense). In general, we conclude

Lemma A.6. *For any weakly differentiable function $f \in L_2(\mathbb{R})$, we have*

$$(f')^{\mathrm{ft}}(t) = -it f^{\mathrm{ft}}(t),$$

for Lebesgue almost all $t \in \mathbb{R}$.

Proof. Consider for any g as in the definition of the weak derivative f' that

$$\langle (f')^{\mathrm{ft}}, g^{\mathrm{ft}} \rangle = 2\pi \langle f', g \rangle = -2\pi \langle f, g' \rangle = -\langle f^{\mathrm{ft}}, (g')^{\mathrm{ft}} \rangle,$$

where we have used the Plancherel isometry (Theorem A.4) twice and the definition of f'. We may apply (A.6) with $l = 1$ as all necessary conditions for g are assumed. Hence,

$$\langle f^{\mathrm{ft}}, (g')^{\mathrm{ft}} \rangle = \langle f^{\mathrm{ft}}, (-\mathrm{i}\cdot) g^{\mathrm{ft}} \rangle = -\int f^{\mathrm{ft}}(t) \mathrm{i}t \overline{g^{\mathrm{ft}}(t)}\, dt = \langle -(\mathrm{i}\cdot) f^{\mathrm{ft}}, g^{\mathrm{ft}} \rangle.$$

Therefore,

$$\langle (f')^{\mathrm{ft}} + (\mathrm{i}\cdot) f^{\mathrm{ft}}, g^{\mathrm{ft}} \rangle = 0$$

for all g satisfying the imposed conditions. It remains to be shown that the set collecting all continuously differentiable functions $g \in L_1(\mathbb{R}) \cap L_2(\mathbb{R})$ with $g' \in L_1(\mathbb{R}) \cap L_2(\mathbb{R})$ and $\lim_{|x| \to \infty} g(x) = 0$ is dense in $L_2(\mathbb{R})$. Then, by Parseval's identity (Theorem A.4), the set consisting of the Fourier transforms of those g is also dense in $L_2(\mathbb{R})$.

Without any loss of generality, we may show that the above set is dense in $L_1(\mathbb{R}) \cap L_2(\mathbb{R})$; denseness of that latter set in $L_2(\mathbb{R})$ follows from Lemma A.3. We define, for any $f \in L_1(\mathbb{R}) \cap L_2(\mathbb{R})$, the function

$$f_a(x) = \frac{1}{\sqrt{2\pi} a} \int \exp\left(-(x-y)^2/(2a^2)\right) f(y)\, dy,$$

thus f_a, $a > 0$, is the convolution of f and the normal density with mean zero and the variance a^2. By Lemma A.5, we derive that

$$f_a^{\mathrm{ft}}(t) = \exp\Big(-\frac{1}{2}a^2t^2\Big) \cdot f^{\mathrm{ft}}(t).$$

By Parseval's identity and dominated convergence, we may show that

$$\lim_{a \downarrow 0} \|f_a - f\|_2 = 0.$$

Therefore, the proof of the lemma is completed by verifying that any f_a, $a > 0$, satisfies the conditions of g. By dominated convergence, f_a is continuously differentiable and its (classical) derivative equals

$$f_a'(x) = -\frac{1}{\sqrt{2\pi a^2}} \int (x-y) \exp\big(-(x-y)^2/(2a^2)\big) f(y)\, \mathrm{d}y.$$

Also, f_a' lies in $L_1(\mathbb{R}) \cap L_2(\mathbb{R})$ as the functions being convolved, that is, $\cdot \exp\big(-\cdot^2/(2a^2)\big)$ and f, are both in $L_1(\mathbb{R}) \cap L_2(\mathbb{R})$. Furthermore, by dominated convergence, we may conclude that $f_a(x)$ tends to zero as $|x| \to \infty$. That completes the proof. ■

Lemma A.6 has been used, for example, to prove Theorem 2.13.

A.3 Fourier Series

In the current section, we are concerned with discrete Fourier approaches; in particular, we consider Fourier series rather than continuous transforms. Therefore, we focus on functions f, which are supported on $[-\pi, \pi]$ and square-integrable. Obviously, those functions lie in $L_1(\mathbb{R}) \cap L_2(\mathbb{R})$, and we have

$$f^{\mathrm{ft}}(t) = \int_{-\pi}^{\pi} \exp(\mathrm{i}tx) f(x)\, \mathrm{d}x.$$

The values that f^{ft} takes at $t = k$, integer k, are of particular interest. They are denoted by the kth (complex) *Fourier coefficient* of f,

$$f_k = f^{\mathrm{ft}}(k).$$

Let us consider the term

$$F_n(x) = \frac{1}{2\pi} \sum_{|k| \le n} \exp(-\mathrm{i}kx) f^{\mathrm{ft}}(k).$$

We derive that, for $x \in [-\pi, \pi]$,

$$\begin{aligned} F_n(x) &= \frac{1}{2\pi} \sum_{|k| \le n} \exp(-\mathrm{i}kx) \int_{-\pi}^{\pi} \exp(\mathrm{i}kt) f(t)\, \mathrm{d}t \\ &= \frac{1}{2\pi} \int_{-\pi}^{\pi} \sum_{|k| \le n} \exp\big(\mathrm{i}k(t-x)\big) f(t)\, \mathrm{d}t \end{aligned}$$

$$= \frac{1}{2\pi} \int_{-\pi}^{\pi} \frac{\exp\big(-\mathrm{i}n(t-x)\big) - \exp\big(\mathrm{i}(n+1)(t-x)\big)}{1 - \exp\big(\mathrm{i}(t-x)\big)} f(t)\,\mathrm{d}t$$

$$= \frac{1}{2\pi} \int_{-\pi}^{\pi} \frac{\sin\big((n+1/2)(x-t)\big)}{\sin\big((x-t)/2\big)} f(t)\,\mathrm{d}t,$$

by the elementary formula for geometric sums. The function

$$D_n(s) = \frac{\sin\big((n+1/2)s\big)}{2\sin(s/2)}$$

for $s \in \mathbb{R}$ is called a Dirichlet kernel. We can show that it integrates to π over $[-\pi, \pi]$ by using the reverse decomposition as derived earlier,

$$\int_{-\pi}^{\pi} D_n(s)\,\mathrm{d}s = \frac{1}{2} \int_{-\pi}^{\pi} \sum_{|k| \le n} \exp(\mathrm{i}ks)\,\mathrm{d}s = \frac{1}{2} \sum_{|k| \le n} \int_{-\pi}^{\pi} \exp(\mathrm{i}ks)\,\mathrm{d}s$$
$$= \pi + \frac{1}{2} \sum_{|k| \le n, k \ne 0} \frac{2\sin(\pi k)}{k} = \pi.$$

Therefore,

$$\big|F_n(x) - f(x)\big| = \Big|\frac{1}{\pi} \int_{-\pi}^{\pi} D_n(x-t)\big(f(t) - f(x)\big)\,\mathrm{d}t\Big|$$
$$\le \frac{1}{\pi} \int_{|t| \le \pi, |x-t| < R} |D_n(x-t)|\big|f(t) - f(x)\big|\,\mathrm{d}t$$
$$+ \frac{1}{\pi}\Big| \int_{|t| \le \pi, |x-t| \ge R} D_n(x-t)\big(f(t) - f(x)\big)\,\mathrm{d}t\Big|$$

for some $R \in (0, \pi)$. With respect to the first integral above, we assume that f is continuously differentiable on the whole real line so that this term is bounded above by $2R\|f'\|_\infty$. Note that f and, hence, f' are supported on $[-\pi, \pi]$ and that $|D_n(s)| \le 1/\big|2\sin(s/2)\big|$ for all $s \in \mathbb{R}$. With respect to the second integral, we consider that the bivariate function

$$\varphi(x,t) = \frac{f(x) - f(t)}{\sin((x-t)/2)},$$

with $(x,t) \in [-\pi, \pi]^2 \backslash \big(\{(x,t) : |x-t| < R\} \cup \{(-\pi, \pi)\} \cup \{(\pi, -\pi)\}\big)$, is continuous and has finite limits as $(x,t) \to (-\pi, \pi)$ or $(x,t) \to (\pi, -\pi)$ since $f(\pi) = f(-\pi) = 0$ and f is continuously differentiable. That also implies partial differentiability of $\varphi(x,t)$ with respect to t on the compact set $\{t : |t| \le \pi, |x-t| \ge R\}$ so that this partial derivative is bounded on that set. Therefore, the function φ is bounded on the whole of its domain, which coincides with the set $\{(x,t) : |t| \le \pi, |x-t| \ge R\}$ up to a Lebesgue zero set. Hence, the second integral is bounded above by

$$
\begin{aligned}
& \Big|\frac{1}{2\pi}\int_{t=-\pi}^{-R+x} \sin\big((n+1/2)(x-t)\big)\varphi(x,t)\,\mathrm{d}t\Big| \\
& \qquad\qquad + \Big|\frac{1}{2\pi}\int_{t=R+x}^{\pi} \sin\big((n+1/2)(x-t)\big)\varphi(x,t)\,\mathrm{d}t\Big| \\
& \le \Big|\frac{1}{2\pi(n+1/2)} \cos\big((n+1/2)(x-t)\big)\varphi(x,t)\Big|_{t=R+x}^{\pi} \\
& \qquad - \int_{t=R+x}^{\pi} \frac{1}{2\pi(n+1/2)} \cos\big((n+1/2)(x-t)\big)\frac{\partial}{\partial t}\varphi(x,t)\,\mathrm{d}t\Big| \\
& + \Big|\frac{1}{2\pi(n+1/2)} \cos\big((n+1/2)(x-t)\big)\varphi(x,t)\Big|_{t=-\pi}^{-R+x} \\
& \qquad - \int_{t=-\pi}^{-R+x} \frac{1}{2\pi(n+1/2)} \cos\big((n+1/2)(x-t)\big)\frac{\partial}{\partial t}\varphi(x,t)\,\mathrm{d}t\Big| \\
& = O(1/n)
\end{aligned}
$$

for any fixed small $R > 0$ where we have used integration by parts. By selecting $R > 0$ arbitrarily small and, then, n arbitrarily large, we may finally verify the pointwise convergence of $F_n(x)$ to $f(x)$ for any $x \in [-\pi, \pi]$ and any function f, which is supported on $[-\pi, \pi]$ and continuously differentiable on the whole real line. We notice that $F_n(x)$ is 2π-periodic. Therefore, when the restriction of f to the domain $[-\pi, \pi]$ is 2π-periodically continued to $\mathbb{R}$, we may fix pointwise convergence for all $x \in \mathbb{R}$.

If, in addition, we assume that f is twofold continuously differentiable on $\mathbb{R}$, we derive $|f^{\mathrm{ft}}(t)| \le O(|t|^{-2})$, for large $|t|$, following from (A.6). Then, we derive that

$$|F_n(x)| \le \frac{1}{2\pi} \sum_{k=-\infty}^{\infty} |f^{\mathrm{ft}}(k)| < \infty,$$

for all $x \in [-\pi, \pi]$ so that

$$\sup_{x\in[-\pi,\pi],\, n>0} \big|F_n(x) - f(x)\big|^2 < \infty.$$

Note that f is continuous and compactly supported and, hence, a bounded function. By dominated convergence, we conclude that

$$\big\|(F_n - f)\cdot\chi_{[-\pi,\pi]}\big\|_2^2 = \int_{-\pi}^{\pi} \big|F_n(x) - f(x)\big|^2\,\mathrm{d}x \stackrel{n\to\infty}{\longrightarrow} 0.$$

Summarizing, we have shown that the class of the trigonometric polynomials,

$$\Big\{\sum_{|k|\le n} c_k \exp(\mathrm{i}k\cdot)\chi_{[-\pi,\pi]} : c_k \in \mathbb{C},\, n \in \mathbb{N}\Big\},$$

lies dense in the set consisting of all those $f \in L_2(\mathbb{R})$, which are supported on $[-\pi, \pi]$ and twofold continuously differentiable on $\mathbb{R}$, with respect to the

$\|\cdot\|_2$-norm. The restriction to twofold continuously differentiable functions can be removed by the following consideration: For any f, which is supported on $[-\pi,\pi]$, consider the function

$$f_a = \left[\vartheta_a * \left(f \cdot \chi_{[-\pi+2/a,\pi-2/a]}\right)\right],$$

where $\vartheta_a = a^{-1}\vartheta(\cdot/a)$, $a > 0$, for some φ, which is twofold continuously differentiable on $\mathbb{R}$ and supported on $[-1,1]$. Then, f_a'' exists and is continuous (again, by dominated convergence) and f_a is supported on $[-\pi,\pi]$. Furthermore, by Parseval's identity and Lemma A.1(e),

$$\begin{aligned}
&\|\vartheta_a * (f\chi_{[-\pi+2/a,\pi-2/a]}) - f\|_2 = \frac{1}{\sqrt{2\pi}}\|\vartheta^{\mathrm{ft}}(a\cdot)(f\chi_{[-\pi+2/a,\pi-2/a]})^{\mathrm{ft}} - f^{\mathrm{ft}}\|_2 \\
&\le \frac{1}{\sqrt{2\pi}}\left\|\vartheta^{\mathrm{ft}}(a\cdot)\left[(f\chi_{[-\pi+2/a,\pi-2/a]})^{\mathrm{ft}} - f^{\mathrm{ft}}\right]\right\|_2 + \frac{1}{\sqrt{2\pi}}\left\|\left[\vartheta^{\mathrm{ft}}(a\cdot) - 1\right]f^{\mathrm{ft}}\right\|_2 \\
&\le \|\vartheta\|_1 \left\|f\chi_{[-\pi+2/a,\pi-2/a]} - f\right\|_2 + \frac{1}{\sqrt{2\pi}}\left\|\left[\vartheta^{\mathrm{ft}}(a\cdot) - 1\right]f^{\mathrm{ft}}\right\|_2,
\end{aligned}$$

as ϑ^{ft} is uniformly bounded by $\|\vartheta\|_1$. Then, dominated convergence allows us to fix that both aforementioned integrals converge to zero as $a \downarrow 0$. Hence, we may approximate any $f \in L_2(\mathbb{R})$, which is supported on $[-\pi,\pi]$ by twofold continuously differentiable functions with respect to $\|\cdot\|_2$.

Now, it is evident that the trigonometric polynomials on $[-\pi,\pi]$ are dense in the class of all $L_2(\mathbb{R})$-functions, which are supported on $[-\pi,\pi]$, with respect to $\|\cdot\|_2$.

We may easily derive orthonormality of the scaled trigonometric polynomials $\frac{1}{\sqrt{2\pi}}\exp(\mathrm{i}k\cdot)$, integer k, as

$$\left\langle \frac{1}{\sqrt{2\pi}}\exp(\mathrm{i}k\cdot), \frac{1}{\sqrt{2\pi}}\exp(\mathrm{i}j\cdot)\right\rangle = \delta_{j,k}$$

can be verified by simple integration. Using the argument as in (2.12), we realize that, for any $f \in L_2(\mathbb{R})$, supported on $[-\pi,\pi]$, the function

$$F_n(x) = \frac{1}{2\pi}\sum_{|k|\le n}\langle f, \exp(-\mathrm{i}k\cdot)\rangle \exp(-\mathrm{i}kx) = \frac{1}{2\pi}\sum_{|k|\le n} f^{\mathrm{ft}}(k)\exp(-\mathrm{i}kx)$$

is the best approximation of f within the class of all trigonometric polynomials on $[-\pi,\pi]$ with degree $\le n$. Combining that with the denseness result for the trigonometric polynomials, we have shown that the trigonometric polynomials represent an orthonormal basis of all $L_2(\mathbb{R})$-functions supported on $[-\pi,\pi]$; and all such functions may be represented by their (complex) Fourier series

$$f(x) = \frac{1}{2\pi}\sum_{k=-\infty}^{\infty}\exp(-\mathrm{i}kx)f^{\mathrm{ft}}(k)\cdot\chi_{[\pi,\pi]}(x),$$

where the infinite sum is to be understood as an $L_2(\mathbb{R})$-limit. That representation also allows us to derive a discrete version of Parseval's identity. Consider that

$$\begin{aligned}
& \Big\| \frac{1}{2\pi} \sum_{|k|\leq n} f^{\mathrm{ft}}(k) \exp(-\mathrm{i}k\cdot)\chi_{[\pi,\pi]} \Big\|_2^2 \\
&= \frac{1}{(2\pi)^2} \sum_{|k|,|k'|\leq n} f^{\mathrm{ft}}(k)\overline{f^{\mathrm{ft}}(k')} \int_{-\pi}^{\pi} \exp\big(-\mathrm{i}(k-k')x\big)\,\mathrm{d}x \\
&= \frac{1}{2\pi} \sum_{|k|,|k'|\leq n} f^{\mathrm{ft}}(k)\overline{f^{\mathrm{ft}}(k')} \cdot \delta_{k,k'} \\
&= \frac{1}{2\pi} \sum_{|k|\leq n} \big|f^{\mathrm{ft}}(k)\big|^2
\end{aligned}$$

holds true for all integer n. Because of the basic inequality $\|f-g\| \geq \big|\|f\|-\|g\|\big|$ for all norms $\|\cdot\|$, we may put $n \to \infty$ in the above equation, leading to

$$\|f\|_2^2 = \frac{1}{2\pi} \sum_k \big|f^{\mathrm{ft}}(k)\big|^2,$$

thus, we have an analogous result for Parseval's identity as in Theorem A.4, referring to functions, which are supported on $[-\pi,\pi]$. The integral is replaced by a sum taken over all integers.

We may summarize our findings in the following theorem.

Theorem A.7. *Assume that $f \in L_2(\mathbb{R})$ is supported on $[-\pi,\pi]$. Then, as $n \to \infty$, we have*

$$\int_{-\pi}^{\pi} \Big| f(x) - \frac{1}{2\pi} \sum_{|k|\leq n} \exp(-\mathrm{i}kx) f^{\mathrm{ft}}(k) \Big|^2 \,\mathrm{d}x \overset{n\to\infty}{\longrightarrow} 0\,.$$

Furthermore, the following discrete version of Parseval's identity holds true

$$\|f\|_2^2 = \frac{1}{2\pi} \sum_k \big|f^{\mathrm{ft}}(k)\big|^2,$$

where the sum is to be taken over all integer k.

Combining the Theorems A.4 and A.7, it is interesting to see that

$$\int \big|f^{\mathrm{ft}}(t)\big|^2 \,\mathrm{d}t = \sum_k \big|f^{\mathrm{ft}}(k)\big|^2$$

holds true for all $f \in L_2(\mathbb{R})$, which are supported on $[-\pi,\pi]$.

Simple rescaling techniques allow to extend the Fourier series approach to those functions f with a more general compact support. For any function

f supported on $[a,b]$ with $a < b$, we may introduce the function $\tilde{f}(x) = f\big((b-a)x/(2\pi) + (a+b)/2\big)$, which vanishes outside the interval $[-\pi,\pi]$. Then, Theorem A.7 gives us

$$\tilde{f} = \frac{1}{2\pi}\sum_k \tilde{f}^{\mathrm{ft}}(k)\,\exp(-\mathrm{i}k\cdot)\,\chi_{[-\pi,\pi]},$$

viewed as an $L_2(\mathbb{R})$-limit. We have

$$\tilde{f}^{\mathrm{ft}}(k) = \frac{2\pi}{b-a}\,\exp\Big(-\mathrm{i}\pi k\frac{a+b}{b-a}\Big)\,f^{\mathrm{ft}}\big(2\pi k/(b-a)\big),$$

for all integer k, according to Lemma A.1(e). It follows from there that

$$\begin{aligned}&f\big((b-a)\cdot/(2\pi)+(a+b)/2\big)\\&=\frac{1}{b-a}\sum_k \exp\big(-\mathrm{i}k\cdot\big)\,\exp\Big(-\mathrm{i}\pi k\frac{a+b}{b-a}\Big)\,f^{\mathrm{ft}}\big(2\pi k/(b-a)\big)\,\chi_{[-\pi,\pi]}.\end{aligned}$$

By simple substitution, we obtain that

$$f = \frac{1}{b-a}\sum_k \exp\Big(-\frac{2\pi\mathrm{i}k}{b-a}\cdot\Big)\,f^{\mathrm{ft}}\big(2\pi k/(b-a)\big)\chi_{[a,b]}\,. \tag{A.9}$$

Also, Parseval's identity may be extended to functions with general support. From Theorem A.7 we derive that

$$\|\tilde{f}\|_2^2 = \frac{1}{2\pi}\sum_k \big|\tilde{f}^{\mathrm{ft}}(k)\big|^2 = \frac{1}{2\pi}\sum_k\Big(\frac{2\pi}{b-a}\Big)^2\big|f^{\mathrm{ft}}\big(2\pi k/(b-a)\big)\big|^2,$$

and, hence,

$$\|f\|_2^2 = \frac{1}{b-a}\sum_k \big|f^{\mathrm{ft}}\big(2\pi k/(b-a)\big)\big|^2 \tag{A.10}$$

holds true for all $f \in L_2(\mathbb{R})$, which are supported on $[a,b]$.

We are also able to define smoothness properties by the Fourier coefficients. Let us consider $L_2(\mathbb{R})$-functions f, which are β-fold continuously differentiable on the whole real line, for integer $\beta > 0$, and which satisfy $f^{(l)}(a) = f^{(l)}(b)$ for all $l = 0,\ldots,\beta-1$ and some $a < b$. As in (A.6), we derive that

$$\begin{aligned}&\big(\chi_{[a,b]}f^{(\beta)}\big)^{\mathrm{ft}}\big(2\pi k/(b-a)\big) = \int_a^b \exp\big(2\pi\mathrm{i}kx/(b-a)\big)f^{(\beta)}(x)\,\mathrm{d}x\\&=\underbrace{\exp\big(2\pi\mathrm{i}kx/(b-a)\big)f^{(\beta-1)}(x)\Big|_{x=a}^{b}}_{=0}\\&\qquad\qquad -\frac{2\pi\mathrm{i}k}{b-a}\int_a^b \exp\big(2\pi\mathrm{i}kx/(b-a)\big)f^{(\beta-1)}(x)\,\mathrm{d}x\end{aligned}$$

$$= -\frac{2\pi \mathrm{i} k}{b-a} \left(\chi_{[a,b]} f^{(\beta-1)}\right)^{\mathrm{ft}} \left(2\pi k/(b-a)\right)$$

$$\vdots$$

$$= \Big(-\frac{2\pi \mathrm{i} k}{b-a}\Big)^{\beta} \left(\chi_{[a,b]} f\right)^{\mathrm{ft}} \left(2\pi k/(b-a)\right),$$

as $\exp\big(2\pi \mathrm{i} k b/(b-a)\big) = \exp\big(2\pi \mathrm{i} k a/(b-a)\big)$ for all integer k, and f is supported on $[a,b]$. Applying (A.10), it follows that

$$\int_a^b \left|f^{(\beta)}(x)\right|^2 \mathrm{d}x = \frac{1}{b-a} \sum_k \Big(\frac{2\pi k}{b-a}\Big)^{2\beta} \left|f^{\mathrm{ft}}\left(2\pi k/(b-a)\right)\right|^2 \tag{A.11}$$

for all functions f, which are supported on $[a,b]$ and whose $(b-a)$-periodic continuation to $\mathbb{R}$ is β-fold continuously differentiable, where the values of each derivative with degree 0 to $\beta-1$ at a and b coincide. Therein, the βth derivative of this continued function is denoted by $f^{(\beta)}(x)$ for all $x \in \mathbb{R}$. Analogously to (2.30), we may define the discrete Sobolev condition by (A.11). More concretely, we may assume that

$$\sum_k \Big(\frac{2\pi k}{b-a}\Big)^{2\beta} \left|f^{\mathrm{ft}}\left(2\pi k/(b-a)\right)\right|^2 \leq C, \tag{A.12}$$

with some uniform constant $C > 0$. Using (A.11), the condition (A.12) may be interpreted as a smoothness condition on f restricted to the interval $[a,b]$. In addition, the coincidence of the derivatives with degree 0 to $\beta - 1$ at the support boundaries must be appreciated. This condition allows us to continue the domain of the restriction of f to $[a,b]$ to the whole of $\mathbb{R}$ $(b-a)$-periodically so that the continuation is l-fold continuously differentiable on the whole real line. Therefore, under certain circumstances, the condition (A.12) allows for jump discontinuities of f at a and b. For instance, (A.12) is still satisfied for $\beta = 1$ as long as those jumps have the same height; and the same one-side derivatives occur at the endpoints of the interval. The endpoints are obviously exempted from the smoothness constraints while the smooth periodic continuability of the restriction of f to the domain $[a,b]$ must be respected. For that consideration, the function $\chi_{[a,b]}$ is particularly interesting. In spite of its discontinuity at a and b, we have

$$\chi_{[a,b]}^{\mathrm{ft}}(t) = \frac{2}{t} \sin\big((b-a)t/2\big) \exp\big(\mathrm{i}t(b+a)/2\big),$$

so that all Fourier coefficients occurring in (A.12) vanish except the coefficient for $k=0$. Hence, there exists some $C > 0$ sufficiently large so that (A.12) is satisfied for any $\beta > 0$. This can be understood as the periodic continuation of $\chi_{[a,b]}$, restricted to $[a,b]$, is equal to the constant function $\tilde{f} \equiv 1$, which is perfectly smooth, so to say.

Also, we mention that the representation (A.12) also allows for noninteger smoothness degrees $\beta > 0$.

Furthermore, we mention that the complex Fourier series (A.9) may be written as a sine–cosine series. We realize that

$$\begin{aligned}
f &= \frac{1}{2\pi}\sum_k \exp(-\mathrm{i}k\cdot)f^{\mathrm{ft}}(k)\chi_{[-\pi,\pi]} \\
&= \frac{1}{2\pi}\sum_k \cos(k\cdot)f^{\mathrm{ft}}(k)\chi_{[-\pi,\pi]} - \mathrm{i}\cdot\frac{1}{2\pi}\sum_k \sin(k\cdot)f^{\mathrm{ft}}(k)\chi_{[-\pi,\pi]} \\
&= \frac{1}{2\pi}\sum_k \cos(k\cdot)\big(\mathrm{Re}\, f^{\mathrm{ft}}(k)\big)\chi_{[-\pi,\pi]} + \mathrm{i}\cdot\frac{1}{2\pi}\sum_k \cos(k\cdot)\big(\mathrm{Im}\, f^{\mathrm{ft}}(k)\big)\chi_{[-\pi,\pi]} \\
&\quad - \mathrm{i}\cdot\frac{1}{2\pi}\sum_k \sin(k\cdot)\big(\mathrm{Re}\, f^{\mathrm{ft}}(k)\big)\chi_{[-\pi,\pi]} + \frac{1}{2\pi}\sum_k \sin(k\cdot)\big(\mathrm{Im}\, f^{\mathrm{ft}}(k)\big)\chi_{[-\pi,\pi]} \\
&= \frac{1}{2\pi}\sum_k \cos(k\cdot)f^{\cos}(k)\chi_{[-\pi,\pi]} + \frac{1}{2\pi}\sum_k \sin(k\cdot)f^{\sin}(k)\chi_{[-\pi,\pi]} \\
&\quad + \mathrm{i}\cdot\Big(\frac{1}{2\pi}\sum_k \cos(k\cdot)f^{\sin}(k)\chi_{[-\pi,\pi]} - \frac{1}{2\pi}\sum_k \sin(k\cdot)f^{\cos}(k)\chi_{[-\pi,\pi]}\Big),
\end{aligned}$$

where, at this stage, we define $f^{\cos}$ and $f^{\sin}$ by the real and imaginary part of f^{ft}, respectively. In most applications, f can be seen as a real-valued function (e.g., f is assumed to be a density or a regression function) so that the imaginary part of the above term may be omitted. That gives us the representation

$$f = \frac{1}{2\pi}\sum_k \cos(k\cdot)f^{\cos}(k)\chi_{[-\pi,\pi]} + \frac{1}{2\pi}\sum_k \sin(k\cdot)f^{\sin}(k)\chi_{[-\pi,\pi]} \tag{A.13}$$

for all real-valued $f \in L_2(\mathbb{R})$, which are supported on $[-\pi,\pi]$. Then we derive that

$$\begin{aligned}
f^{\cos}(t) &= \int \cos(tx)f(x)\,\mathrm{d}x, \\
f^{\sin}(t) &= \int \sin(tx)f(x)\mathrm{d}x,
\end{aligned}$$

so that the notation is justified. Let us study how convolution affects the sine–cosine transforms. We obtain that

$$\begin{aligned}
&(f*g)^{\cos}(t) \\
&= \mathrm{Re}\,\big(f^{\mathrm{ft}}(t)g^{\mathrm{ft}}(t)\big) = \mathrm{Re}\,\big[\big(f^{\cos}(t)+\mathrm{i}f^{\sin}(t)\big)\big(g^{\cos}(t)+\mathrm{i}g^{\sin}(t)\big)\big] \\
&= \mathrm{Re}\,\big[f^{\cos}(t)g^{\cos}(t) - f^{\sin}(t)g^{\sin}(t) + \mathrm{i}f^{\cos}(t)g^{\sin}(t) + \mathrm{i}f^{\sin}(t)g^{\cos}(t)\big] \\
&= f^{\cos}(t)g^{\cos}(t) - f^{\sin}(t)g^{\sin}(t).
\end{aligned}$$

In the same way, we derive that

$$(f*g)^{\sin}(t) = \mathrm{Im}\,\big(f^{\mathrm{ft}}(t)g^{\mathrm{ft}}(t)\big) = f^{\cos}(t)g^{\sin}(t) + f^{\sin}(t)g^{\cos}(t).$$

Now assume a deconvolution problem where we shall reconstruct the sine–cosine coefficients of f from those of $f * g$ when the coefficients of g are known. Then, we have to solve the following system of linear equations.

$$\begin{pmatrix} g^{\cos}(t) & -g^{\sin}(t) \\ g^{\sin}(t) & g^{\cos}(t) \end{pmatrix} \begin{pmatrix} f^{\cos}(t) \\ f^{\sin}(t) \end{pmatrix} = \begin{pmatrix} (f * g)^{\cos}(t) \\ (f * g)^{\sin}(t) \end{pmatrix}. \tag{A.14}$$

Hence, we realize that the sine–cosine approach is also applicable for deconvolution topics; it is used in [59] and [30], for instance.

A.4 Multivariate Case

The concept of Fourier transforms can be extended to the multivariate setting. Now consider d-variate functions f mapping $\mathbb{R}^d$ to $\mathbb{R}$. Then the (d-variate) Fourier transform of f is defined by

$$\begin{aligned} f^{\mathrm{ft}}(t) &= \int \exp\left(\mathrm{i}t \cdot x\right) f(x)\,\mathrm{d}x \\ &= \int \cdots \int \exp\Big(\mathrm{i}\sum_{j=1}^{d} t_j x_j\Big) f(x_1, \ldots, x_d)\,\mathrm{d}x_1 \cdots \mathrm{d}x_d\,. \end{aligned}$$

Thus, $t \cdot x$ denotes the standard inner product (dot-product) in $\mathbb{R}^d$ in the above equation. Hence we write $t = (t_1, \ldots, t_d)$ and $x = (x_1, \ldots, x_d)$. Note that the d-variate Fourier transform maps $\mathbb{R}^d$ to $\mathbb{C}$.

Most results derived for univariate functions can be extended to the multivariate setting, in particular Lemma A.1(a)–(d). We will prove the important deconvolution formula Lemma A.1(b) or, in the $L_2(\mathbb{R}^d)$ setting, Lemma A.5 by extension from the corresponding univariate results. We have, by Fubini's theorem,

$$\begin{aligned} (f * g)^{\mathrm{ft}}(t) = \int \cdots \int \exp\Big(i\sum_{j=1}^{d} x_j t_j\Big) \int \cdots \int f(x_1 - y_1, \ldots, x_d - y_d) \\ g(y_1, \ldots, y_d)\,\mathrm{d}y_1 \cdots \mathrm{d}y_d \mathrm{d}x_1 \cdots \mathrm{d}x_d \end{aligned}$$

$$\begin{aligned} &= \int \cdots \int \exp\Big(i\sum_{j=1}^{d} y_j t_j\Big) g(y_1, \ldots, y_d) \\ &\quad \int \cdots \int \exp\Big(i\sum_{j=1}^{d} (x_j - y_j) t_j\Big) f(x_1 - y_1, \ldots, x_d - y_d)\,\mathrm{d}x_1 \cdots \mathrm{d}x_d \mathrm{d}y_1 \cdots \mathrm{d}y_d \end{aligned}$$

$$= \int \cdots \int \exp\Big(i \sum_{j=1}^{d} y_j t_j\Big) g(y_1, \ldots, y_d)\, \mathrm{d}y_1 \cdots \mathrm{d}y_d$$

$$\cdot \int \cdots \int \exp\Big(i \sum_{j=1}^{d} x_j t_j\Big) f(x_1, \ldots, x_d) \mathrm{d}x_1 \cdots \mathrm{d}x_d$$

$$= g^{\mathrm{ft}}(t) \cdot f^{\mathrm{ft}}(t). \tag{A.15}$$

With respect to Fourier inversion as in Theorem A.2, we consider a bounded and continuous d-variate function $f \in L_1(\mathbb{R}^d)$ with integrable d-variate Fourier transform. Then,

$$\int \exp(-\mathrm{i}t \cdot x) f^{\mathrm{ft}}(t)\, \mathrm{d}t = \int \cdots \int \exp\Big(-i \sum_{j=1}^{d} t_j x_j\Big) f^{\mathrm{ft}}(t_1, \ldots, t_d)\, \mathrm{d}t_1 \cdots \mathrm{d}t_d$$

$$= \int \exp(-\mathrm{i}t_1 x_1) \cdots \int \exp(-\mathrm{i}t_d x_d) f^{\mathrm{ft}}(t_1, \ldots, t_d)\, \mathrm{d}t_d \cdots \mathrm{d}t_1, \tag{A.16}$$

by Fubini's theorem. Now we focus on the inner integral first. Consider $f^{\mathrm{ft}}(t_1, \ldots, t_{d-1}, \cdot)$ as a univariate function for arbitrary but fixed $t_1, \ldots, t_d \in \mathbb{R}$. Then, introducing the function

$$\varphi_{(t_1,\ldots,t_{d-1})}(x_d) = \int \cdots \int \exp\Big(i \sum_{j=1}^{d-1} x_j t_j\Big) f(x_1, \ldots, x_d)\, \mathrm{d}x_1 \cdots \mathrm{d}x_{d-1}$$

$$= \phi^{\mathrm{ft}}_{x_d}(t_1, \ldots, t_{d-1}),$$

with $\phi_{x_d}(x_1, \ldots, x_{d-1}) = f(x_1, \ldots, x_d)$, we have

$$f^{\mathrm{ft}}(t_1, \ldots, t_{d-1}, t_d) = \int \exp(\mathrm{i}x_d t_d) \varphi_{(t_1,\ldots,t_{d-1})}(x_d)\, \mathrm{d}x_d = \varphi^{\mathrm{ft}}_{(t_1,\ldots,t_{d-1})}(t_d).$$

Then, we may apply Theorem A.2 to the inner integral in (A.16) so that

$$\int \exp(-\mathrm{i}t \cdot x) f^{\mathrm{ft}}(t)\, \mathrm{d}t$$

$$= \int \exp(-\mathrm{i}t_1 x_1) \cdots \int \exp(-\mathrm{i}t_{d-1} x_{d-1})\, 2\pi \varphi_{(t_1,\ldots,t_{d-1})}(x_d)\, \mathrm{d}t_{d-1} \cdots \mathrm{d}t_1$$

$$= 2\pi \int \exp(-\mathrm{i}t_1 x_1) \cdots \int \exp(-\mathrm{i}t_{d-1} x_{d-1})\, \phi^{\mathrm{ft}}_{x_d}(t_1, \ldots, t_{d-1})\, \mathrm{d}t_{d-1} \cdots \mathrm{d}t_1. \tag{A.17}$$

Then, we may apply the same procedure again when the function ϕ_{x_d} replaces f. After that step, we obtain that

$$\int \exp(-\mathrm{i}t \cdot x) f^{\mathrm{ft}}(t)\, \mathrm{d}t$$

$$= (2\pi)^2 \int \exp(-\mathrm{i}t_{d-1} x_{d-1})\, \phi^{\mathrm{ft}}_{x_d, x_{d-1}}(t_1, \ldots, t_{d-2})\, \mathrm{d}t_{d-2} \cdots \mathrm{d}t_1,$$

where $\phi_{x_d,x_{d-1}}(x_1,\ldots,x_{d-2}) = f(x_1,\ldots,x_d)$. Then, after applying $d-2$ in further steps, we obtain that

$$f(x_1,\ldots,x_d) = (2\pi)^{-d} \int \exp(-\mathrm{i}t \cdot x) f^{\mathrm{ft}}(t)\,\mathrm{d}t. \tag{A.18}$$

Therefore, Theorem A.2 can be extended to the d-variate case by d-fold iteration of its univariate version. As an important difference, we notice that the scaling factor of the inverse Fourier integral has changed from $(2\pi)^{-1}$ to $(2\pi)^{-d}$.

Also, the Plancherel isometry (Theorem A.4) is extendable to the d-variate setting. We consider two d-variate functions $f, g \in L_1(\mathbb{R}^d) \cap L_2(\mathbb{R}^d)$. Then, by Fubini's theorem,

$$\begin{aligned}
&\langle f^{\mathrm{ft}}, g^{\mathrm{ft}} \rangle \\
&= \underbrace{\int \cdots \int}_{t_1,\ldots,t_d} \underbrace{\int \cdots \int}_{x_1,\ldots,x_d} \exp(\mathrm{i}t_1 x_1) \cdot \ldots \cdot \exp(\mathrm{i}t_d x_d) f(x_1,\ldots,x_d)\,\mathrm{d}x_1 \cdots \mathrm{d}x_d \\
&\quad \cdot \underbrace{\int \cdots \int}_{y_1,\ldots,y_d} \exp(-\mathrm{i}t_1 y_1) \cdot \ldots \cdot \exp(-\mathrm{i}t_d y_d) \overline{g(y_1,\ldots,y_d)}\,\mathrm{d}y_1 \cdots \mathrm{d}y_d \mathrm{d}t_1 \cdots \mathrm{d}t_d \\
&= \int_{t_1} \int_{x_1} \exp(\mathrm{i}t_1 x_1) \int_{y_1} \exp(-\mathrm{i}t_1 y_1) \cdots \Big(\int_{t_d} \int_{x_d} \exp(\mathrm{i}t_d x_d) f(x_1,\ldots,x_d)\,\mathrm{d}x_d \\
&\qquad \int_{y_d} \exp(-\mathrm{i}t_d y_d) \overline{g(y_1,\ldots,y_d)}\,\mathrm{d}y_d \mathrm{d}t_d \Big) \mathrm{d}t_1 \cdots \mathrm{d}t_{d-1} \\
&= 2\pi \int_{t_1} \int_{x_1} \exp(\mathrm{i}t_1 x_1) \int_{y_1} \exp(\mathrm{i}t_1 y_1) \\
&\qquad \cdots \underbrace{\int f(x_1,\ldots,x_d) \overline{g(x_1,\ldots,x_d)} \mathrm{d}x_d}_{=\varphi(x_1,\ldots,x_{d-1})} \mathrm{d}t_1 \cdots \mathrm{d}t_{d-1},
\end{aligned}$$

where Theorem A.4 has been used in the last step. We may utilize the same technique again when φ replaces f. Then, repeated application of that procedure and, finally, extension to all $f, g \in L_2(\mathbb{R}^d)$ by the same arguments as in the univariate case lead to

$$\langle f^{\mathrm{ft}}, g^{\mathrm{ft}} \rangle = (2\pi)^d \langle f, g \rangle, \tag{A.19}$$

and Parseval's identity follows the line by putting $f = g$.

Finally, we extend the Fourier series to the multivariate setting. Therefore, assume a function f, which is supported on the d-dimensional cube $[-\pi,\pi]^d$ and square integrable. Consider the series

$$\sum_{k_1=-K_1}^{K_1} \cdots \sum_{k_d=-K_d}^{K_d} f^{\text{ft}}(k_1, \ldots, k_d) \exp(\mathrm{i}k_1x_1) \cdots \exp(\mathrm{i}k_dx_d)$$

$$= \sum_{k_1=-K_1}^{K_1} \exp(\mathrm{i}k_1x_1) \cdots \sum_{k_d=-K_d}^{K_d} \exp(\mathrm{i}k_dx_d) \int \exp(\mathrm{i}k_dy_d)$$

$$\Big[\int \cdots \int \exp\Big(\mathrm{i}\sum_{j=1}^{d-1} k_jy_j\Big) f(y_1, \ldots, y_d)\, \mathrm{d}y_1 \cdots \mathrm{d}y_{d-1}\Big] \mathrm{d}y_d.$$

Putting $K_d \to \infty$, we may apply Theorem A.7 to the inner sum; and we obtain equality to

$$2\pi \sum_{k_1=-K_1}^{K_1} \exp(\mathrm{i}k_1x_1) \cdots \sum_{k_{d-1}=-K_{d-1}}^{K_{d-1}} \exp(\mathrm{i}k_{d-1}x_{d-1})$$

$$\int \cdots \int \exp\Big(\mathrm{i}\sum_{j=1}^{d-1} k_jy_j\Big) f(y_1, \ldots, y_d)\, \mathrm{d}y_1 \cdots \mathrm{d}y_{d-1},$$

so that, again, repeated use of Theorem A.7 leads to the representation

$$f(x) = (2\pi)^{-d} \sum_{k \in \mathbb{Z}^d} \exp\big(\mathrm{i}x \cdot k\big) f^{\text{ft}}(k) \cdot \chi_{[-\pi,\pi]^d}(x) \tag{A.20}$$

as a generalization of Theorem A.7 in the multivariate case; where the infinite sum is to be understood as the $L_2(\mathbb{R}^d)$-limit. Also, the discrete version of Parseval's identity has its multivariate analogue, namely

$$\int_{x_1=-\pi}^{\pi} \cdots \int_{x_d=-\pi}^{\pi} \big|f(x_1, \ldots, x_d)\big|^2 \mathrm{d}x_1 \cdots \mathrm{d}x_d$$
$$= (2\pi)^{-d} \sum_{(k_1,\ldots,k_d) \in \mathbb{Z}^d} \big|f^{\text{ft}}(k_1, \ldots, k_d)\big|^2. \tag{A.21}$$

Thus, (A.20) and (A.21) are valid for all square integrable functions f and g, which are supported on the d-dimensional cube $[-\pi, \pi]^d$. Extensions to more general supports in the spirit of (A.9) are also possible.

For more topics about Fourier analysis, we refer to the books of [49, 77, 127].

B

List of Symbols

Symbol	Definition/Description
$\mathbb{N}$	Set consisting of all positive integers
$\mathbb{Z}$	Set consisting of all integers
$\mathbb{Q}$	Set consisting of all rational numbers
$\mathbb{R}$	Set consisting of all real numbers
$\mathbb{C}$	Set consisting of all complex numbers
$A \times B$	Set product
A^n	Set power, that is, $A = \underbrace{A \times \cdots \times A}_{n\text{–fold}}$
$A \cap B$	Intersection of the sets A and B
$A \cup B$	Union of the sets A and B
$\bigcap_{i \in I} A_i$	Intersection of all the sets A_i where $i \in I$
$\bigcup_{i \in I} A_i$	Union of all the sets A_i where $i \in I$
$A \backslash B$	Set consisting of those elements of A which are not in B
$A \subseteq B$	The assertion: A is a subset of B
$\#A$	The number of elements contained in the set A
$\lfloor x \rfloor$	The largest integer, which is smaller or equal to $x \in \mathbb{R}$
$\lceil x \rceil$	The smallest integer, which is larger or equal to $x \in \mathbb{R}$
$\operatorname{sign} x$	Sign of x
$A \wedge B$	The assertion: A and B hold true
$A \vee B$	The assertion: A or B hold true
$K : A \rightarrow B$	Function K mapping the set A to B,
$a \mapsto b$	and mapping the element a to b
$[a, b]$	The interval $\{x \in \mathbb{R} : a \leq x \leq b\}$
(a, b)	The interval $\{x \in \mathbb{R} : a < x < b\}$
$[a, b)$	The interval $\{x \in \mathbb{R} : a \leq x < b\}$
$(a, b]$	The interval $\{x \in \mathbb{R} : a < x \leq b\}$
$B_\varepsilon(x)$	Ball around x with the radius ε
$\Longrightarrow$	Implies
$\equiv$	Is equal to the constant

$\approx$	Approximately equal to
$\binom{n}{k}$	Binomial coefficient $n!/[k!(n-k)!]$
sup	Supremum
inf	Infimum
$f(x)\Big\vert_{x=a}^{b}$	Equal to $f(b)-f(a)$
v^t	The transposed version of the vector or matrix v
$\operatorname{supp} f$	The support of the function f
i.i.d.	Independent identically distributed
a.s.	Almost sure(ly)
$P(A)$	Probability measure of the set A
$P(A\vert B)$	Conditional probability measure of A given B
$P[A]$	Probability that the assertion A is true
$\mathrm{d}P$	Differential for the probability measure P
$\lambda(A)$	Lebesgue measure of the set A
$\mathfrak{B}(\mathbb{R}^d)$	Borel algebra of $\mathbb{R}^d$
$E\,X$	Expectation of the random variable X
$E(X\vert Y)$	Conditional expectation of X given Y
$\operatorname{var} X$	Variance of the random variable X
$\operatorname{var}(X\vert Y)$	Conditional variance of the random variable X given Y
$(a_n)_n$	Sequence $(a_n)_n$
∞	Infinity
lim	Limit
lim sup	Limit superior
lim inf	Limit inferior
$\stackrel{x\to y}{\longrightarrow}$	Tends to as x tends to y
const	A generic positive constant, which may change its value in a calculation
$b_n = O(a_n)$	$b_n \le \text{const.} \cdot a_n$ for all n
$b_n = o(a_n)$	$\lim_{n\to\infty} b_n/a_n = 0$
$a_n \asymp b_n$	$a_n = O(b_n)$ and $b_n = O(a_n)$
$\delta_{j,k}$	Kronecker-symbol: equal to 1 if $j=k$; and 0, otherwise
i	Imaginary unit
e	Euler's number
π	The number π
Re	Real part
Im	Imaginary part
$\overline{z}$	Conjugate of the complex number z
$\frac{\mathrm{d}}{\mathrm{d}x}$	Derivative with respect to x
$\frac{\partial}{\partial x}$	Partial derivative with respect to x
f'	(Weak) derivative of the function f
$f^{(\beta)}$	The βth derivative of f
$N(\mu,\sigma^2)$	Normal density with mean μ and variance σ^2
$f \times g$	Convolution of the functions f and g
f^{ft}	Fourier transform of a function f

χ_I	Indicator function of a set I
$f(\cdot)$	Function f
$\int$	Integral to be taken over the whole measure space (unless something else is said)
$L_p(I)$	Banach space consisting of all functions mapping I to $\mathbb{C}$ where $\int_I \lvert f(x)\rvert^p \, \mathrm{d}x$ exists; and is finite
$\lVert \cdot \rVert_p$	$L_p(I)$-norm of a function, that is, $\lVert f\rVert_p = \left(\int_I \lvert f(x)\rvert^p \, \mathrm{d}x \right)^{1/p}$
$\lVert \cdot \rVert_\infty$	Supremum-norm, that is, $\lVert f\rVert_\infty = \sup_{x\in\mathbb{R}} \lvert f(x)\rvert$
$\langle \cdot, \cdot \rangle$	$L_2(\mathbb{R})$-inner product
CV	Cross-validation
MSE	Mean squared error
MISE	Mean integrated squared error
MSSE	Mean summed squared error
$\mathrm{med}\{a,b,c\}$	Equal to b if $a \leq b \leq c$
f_+	Equal to $\max\{f,0\}$

References

1. Berkson, J.: Are there two regression problems? J. Am. Stat. Assoc. **45**, 164–180 (1950)
2. Bissantz, N., Dümbgen, L., Holzmann, H. and Munk, A.: Nonparametric confidence bands in deconvolution density estimation. J. Roy. Stat. Soc. Ser. B **69**, 483–506 (2007)
3. Bland, J.M. and Altman, D.G.: Statistical methods for assessing agreement between two methods of clinical measurement. Lancet **i**, 307–310 (1986)
4. Bretagnolle, J. and Huber, C.: Estimation des densités: risque minimax (in French), Z. Wahrsch. Verw. Gebiete **47**, 119–137 (1979)
5. Brown, L.D. and Low, M.G.: Asymptotic equivalence of nonparametric regression and white noise. Ann. Stat. **24**, 2384–2398 (1996)
6. Buonaccorsi, J.P. and Lin, C.-D.: Berkson measurement error in designed repeated measures studies with random coefficients. J. Stat. Plan. Inf. **104**, 53–72 (2002)
7. Butucea, C.: Goodness-of-fit testing and quadratic functional estimation from indirect observations. Ann. Stat. **35**, 1907–1930 (2007)
8. Butucea, C. and Matias, C.: Minimax estimation of the noise level and of the signal density in a semiparametric convolution model. Bernoulli **11**, 309–340 (2005)
9. Butucea, C. and Tsybakov, A.B.: Sharp optimality for density deconvolution with dominating bias. I. Theo. Probab. Appl. **52**, 111–128 (2008)
10. Butucea, C. and Tsybakov, A.B.: Sharp optimality for density deconvolution with dominating bias. II. Theo. Probab. Appl. **52**, 237–249 (2008)
11. Carasso, A.S.: Linear and nonlinear image deblurring – a documented study. SIAM J. Num. Anal. **36**, 1659–1689 (1999)
12. Carasso, A.S.: Direct blind deconvolution. SIAM J. Appl. Math. **61**, 1980–2007 (2001)
13. Carroll, R.J., Delaigle, A. and Hall, P.: Nonparametric regression estimation from data contaminated by a mixture of Berkson and classical errors. J. Roy. Stat. Soc. Ser. B. **69**, 859–878 (2007)
14. Carroll, R.J. and Hall, P.: Optimal rates of convergence for deconvolving a density. J. Am. Stat. Assoc. **83**, 1184–1186 (1988)
15. Carroll, R.J. and Hall, P.: Low-order approximations in deconvolution and regression with errors in variables. J. Roy. Stat. Soc. Ser. B **66**, 31–46 (2004)
16. Carroll, R.J., Ruppert, D., Stefanski, L.A. and Crainiceanu, C. M.: Measurement Error in Nonlinear Models, 2nd Edn. Chapman and Hall CRC Press, Boca Raton (2006)
17. Cavalier, L. and Raimondo, M.: Wavelet deconvolution with noisy eigen-values. IEEE Trans. Sign. Proc. **55**, 2414–2424 (2007)
18. Cavalier, L. and Tsybakov, A.B.: Sharp adaption for inverse problems with random noise. Prob. Theo. Rel. Fields **123**, 323–354 (2002)

19. Clayton, D.G.: Models for the analysis of cohort and case control studies with inaccurately measured exposures. In: Dwyer, J.H., Feinleib, M., Lippert, P. and Hoffmeister, H. (ed) Statistical Models for Longitudinal Studies of Health, Oxford University Press, New York, pp. 301–331. (1992)
20. Comte, F., Rozenholc, Y. and Taupin, M.-L.: Penalized contrast estimator for adaptive density deconvolution. Can. J. Stat. **34**, 431–452 (2006)
21. Comte, F., Rozenholc, Y. and Taupin, M.-L.: Finite sample penalization in adaptive density deconvolution. J. Stat. Comp. Sim. **77**, 977–1000 (2007)
22. Cook, J.R. and Stefanski, L.A.: Simulation-extrapolation estimation in parametric measurement error models. J. Am. Stat. Assoc. **89**, 1314–1328 (1994)
23. Delaigle, A. and Gijbels, I.: Estimation of integrated squared density derivatives from a contaminated sample. J. Roy. Stat. Soc. Ser. B **64**, 869–886 (2002)
24. Delaigle, A. and Gijbels, I.: Bootstrap bandwidth selection in kernel density estimation from a contaminated sample. Ann. Inst. Stat. Math. **56**, 19–47 (2004a)
25. Delaigle, A. and Gijbels, I.: Comparison of data-driven bandwidth selection procedures in deconvolution kernel density estimation. Comp. Stat. Data Anal. **45**, 249–267 (2004b)
26. Delaigle, A. and Gijbels, I.: Estimation of boundary and discontinuity points in deconvolution problems. Stat. Sinica **16**, 773–788 (2006)
27. Delaigle, A. and Hall, P.: On the optimal kernel choice for deconvolution. Stat. Probab. Lett. **76**, 1594–1602 (2006)
28. Delaigle, A. and Hall, P.: Using SIMEX for smoothing-parameter choice in errors-in-variables problems. J. Am. Stat. Assoc. **103**, 280–287 (2008)
29. Delaigle, A., Hall, P. and Meister, A.: On deconvolution with repeated measurements. Ann. Stat. **36**, 665–685 (2008)
30. Delaigle, A. and Meister, A.: Nonparametric regression estimation in the heteroscedastic errors-in-variables problem. J. Am. Stat. Assoc. **102**, 1416–1426 (2007)
31. Delaigle, A. and Meister, A.: Density estimation with heteroscedastic error. Bernoulli **14**, 562–579 (2008)
32. Delaigle, A., Hall, P. and Qiu, P.: Nonparametric methods for solving the Berkson errors-in-variables problem. J. Roy. Stat. Soc. Ser. B **68**, 201–220 (2006)
33. Devroye, L.: A Course in Density Estimation. Birkhäuser, Boston (1987)
34. Devroye, L.: Consistent deconvolution in density estimation. Can. J. Stat. **17**, 235–239 (1989)
35. Devroye, L. and Györfi, L.: Nonparametric Density Estimation: The L_1 View. Wiley, New York (1985)
36. Devroye, L. and Györfi, L.: No empirical measure can converge in the total variation sense for all distributions. Ann. Stat. **18**, 1496–1499 (1990)
37. Diggle, P. and Hall, P.: A Fourier approach to nonparametric deconvolution of a density estimate. J. Roy. Stat. Soc. Ser. B **55**, 523–531 (1993)
38. Donoho, D.L.: Statistical estimation and optimal recovery. Ann. Stat. **22**, 238–270 (1994)
39. Dunn, G.: Statistical Evaluation of Measurement Errors, Design and Analysis of Reliability Studies, 2nd Edn. Arnold, London (2004)
40. Efromovich, S.: Density estimation for the case of supersmooth measurement error. J. Am. Stat. Assoc. **92**, 526–535 (1997)
41. Efromovich, S.: Nonparametric Curve Estimation: Methods, Theory and Applications. Springer, Berlin Heidelberg New York (1999)
42. Efromovich, S. and Pinsker, M.S.: Estimation of square-integrable probability density of a random variable. Probl. Inf. Transm. **18**, 175–189 (1983)
43. Fan, J.: On the optimal rates of convergence for non-parametric deconvolution problems. Ann. Stat. **19**, 1257–1272 (1991a)
44. Fan, J.: Asymptotic normality for deconvolution kernel density estimators. Sankhya A **53**, 97–110 (1991b)

45. Fan, J.: Deconvolution with supersmooth distributions. Can. J. Stat. **20**, 155–169 (1992)
46. Fan, J.: Adaptively local one-dimensional subproblems with application to a deconvolution problem. Ann. Stat. **21**, 600–610 (1993)
47. Fan, J. and Truong, Y.: Nonparametric regression with errors in variables. Ann. Stat. **21**, 1900–1925 (1993)
48. Fan, J. and Koo, J.-Y.: Wavelet deconvolution. IEEE Trans. Inform. Theory **48**, 734–747 (2002)
49. Folland, G.B.: Fourier analysis and its applications. Brooks/Cole, Pacific Grove, CA (1992)
50. Goldenshluger, A.: Density deconvolution in the circular structural model. J. Multivar. Anal. **84**, 350–375 (2002)
51. Goldenshluger, A. and Tsybakov, A.: Estimating the endpoint of a distribution in presence of additive observation errors. Stat. Probab. Lett. **68**, 39–49 (2004)
52. Goldenshluger, A. and Zeevi, A.: Recovering convex boundaries from blurred and noisy observations. Ann. Stat. **34**, 1375–1394 (2006)
53. Goldenshluger, A., Tsybakov, A. and Zeevi, A.: Optimal change-point estimation from indirect observations. Ann. Stat. **34**, 350–372 (2006)
54. Groeneboom, P. and Jongbloed, G.: Density estimation in the uniform deconvolution model. Stat. Neerlandica **57**, 136–157 (2003)
55. Hall, P.: Large sample optimality of least squares cross-validation in density estimation. Ann. Stat. **11**, 1156–1174 (1983)
56. Hall, P. and Lahiri, S.N.: Estimation of distributions, moments and quantiles in deconvolution problems. Ann. Stat., to appear
57. Hall, P. and Marron, J.S.: Extent to which least-squares cross-validation minimises integrated square error in nonparametric density estimation. Prob. Theo. Rel. Fields **74**, 567–581 (1987)
58. Hall, P. and Meister, A.: A ridge-parameter approach to deconvolution. Ann. Stat. **35**, 1535–1558 (2007)
59. Hall, P. and Qiu, P.: Discrete-transform approach to deconvolution problems. Biometrika **92**, 135–148 (2005)
60. Hall, P. and Qiu, P.: Nonparametric estimation of a point-spread function in multivariate problems. Ann. Stat. **35**, 1512–1534 (2007)
61. Hall, P. and Simar, L.: Estimating a changepoint, boundary or frontier in the presence of observation error. J. Am. Stat. Assoc. **97**, 523–534 (2002)
62. Hall, P. and Yao, Q.: Inference in components of variance models with low replication. Ann. Stat. **31**, 414–441 (2003)
63. Hesse, C.H.: Distribution function estimation from noisy observations. Publ. Inst. Stat. Paris Sud **39**, 21–35 (1995)
64. Hesse, C.H.: Data-driven deconvolution. J. Nonparametric Stat. **10**, 343–373 (1999)
65. Hesse, C.H. and Meister, A.: Optimal iterative density deconvolution. J. Nonparametric Stat. **16**, 879–900 (2004)
66. Hoeffding, W.: Probability inequalities for sums of bounded random variables. J. Am. Stat. Assoc. **58**, 13–30 (1963)
67. Holzmann, H., Bissantz, N. and Munk, A.: Density testing in a contaminated sample. J. Multivar. Anal. **98**, 57–75 (2007)
68. Holzmann, H. and Boysen, L.: Integrated square error asymptotics for supersmooth deconvolution. Scand. J. Stat. **33**, 849–860 (2006)
69. Horowitz, J.L.: Semiparametric Methods in Econometrics. Springer, Berlin Heidelberg New York (1998)
70. Horowitz, J.L. and Markatou, M.: Semiparametric estimation of regression models for panel data. Rev. Econom. Stud. **63**, 145–168 (1996)
71. Huwang, L. and Huang, H.Y.S.: On errors-in-variables in polynomial regression – Berkson case. Stat. Sinica **10**, 923–936 (2000)

72. Johnstone, I., Kerkyacharian, G., Picard, D. and Raimondo, M.: Wavelet deconvolution in a periodic setting. J. Roy. Stat. Soc., Ser. B **66**, 547–573 (2004). with discussion pp. 627–652
73. Johnstone, I.M. and Raimondo, M.: Periodic boxcar deconvolution and diophantine approximation. Ann. Stat. **32**, 1781–1804 (2004)
74. Karatzas, I. and Shreve, S.S.: Brownian motion and stochastic calculus. 2nd ed. Springer, Berlin Heidelberg New York (1999)
75. Kerkyacharian, G., Picard, D. and Raimondo, M.: Adaptive boxcar deconvolution on full Lebesgue measure sets. Stat. Sinica **17**, 317–340 (2007)
76. Komlos, J., Major, P. and Tusnady, G.: An approximation of partial sums of independent rv's and the sample df. Z. Wahrsch. verw. Gebiete **32**, 111–131 (1975)
77. Körner, T.W.: Fourier Analysis. Cambridge University Press, UK (1988)
78. Korostelev, A.P. and Tsybakov, A.B.: Minimax Theory of Image Reconstruction. Lecture Notes in Statistics 82, Springer, Berlin Heidelberg New York (1993)
79. Kundur, D. and Hatzinakos, D.: A novel blind deconvolution scheme for image restoration using recursive filtering. IEEE Trans. Sig. Process. **46**, 375–390 (1998)
80. Lenglart, E.: Transformation de martingales locales par changement absolue continu de probabilités (in French). Z. Wahrsch. **39**, 65–77 (1977)
81. Lepski, O.V.: Asymptotically minimax adaptive estimation. I: Upper bounds. Optimally adaptive estimates. Theo. Probab. Appl. **36**, 682–697 (1991)
82. Lepski, O.V.: Asymptotically minimax adaptive estimation. II: Schemes without optimal adaption. Adaptive estimates. Theo. Probab. Appl. **37**, 433–448 (1992)
83. Li, T. and Vuong, Q.: Nonparametric estimation of the measurement error model using multiple indicators. J. Multivar. Anal. **65**, 139–165 (1998)
84. Linton, O. and Whang, Y.L.: Nonparametric estimation with aggregated data. Econometric Theo. **18** 420–468 (2002)
85. Liu, M.C. and Taylor, R.C.: A consistent nonparametric density estimator for the deconvolution problem. Can. J. Stat. **17**, 427–438 (1989)
86. Low, M.G.: Non-existence of an adaptive estimator for the value of an unknown probability density. Ann. Stat. **20**, 598–602 (1992)
87. Meister, A.: On the effect of misspecifying the error density in a deconvolution problem. Can. J. Stat. **32**, 439–449 (2004)
88. Meister, A.: Non-estimability in spite of identifiability in density deconvolution. Math. Meth. Stat. **14**, 479–487 (2005)
89. Meister, A.: Density estimation with normal measurement error with unknown variance. Stat. Sinica **16**, 195–211 (2006a)
90. Meister, A.: Estimating the support of multivariate densities under measurement error. J. Multivar. Anal. **97**, 1702–1717 (2006b); [erratum: (2008), 99, 308]
91. Meister, A.: Support estimation via moment estimation in presence of noise. Statistics **40**, 259–275 (2006c)
92. Meister, A.: Deconvolving compactly supported densities. Math. Meth. Stat. **16**, 63–76 (2007a)
93. Meister, A.: Optimal convergence rates for density estimation from grouped data. Stat. Probab. Lett **77**, 1091–1097 (2007b)
94. Meister, A.: Deconvolution from Fourier-oscillating error densities under decay and smoothness restrictions. Inverse Problems **24**, 015003 (2008)
95. Meister, A. and Neumann, M.H.: Deconvolution from non-standard error densities under replicated measurements. submitted preprint (2008)
96. Meyer, Y.: Ondelettes et oprateurs, Tome I, Hermann (1990)
97. Morris, J.N., Marr, J.W. and Clayton, D.G.: Diet and heart: a postscript. Br. Med. J. **2**, 1307–1314 (1977)
98. Nadaraya, E.A.: On estimating regression. Theo. Probab. Appl. **10**, 186–190 (1964)
99. Neelamani, R., Choi, H. and Baraniuk, R.: Forward: Fourier-wavelet regularized deconvolution for ill-conditioned systems. IEEE Trans. Sig. Proc. **52**, 418–433 (2004)

100. Nelson, J.A.: Consumer Expenditure Surveys 1980–1989: Interview Surveys, for Household-Level Analysis, Computer file, United States Department of Labor, Bureau of Labor Statistics, Washington, DC. Distributed by: Inter-University Consortium for Political and Social Research, Ann Arbor, MI (1994)
101. Neumann, M.H.: On the effect of estimating the error density in nonparametric deconvolution. J. Nonparametric Stat. **7**, 307–330 (1997a)
102. Neumann, M.H.: Optimal change-point estimation in inverse problems. Scan. J. Stat. **24**, 503–521 (1997b)
103. Neumann, M.H.: Deconvolution from panel data with unknown error distribution. J. Multivar. Anal. **98**, 1955–1968 (2007)
104. Nussbaum, M.: Asymptotic equivalence of density estimation and Gaussian white noise. Ann. Stat. **24**, 2399–2430 (1996)
105. Parzen, E.: On estimation of a probability density function and mode. Ann. Math. Stat. **33**, 1065–1076 (1962)
106. Oksendal B.: Stochastic Differential Equations: An Introduction with Applications, 5th ed. Springer, Berlin Heidelberg New York (1998)
107. Pensky, M. and Vidakovic, B.: Adaptive wavelet estimator for nonparametric density deconvolution. Ann. Stat. **27**, 2033–2053 (1999)
108. Qiu, P.: Image Processing and Jump Regression Analysis. Wiley, New York (2005)
109. Rachdi, M. and Sabre, R.: Consistent estimates of the mode of the probability density function in nonparametric deconvolution problems. Stat. Probab. Lett. **47**, 105–114 (2000)
110. Reeves, G.K., Cox, D.R., Darby, S.C. and Whitley, E.: Some aspects of measurement error in explanatory variables for continuous and binary regression models. Stat. Med. **17**, 2157–2177 (1998)
111. Rice, J.: Bandwidth choice for nonparametric regression. Ann. Stat. **12**, 1215–1230 (1984)
112. Rosenblatt, M.: Remarks on some nonparametric estimates of a density function. Ann. Math. Stat. **27**, 832–837 (1956)
113. Schennach, S. M.: Estimation of nonlinear models with measurement error. Econometrica **72**, 33–75 (2004a)
114. Schennach, S. M.: Nonparametric regression in the presence of measurement error. Econometric Theo. **20**, 1046–1093 (2004b)
115. Schipper, M.: Optimal rates and constants in L2 -minimax estimation of proba- bility density functions. Math. Meth. Stat. **5**, 253–274 (1996)
116. Stark, H. and Woods, J.W.: Probability and Random Processes with Applications to Signal Processing, 3rd ed. Prentice Hall, New Jersey (2002)
117. Stefanski, L.A. and Carroll, R.J.: Deconvoluting kernel density estimators. Statistics **21**, 169–184 (1990)
118. Stein, C.M.: Estimation of the mean of a multivariate normal distribution. Ann. Stat. **9**, 1135–1151 (1981)
119. Stone, C.J.: Optimal rates of convergence for nonparametric estimators. Ann. Stat. **8**, 1348–1360 (1980)
120. Stone, C.J.: An asymptotically optimal window selection rule for kernel density estimates. Ann. Stat. **12**, 1285–1297 (1984)
121. Tournier, J.D., Calamante, F., Gadian, D.G. and Connelly, A.: Direct estimation of the fiber orientation density function from diffusion-weighted MRI data using spherical deconvolution. NeuroImage **23**, 1176–1185 (2004)
122. Turner, S.W., Tonne, B.K. and Brett-Jones, J.R.: Computerized tomographic scan changes in early schizophrenia – preliminary findings. Psychol. Med. **16**, 219–225 (1986)
123. van Es, A.J. and Uh, H.-W.: Asymptotic normality of nonparametric kernel type deconvolution density estimators: crossing the Cauchy boundary. J. Nonparametric Stat. **16**, 261–277 (2004)

124. van Es, A.J. and Uh, H.-W.: Asymptotic normality of kernel type deconvolution estimators. Scand. J. Stat. **32**, 467–483 (2005)
125. Watson, G.S.: Smooth regression analysis. Sankhya, Ser. A **26**, 359–372 (1964)
126. Watson, G.S. and Leadbetter, M.R.: On the estimation of probability density, I. Ann. Math. Stat. **34**, 480–491 (1963)
127. Werner, D.: Funktionalanalysis (in German). Springer, New York (1997)
128. Yang, Y., Galatsanos, N.P. and Stark, H.: Projection-based blind deconvolution. J. Opt. Soc. Am. Ser. A **11**, 2401–2409 (1994)
129. Yazici, B.: Stochastic deconvolution over groups. IEEE Trans. Inf. Theo. **50**, 491–510 (2004)
130. Yosida, K.: Functional Analysis, 6th ed. Springer, New York (1980)
131. Zhang, C.H.: Fourier methods for estimating mixing densities and distributions. Ann. Stat. **18**, 806–830 (1990)